Insects and Diseases of Woody Plants of the Central Rockies

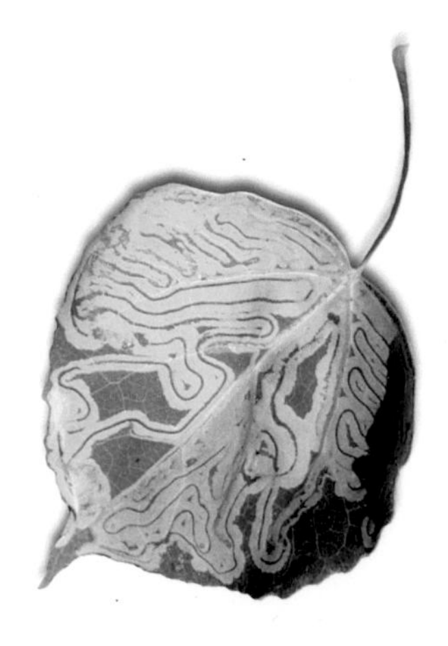

Bulletin 506A

Colorado State University Cooperative Extension

Putting Knowledge to Work

acknowledgments

This publication has been an evolving project, with many predecessors. The 1980 edition, *Insect Pests of Colorado Trees and Shrubs,* was written by Wayne Brewer, William Hantsbarger and Susan Taylor. In 1993, this publication was revised by Whitney Cranshaw, David Leatherman, Boris Kondratieff, Robert Stevens, and Robert Wawrzynski, as *Insects That Feed on Colorado Trees and Shrubs.*

This latest version is a major revision and expansion, most notably in its incorporation of woody plant diseases. Diagnostic keys were greatly expanded as well as sections on vertebrate damage. This latest edition is our attempt as a more comprehensive and useful publication for an individual who wants to identify an organism and injury associated with woody plants in the central Rockies region. This printing also includes an addendum produced in the summer of 2004 (pages 237-242).

In this edition, information on insects was organized and developed by Whitney Cranshaw and David Leatherman. Information involving insects on woody plants were contributed by Boris Kondratieff, Paul Opler and Casey Sclar.

The new plant disease sections were coordinated by William Jacobi and Loretta Mannix. This involved numerous contributors who wrote individual sections including Loretta Mannix, Laura Pottorff, Linnea Skoglund, Holly Kearns, Sam Harrison, Mary Small, Gary Franc, and Veva McCaig.

Sections involving vertebrate injuries were written by David Leatherman.

This publication was published by Colorado State University Cooperative Extension and designed and printed by Publications and Printing at Colorado State University.

Insects and Diseases of Woody Plants of the Central Rockies is a project of Colorado State Cooperative Extension and the Colorado State Forest Service. Financial assistance in its development has been received from the Colorado IPM Program.

how to use this publication

Content

Insects and Diseases of Woody Plants of the Central Rockies was an effort to pull together information on most of the organisms associated with woody plants in Colorado and adjacent areas of the Central Rockies. This includes insects and mites, viruses, bacteria, fungi, parasitic plants, and vertebrates.

For easy browsing, this publication includes several ways to find information. In the beginning there is a **Table of Contents** describing the order of entries. An effort was made to try and group the organisms according to the areas of the plants where they most commonly are associated. For example, the first section includes organisms found on foliage. Subsequent sections discuss what one might find on twigs or small branches, trunks and large branches, or roots. There is overlap in these sites as insects or fungi, for example, may have life stages that develop on both foliage and twigs.

The section **Diagnostic Key to Common Woody Plant Disorders** is a unique development designed to assist in identifications. Major sections of this key are organized by plant species. These are then subdivided by plant parts and ultimately by symptoms. It is hoped that this will prove a useful means for users to identify an organism associated with the plant. There is also a reference to a page where one may find additional information.

Finally, there is a standard **Index** organized by the names of the various insects, diseases, and vertebrates covered in the publication.

Throughout this publication there are sections related to the management of the various insects, diseases, and vertebrates - where they occur as significant species associated with woody plants. These sections are written in a generally descriptive format, because specific recommendations can change rapidly due to altered registrations.

Additional References

Many fact sheets have been written and produced by Colorado State University Cooperative Extension and the Colorado State Forest Service as a source for specific updated information of management recommendations on these disorders. The publication, *Insect Management Recommendations for Turf and Ornamentals*, XCM-38, was written as a companion to this publication in 2004. Copies of all these publications are available from: the Cooperative Extension Resource Center, 115 General Services Building, Colorado State University, Fort Collins, CO 80523-4061, (877) 692-9358, http://www.cerc.colostate.edu.

table of contents

conditions generally associated with smaller branches and twigs

Abiotic Problems of Unknown Origin

Pathogen-Associated Twig Injuries

Scale Insects

Insects That Oviposit in Twigs

Tip Moths, Twig and Terminal Feeding Insects

conditions generally associated with larger branches and stems

Vascular Wilts

Cankers

Decay Fungi

Bark Beetles

Borers of Trunks and Larger Branches

IO MOTH

Automeris io (F.)
Lepidoptera: Saturniidae

Hosts: Many varieties of deciduous trees and shrubs

Damage and Diagnosis: The primary injury results from the stinging hairs on the larvae and pupae which can cause painful rashes. These are large pale green caterpillars with a lateral stripe of pink and creamy white down each side, covered with clusters of branching spines. About 3 inches long at maturity, larvae feed on leaves late in the summer but do not cause noticeable defoliation to trees and shrubs. Most outbreaks have historically been around the Arkansas River Basin area (Fremont/El Paso counties).

Large moths with a wing-span of two and one half to three inches, the females with lightly patterned purplish-red wings, the males smaller with yellowish wings, both with a large circular black eyespot on the hind wings.

Life History and Habits: Io moth spends the winter as a pupa, within a tough brown cocoon covered with bits of hair and debris. (The pupa, containing the stinging hairs, also can produce painful reactions on contact.) Adults fly in the late spring and early summer and females lay eggs. Caterpillars are present from July to September. When full grown, they wander from the plant and search for a site to pupate. There is one generation per year.

Management: This species is never abundant enough to seriously damage trees. However, control of outbreaks may be considered to limit injury by the stinging hairs of the caterpillars. Controls outlined under the section for **Leaf feeding caterpillars** in the control supplement should be effective.

TOP: Io moth caterpillar. (J. Capinera)
BOTTOM: Buck moth caterpillar.

NEVADA BUCK MOTH

Hemileuca nevadensis Stretch
Lepidoptera: Saturniidae

Hosts: Cottonwood, poplars and willow, occasionally aspen

Damage: Larvae feed on foliage and during outbreaks can defoliate trees. The spines can cause skin irritation. Caterpillars are black when young, turning yellow as they mature. Older caterpillars have red heads and branched, black and white lateral spines, plus two rows of yellow dorsal spines.

Life History and Habits: Egg masses that ring the twigs are the overwintering stage. These hatch in late April or early May. Larvae first feed gregariously, later dispersing. They grow and molt five times before dropping to the ground to pupate. Most adults emerge in fall, although some may overwinter as pupae. The name "buck moth" was given because adults can be observed during fall deer hunting season.

Related Species: Several other *Hemileuca* species also occur in the region. These include *H. diana* Packard, on gambel oak; *H. magnifica* (Rotger) on *Artemisia tridentata; H. eglanterina annulata* Ferguson on many plants including *Amelanchier*, rose, *Prunus, Purshia tridentata*, and *Symphoricarpos; H. nuttalli nuttalli* (Strecker) which primarily feeds on Antelope bitterbrush (*P. tridentata*) and snowberry; and *H. neumoegeni* Hy. Edwards which feeds on desert almond, skunkbrush, sumac and Apache plum. None of these species are reported to cause significant injury.

TOP: Pandora moth caterpillar.
BOTTOM: Redhumped caterpillar. (H. Evans)

PANDORA MOTH

Coloradia pandora (Blake)
Lepidoptera: Saturniidae

Hosts: Lodgepole pine, ponderosa pine

Damage and Diagnosis: Larvae feed on older foliage, leaving the tree with a tufted appearance. Infested trees have reddish tops with foliage only at branch tips. Heavy damage is not common and usually occurs only where soil is loose enough to allow caterpillars to burrow and pupate. Because of its life history, defoliation episodes occur in alternate years during outbreaks. After outbreaks subside, damaged trees often experience "compensatory growth" and can completely recover.

Pandora moth larvae are from two and one half to three inches long, a gray caterpillar with white stripe down the back. A few stout, branched spines appear on each segment which can produce a stinging sensation on contact. Adults are gray, thick-bodied moths with a wingspan of three to five and one half inches. Forewings are brownish gray, hind wings pinkish gray, with black dot and dark wavy line on each wing.

Life History and Habits: Pandora moth most often has a two-year life cycle. They spend the winter of the first year as young caterpillars on the tree and resume feeding with the return of warm weather. In June or July, mature larvae crawl down tree and pupate in loose soil. Pupation typically lasts for about a year, but may occasionally may extend up to four years. Adults emerge in June or July and mate. Females begin to produce eggs one to two days later, laying clusters on needles, bark, and occasionally in ground litter or brush. Eggs hatch in August, and young larvae feed on needles until fall, then overwinter.

Related Species: A related, uncommon species, *Coloradia doris* Barnes, feeds on ponderosa pine and has an annual life cycle.

REDHUMPED CATERPILLAR

Schizura concinna (J.E. Smith)
Lepidoptera: Notodontidae

Hosts: Larvae feed on foliage of a wide variety of broadleaf trees, most commonly honeylocust, willow and various fruit trees such as crabapple, apple and plum.

Damage and Diagnosis: These unusual caterpillars are marked with a pronounced reddish hump behind the head and during early development they feed gregariously. Newly hatched caterpillars feed first on the underside of leaves, producing skeletonizing injuries, but later become more general feeders. Serious defoliation by this insect is very rare, although individual branches may be stripped of leaves.

Life History and Habits: Winter is spent within a cocoon mixed with leaf litter and other cover around the base of previously infested plants. The adult moths begin to emerge during late spring, but emergence of new moths extends over a couple of months. Females lay eggs on the underside of host plant foliage, in masses of 50 to 100. The newly hatched caterpillars feed as a group, skeletonizing the underside of the leaves. They continue to feed gregariously for much of their life, moving from branch to branch and chewing entire leaves except for the largest vein. Nearly mature caterpillars tend to disperse and when full grown crawl away from the plant to pupate. There is one generation per year.

Related Species: The unicorn caterpillar, *Shizura unicornis* (J.E. Smith), is an unusual brown and green caterpillar that feeds on many broadleaf trees and shrubs.

Management: This species is common, but rarely abundant enough to cause serious injury. It can be readily controlled by several insecticides including carbaryl (Sevin), *Bacillus thuringiensis* products, spinosad (Conserve), and pyrethroids. (See section in control supplement on **Leaf feeding caterpillars**.)

WESTERN SPRUCE BUDWORM

Choristoneura occidentalis (Freeman)
Lepidoptera: Tortricidae

Hosts: Douglas-fir, spruce, and true firs

Damage and Diagnosis: Larvae tunnel buds and chew the needles, concentrating on the newly emerged current season growth. During outbreaks trees may be extensively defoliated. These injuries can cause serious stress that can directly kill the tree or make it susceptible to secondary pests, such as Douglas-fir beetle. Western spruce budworm is the most serious defoliator of native forests in the region.

Larvae are olive-brown or reddish-brown caterpillars, which can reach one to one and one half inches long at maturity. They have ivory-colored paired spots on each body segment and a chestnut-brown head and collar. The adult moths have a wingspan of about one inch. They are highly variable in color and have mottled medium-brown, dark-brown to orange-brown forewings and tan hind wings.

Life History and Habits: Winter is spent as a minute caterpillar protected within a silken hibernacula under bark flakes or among lichen. In late May or June, they begin to feed, mining old needles, tunneling into buds or developing cones. Following bud break the larvae move to the new growth. When this is consumed they will feed on older needles.

Larvae mature about 30 to 40 days after feeding begins in spring. Pupation occurs among foliage and adults emerge in July and early August to mate and lay eggs. Eggs are laid in shingle-like masses on the underside of needles. Eggs hatch in about 10 days and the resulting larvae immediately spin small overwintering cocoons (hibernaculae) on the bark. There is one generation per year.

Related Species: Similar injury is produced to ponderosa pine by *Choristoneura lambertiana*, the pine budworm (sugar pine tortrix).

Management: Western spruce budworm is attacked by many natural enemies including parasitic insects, predators such as ants, spiders and birds, as well as suffering from adverse effects of late spring frosts. Several insecticides are also registered for western spruce budworm control, which are best applied shortly after the flush of new growth is produced in late spring. Specific controls for protection of landscape trees are included in the supplement under the heading **Leaf feeding caterpillars**. Sprays used in forest protection are more limited but include certain carbaryl and Bt products.

PINE BUDWORM (Sugar Pine Tortrix)

Choristoneura lambertiana (Busck)
Lepidoptera: Tortricidae

Hosts: Ponderosa pine and occasionally other pines

Damage and Diagnosis: Larvae feed on developing buds, mine old needles, and chew newly produced needles. Buds are littered with frass and webbing. Repeated annual loss of new growth can leave branch ends sparse and reduce overall tree growth and vigor, although the trees are rarely killed. However, defoliated trees appear to be more susceptible to mountain pine beetle and other bark beetles. For reasons probably related to stress, infested trees are often near roadways.

The larvae are 1-inch long caterpillars when mature, with a dark head capsule and a smooth brown-green body with paired white spots on each body segment. They resemble the western spruce budworm, a close relative associated with Douglas-fir and spruce. Adult moths have a wingspan of about 1 inch, with dull grayish-brown forewings and light tan hindwings.

Life History and Habits: Moths fly and mate in July and August, laying clusters of 25 to 50 eggs in shingle-like patches on young needles. Eggs hatch after two to three

TOP: Defoliation of Douglas-fir new growth by western spruce budworm.
SECOND: Western spruce budworm larva feeding.
THIRD: Pine budworm, middle instar larva.
BOTTOM: Pine budworm, injury to terminal growth of ponderosa pine.

weeks and the larvae, without feeding, seek shelter and spin tiny cocoons, called hibernaculae, in which they spend the winter. In the spring, they emerge and begin feeding on the developing buds and needles. They later feed on older needles and pupate among the damaged foliage. There is one generation per year.

Management: Pine budworm populations are naturally regulated by a great many control agents, including parasitic insects, predation by birds, and effects of temperature extremes. Following sustained outbreaks, starvation can also be important. Where chemical controls are necessary to prevent unacceptable injury, sprays should be timed for periods when larvae resume feeding in the spring. Typically this would be after bud break in May or June.

ABOVE: Aspen leaf tying characteristic of various leaf rollers (aspen tortrix on left).

LARGE ASPEN TORTRIX

Choristoneura conflictana (Walker)
Lepidoptera: Tortricidae

Hosts: Aspen and occasionally other *Populus* species

Damage and Diagnosis: Caterpillars of the large aspen tortix feed on the buds and leaves of aspen early in the season. Outbreaks, although rare, can seriously defoliate native stands of aspen; significant tree mortality is even less common. Later stages of the caterpillars often feed as a leafroller, folding over the edge of leaves and tying it with silk. Within the folded leaves may be found the larvae with are olive-green to black caterpillars, with a dark head and first body segment, that may reach over an inch in length.

Life History and Habits: Large aspen tortrix spends the winter as a second instar larvae, usually in bark crevices. In spring the larvae first feed on developing buds, later moving to newly-formed leaves. During feeding, they commonly tie together two leaves with the larvae feeding and later pupating within this structure.

The dull, gray adult moths appear in late June and July and lay flat, shingle-like egg masses on the leaves. Eggs hatch and larvae move to the tree trunk to overwinter. There is one generation per year.

Related and Associated Insects: The obliquebanded leafroller, *C. rosaceana* (Harris), is sometimes associated with birch, *Populus*, willow and various fruit trees. It develops as a leafroller and may have two generations in a season. The overwintering stage are as eggs, laid in masses on twigs.

There are also numerous other caterpillars associated with aspen that roll, tie, or fold leaves. Most of these are also in the family Torticidae and all are of minor consequence.

Management: Large aspen tortrix is attacked by numerous natural enemies, notably tachinid flies and parasitic wasps. Large numbers may also starve during very high populations. Because of these natural controls outbreaks are sporadic and aspen is quite tolerant of defoliation. In protection of ornamental aspen during outbreaks, insecticides recommended for control of **Leaffeeding caterpillars** in the supplement should be effective. *Bacillus thuringiensis* is sometimes used to control outbreaks in forested areas.

LEAFROLLERS

FRUITTREE LEAFROLLER: *Archips argyrospila* (Walker)
OAK LEAFROLLER: *Archips semiferana* (Walker)
BOXELDER LEAFROLLER: *Archips negundanus* (Dyar)

Lepidoptera: Tortricidae

Hosts: Fruittree leafroller feeds on a wide variety of deciduous trees and shrubs are hosts, including apple, crabapple, honeylocust, ash and linden. Oak leafroller is associated with Gambel oak; boxelder leafroller with boxelder.

Damage and Diagnosis: Larvae are active early in the season and chew leaves of a wide variety of plants. Older larvae have the habit of curling over the edge of leaves and fastening with silk to create a rolled leaf shelter. Damage by leafrollers is usually transitory and mostly cosmetic. However, oak leafroller, often in combination with species such as the oak looper and speckled green fruitworm, have caused episodes of extensive defoliation to native Gambel oak stands.

The caterpillars, which are range from pale to dark green, have a black head. They are usually found within the folded leaves where they feed, chewing in a skeletonizing manner.

When disturbed they can move vigorously and often will drop out of the leaf on a strand of silk.

On fruit trees larvae of the fruittree leafroller may chew pits in developing fruit causing them to prematurely drop or grow in a distorted manner.

Life History and Habits: Leafrollers overwinter as eggs within a flat gray-brown mass, typically containing over 100 eggs. Egg masses are glued to twigs and hatch in spring shortly after leaves emerge. Young larvae usually first feed around the tips and as they get older they begin to tie up leaves with webbing and feed inside this shelter. When mature, larvae usually pupate within the rolled leaves but may disperse to bark cracks of the trunk and large branches. Adult moths appear two weeks later, mate and lay the overwintering egg masses. There is one generation per year.

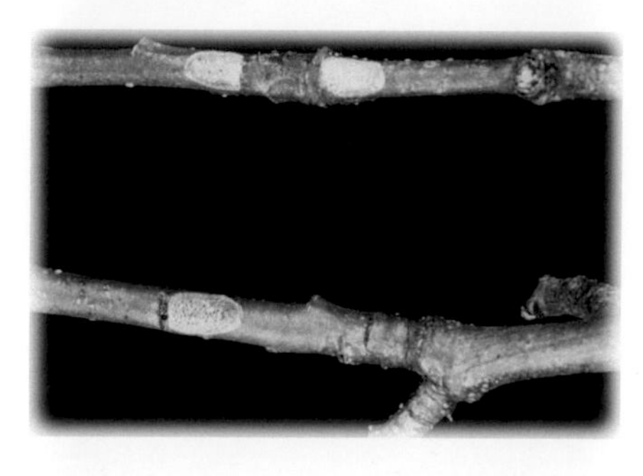

Management: Leafroller outbreaks are often highly sporadic, and numerous natural controls usually prevent high, damaging populations. Supplemental controls are rarely needed for plant protection.

Fruittree leafroller can be controlled by most of the insecticides listed in the supplement under the heading **Leaf feeding caterpillars**. However, coverage can be difficult after leaves are curled and systemic insecticides are more likely to be effective in those situations.

Horticultural oils applied during the dormant season are effective at controlling overwintering egg stages of this insect. Applications should target the twigs, where egg masses are laid.

UGLYNEST CATERPILLAR

Archips cerasivorana (Fitch)

Lepidoptera: Tortricidae

Hosts: Most common on chokecherry, but may also be found on other deciduous trees and shrubs

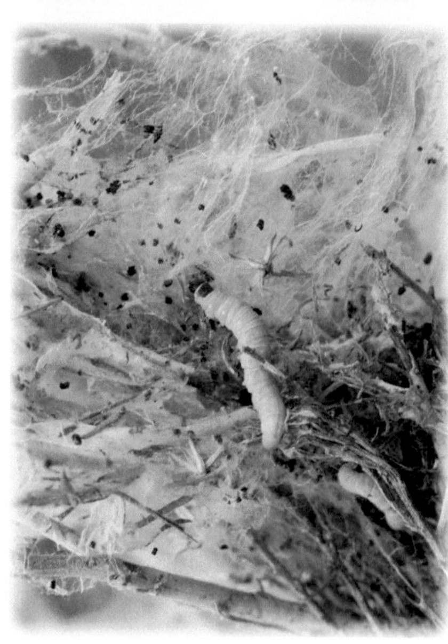

TOP: Fruittree leafroller, larva spinning. (J. Capinera)
SECOND: Oak leafroller, larva on leaf.
THIRD: Egg masses of fruittree leafroller.
BOTTOM: Uglynest caterpillar colony.

Damage and Diagnosis: Larvae chew on foliage and produce loose silken shelters of webbing and leaves. Accumulations of frass and bits of leaves inside the webbing are unsightly. The caterpillars are olive green, reaching about 3/4 inch long when mature.

Life History and Habits: The uglynest caterpillar spends the winter in the egg stage on twigs. Eggs hatch in May or June and the larvae, often working together in small groups, construct a tightly woven silk and leaf nest within which they feed. The caterpillars become full-grown in four to six weeks and pupate around the nest. Adult moths appear from July to September, and eggs are laid in late summer or early fall. There is one generation produced per year.

Management: Although conspicuous, infestations of uglynest caterpillar typically occur on low-value plants for which treatment is not economical. In landscape settings, the tents can be easily removed by hand.

TOP: Rabbitbrush webbing moth, *Synnoma lynosyrana*, larva.
BOTTOM: Juniper webworm.

RABBITBRUSH WEBBING MOTH

Synnoma lynosyrana Walsingham
Lepidoptera: Tortricidae

Hosts: Rabbitbrush and snakeweed

Damage and Diagnosis: Larvae feed on rabbitbrush and can cause serious defoliation. They are grayish green caterpillars that create dense tents of silk mixed with food fragments, frass, and cast skins. Adults are moderate sized moths with a wingspan of about 3/4-inch. The wings are generally gray-brown, speckled with dark spots and often some whitish patches.

Life History and Habits: Life history is generally similar to that of the uglynest caterpillar. However, females of the rabbitbrush webbing moth do not fly, resulting in very localized infestations that spread slowly. Peak period of mating and egg laying occurs from mid-August through early October, with overwintering eggs laid on the host plant. There is one generation produced per season.

JUNIPER WEBWORM

Dichomeris marginella (F.)
Lepidoptera: Gelechiidae

Hosts: Many juniper species. Rocky Mountain juniper is the most common host.

Damage and Diagnosis: The larvae feed on the needles of junipers during late summer, causing damaged foliage to turn brown. The larvae are brown caterpillars with indistinct stripes and a dark head, reaching about one half inch when full grown. Older larvae are almost always found in silken tubes they produce amongst the juniper foliage. Adults are small (1/2-inch wingspan) moths with copper-colored wings bordered with white. This insect has caused significant injury primarily in the Eastern plains areas.

Life History and Habits: The juniper webworm spends the winter as a large larva within the webbed needles. They continue feeding in spring and later pupate. Adults emerge in June and July, mate and lay eggs. The tiny, newly hatched larvae first feed by tunneling needles. They later emerge and form silken tubes, becoming surface feeders. During outbreaks, silk production may be extensive and several larvae may feed together in "nests" on the foliage. There is one generation per year.

Management: Outbreaks of juniper webworm are uncommon in the region and several predators and parasites are reported to attack this insect. During outbreaks systemic insecticides such as acephate (Orthene) are likely to provide best control. The addition of a wetting agent, such as soap, can improve coverage and penetration of webbing.

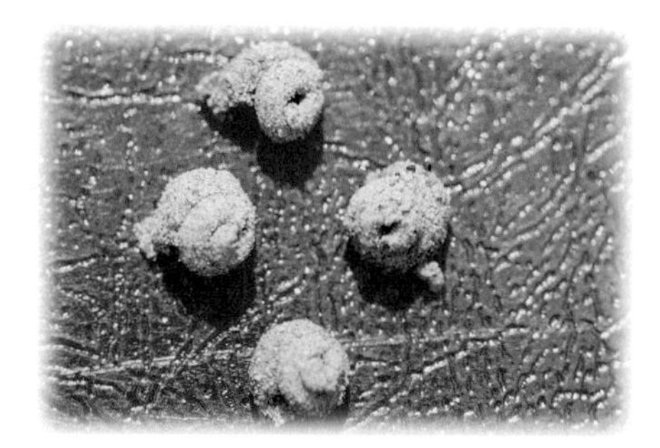

ABOVE: Snailcase bagworm.

SNAILCASE BAGWORM

Apterona (=crenulella) helix (Siebold)
Lepidoptera: Psychidae

Hosts: Sagebrush, saltbush, rabbitbrush, willow, mountain-mahogany, various fruit trees, knapweed, and a variety of herbaceous weeds, vegetable and other garden plants.

Damage and Diagnosis: The snailcase bagworm is an unusual moth that has been introduced into Colorado and is spreading through the state. Currently it has been known to occur in all of the West Slope counties and has been reported from Fremont and Boulder counties in eastern Colorado. Although the individual insects are wingless and can not fly, further spread can be expected to continue as new infestations can develop wherever an insect is moved.

The developing insects feed on a wide variety of plants, but rarely cause any significant plant injury. Instead problems occur as the full grown larvae migrate for sites to pupate. During this time large numbers may firmly attach themselves to sides of buildings, fences, mailboxes and other surfaces, creating a nuisance.

Life History and Habits: All stages of this insect take place within a coiled snail-like case. The caterpillars are greenish or reddish-gray with a black head. Adults are wingless, and nearly legless moths. Only females are known to occur.

Snailcase bagworms survive winter as young caterpillars protected within the case of the mother insect. They become active in midspring and feed on the leaves of a wide variety of native and cultivated plants, including sagebrush, saltbush, rabbitbrush, willow, mountain-mahogany, various fruit trees, squash, and alfalfa. The feeding injuries appear as small areas progressively gouged out of the leaf surface. Serious plant injury is very rare.

As the larvae grow and develop they produce a snail-like case of silk and soil particles. Later they push their fecal matter out of an opening in the center of the case, allowing it to pile up on top of the insect. The larval insects are mobile and can carry the case upright. As they become full-grown, typically in late spring and early summer, snailcase bagworms migrate to high, shaded points. There they firmly attached themselves and transform to the pupal stage.

Transition to the adult moth takes place in the pupal covering over a couple of weeks. The moths are wingless, nearly legless, and do not feed. Females only are produced, but they can fertilize eggs asexually. About one to dozen eggs are produced by the female and during midsummer these eggs hatch. The young larvae remain within the pupal covering throughout the winter until becoming active the following spring.

Management: Controls for snailcase bagworm have not been developed. Developing larvae on plants should be able to be controlled insecticides such as Orthene and pyrethroids (e.g., Tempo, Talstar, Astro), which are known to be effective against other types of bagworms.

However, full-grown migrating larvae can not be controlled with insecticides. To prevent the nuisance of migrating caterpillars attaching themselves to the sides of homes, temporary barriers (flanges, sticky tapes) may be feasible. Snailcase bagworms also be dislodged with a vigorous jet of water (preferably soapy water), *before* they have attached themselves to the surface in preparation to pupation. After attachment, these insects are not readily removed.

TOP: Casebearer.
MIDDLE: Western tent caterpillar colony on mountain mahogany.
BOTTOM: A tent caterpillar, *Malacosoma incurvatum discoloratum*, larva.

CASEBEARERS

Coleophora spp.
Lepidoptera: Coleophoridae

Hosts: Apple, pear, hawthorn, plum, alder, birch, cherry and elm are reported hosts.

Damage and Diagnosis: Caterpillars of these unusual and fairly uncommon moths produce a cylindrical cigar-shaped tube of leaf fragments that it carries for much of its life. Leaf feeding damage is minor.

Life History and Habits: The following is based on the biology of the birch casebearer, also known as the cigar casebearer, *Coleophora serratella* (L.). Adults are minute moths that are active in early summer and lay eggs on the underside of host plant leaves. The newly hatched caterpillars feed briefly along larger veins then enter the leave and feed for a brief period as a leafminer. The larger larvae then emerge and construct the characteristic case that is made from fragments of the mined leaves. Winter is spent as a developing larva, firmly attached to bark or nearby surfaces. They resume feeding in spring and can cause most damage at this time, tunneling buds and chewing on young leaves. Pupation occurs in late spring. There is one generation produced per year.

WESTERN TENT CATERPILLAR

Malacosoma californicum (Packard)
Lepidoptera: Lasiocampidae

Hosts: Mountain-mahogany, wax currant, aspen, plum and other fruit trees, willow and some other hardwoods

Damage and Diagnosis: Western tent caterpillar is an early season defoliator of many deciduous trees and shrubs and is capable of completely defoliating plants. Outbreaks lasting two or more years that extensively defoliate plants can seriously weaken plants. However, high populations almost always within a year or so due to effects of natural controls.

Western tent caterpillars develop as colonies that produce a dense mat of silk in branch crotches. It is only one of the tent caterpillars (*Malacosoma* species) found in the region. The caterpillars can be distinguished from other species by being slightly hairy, having a typically light brown general coloration with powdery blue markings along the sides and a blue head.

Life History and Habits: The western tent caterpillars overwinter as young larvae (first instar) inside eggs. These eggs are laid in masses, glued around twigs and are chocolate brown. Larvae emerge from the egg masses in spring, sometimes by late may and immediately begin construction of a communal "tent" of densely spun silk in branch crotches. This tent expands as the caterpillars develop and is used primarily as a resting area during the day, for shelter, and for molting. For much of their life the caterpillars feed primarily at night, foraging over the plants and returning to the protective tent daily. Larval development typically takes about a month and a half, but can vary widely depending on spring temperatures. During cool, wet seasons caterpillars may be present into July, particularly at upper elevations.

Last instar larvae tend to disperse and when full grown most leave the plant for pupation. They produce a white or pale yellow cocoon within which they pupate. Adults emerge about two weeks later. The moths are generally light brown with forewings divided into three bands by two white lines and a wingspan of about one inch. Females glue their eggs in masses to live twigs that are less than 3/4 inch in diameter. Eggs develop for three to four weeks but do not hatch; young larvae overwinter in the eggs until spring. There is one generation per year.

Related Species: The species *M. incurvatum discoloratum* (Neumogen) can be an important pest in parts of the West Slope, feeding on poplars and cottonwood along riverways. Eggs hatch very early, late March or early April, and feeding is completed by mid-May. There is one generation per year.

The eastern tent caterpillar, *Malacosoma americanum* (Fabricius), is an uncommon species within the region, confined to areas of northeastern Colorado and immediately adjacent areas. It is primarily associated with fruit trees and related species (e.g., hawthorn). This is the most common tent caterpillar in the midwestern United States.

Management: Western tent caterpillars are subject to a great many natural controls. They are heavily parasitized by tachinid flies and parasitic wasps. Outbreaks of viral disease also are commonly devastating when high populations develop. They are also susceptible to fungal diseases such as *Beauveria bassiana*, particularly during wet springs. Birds will feed on some larvae. Together these natural controls keep western tent caterpillar under a high level of natural control.

Additional controls are rarely warranted. Western tent caterpillars are very susceptible to most insecticides including those containing *Bacillus thuringiensis*, spinosad, neem, carbaryl, chlorpyrifos and all pyrethroids. Dormant season sprays of horticultural oil can kill many of the developing larvae within eggs. On small trees they are perhaps best controlled by physically removing the silk tent containing the resting larvae during the day.

SONORAN TENT CATERPILLAR

Malacosoma tigris (Dyar)
Lepidoptera: Lasiocampidae

Hosts: Oak, especially Gambel oak

Damage and Diagnosis: Areas of defoliation surround small silken tents on infested trees. Damaging outbreaks are uncommon. The caterpillars of this tent caterpillar species are marked above with longitudinal black and orange stripes and crossed on each body segment by a blue line.

Life History and Habits: Larvae emerge in early spring when the first leaves appear. They feed in groups and together construct a small tent which is used for protection while molting. Pupation occurs in the leaf litter within chalky white cocoons. Adults appear in midsummer and lay masses of 150 to 200 eggs in spiral bands around slender twigs. Within two weeks the larvae are fully developed inside their eggs, but do not emerge until the following spring.

Management: Sonoran tent caterpillars are readily controlled by several insecticides, including *Bacillus thuringiensis* products. For specific recommendations see section in control supplement on **Leaf feeding caterpillars**.

FOREST TENT CATERPILLAR

Malacosoma disstria (Hubner)
Lepidoptera: Lasiocampidae

Hosts: Ash, aspen, various fruit trees, poplars, willow, are among the many deciduous trees that this insect may feed on.

Damage and Diagnosis: Caterpillars chew on leaves of many trees during spring. Although serious defoliation is rare, outbreaks in stands of native aspen have historically occurred.

Although this is the most common of the tent caterpillars (*Malacosoma* spp.) it does not may a permanent silken tent in branch crotches. Instead colonies will make several resting mats of lightly spun silk during a season. Forest tent caterpillars are also

TOP: Eastern tent caterpillar, larva.
CENTER: Sonoran tent caterpillar, *Malacosoma tigris*, larva.
BOTTOM: Forest tent caterpillar.

TOP: Forest tent caterpillar, mass of larvae on trunk.
BOTTOM: Forest tent caterpillar, egg mass and newly hatched larvae.

distinctive by having an unusual electric blue color and a series of yellow "keyhole" patterns along their back.

Life History and Habits: Winter is spent within light gray egg masses glued to twigs. Caterpillars emerge from the eggs in mid-Spring and feed on the young leaves. Throughout their early development forest tent caterpillars feed gregariously, also resting during the day in groups on a light silken pad spun on trunks or large branches. Such resting sites may be used for a few consecutive days but the colonies move about the tree and produce several such resting mats as they develop. Caterpillars nearing maturity disperse and when full-grown most wander away from the host tree. They pupate within a white cocoon and adults emerge in early summer. Adults are moderate sized light brown moths with a wavy light markings on the wings. After mating the females lay the eggs that survive winter.

Management: Populations of forest tent caterpillar are usually well suppressed by numerous natural enemies including tachinind flies, parasitic wasps, spiders, birds and other general predators. Supplemental controls are rarely warranted. Treatments in the supplement for control of **Leaf feeding caterpillars** should be effective for this insect.

Tent-making Caterpillars and Sawflies of Colorado

Several tree and shrub defoliators, particularly various caterpillars, feed together in groups and produce tents or other structures out of webbing. These tents can be very characteristic, and are useful for diagnosing the pest species. Some of the more common tent-making species are listed below.

INSECT	HOSTS	DESCRIPTION OF TENT
Fall webworm	Cottonwoods, poplars, fruit trees, etc.	Caterpillars construct a large, fairly loose tent that encloses much of the foliage the caterpillars feed on. Tents may be produced from late June through early September.
Sonoran tent caterpillar	Oak	Tents are constructed of dense silk in the crotches of larger branches. Tents are formed in spring and abandoned in early summer.
Western tent caterpillar	Aspen, fruit trees, mountain-mahogany, willow, etc.	Tents are constructed of dense silk in the crotches of larger branches. Tents are ormed in spring and abandoned early summer.
Eastern tent caterpillar	Hawthorn, fruit trees, willow	Tents are constructed of dense silk in the crotches of larger branches. Tents are formed in spring and abandoned in early summer.
Tiger moth (*Lophocampa ingens*)	Pines, juniper, Douglas-fir, fir	Large tents are constructed in the tops of trees and include the leader. Caterpillar activity and tent building occurs during winter and early spring.
Tiger moth (*Lophocampa argentata*)	Pinyon, juniper	Large tents are constructed in the tops of trees and include the leader. Caterpillar activity and tent building occurs during winter and early spring.
Juniper webworm	Juniper	Individual larvae web together needles, but during outbreaks small "nests" may be formed when several caterpillars feed in close proximity. Webbing is produced during late summer and the caterpillars continue to feed in spring.
Uglynest caterpillar	Cherry, various other shrubs	Nests are tightly constructed forming a mat. Dark fecal pellets are mixed with the silk. Nests are formed during late spring and early summer.
Rabbitbrush webbing moth	Rabbitbrush	Nests are constructed of a tight mat of webbing amongst the smaller twigs and branches during late spring and early summer.
Pine webworm (*Tetralopha sp.*)	Ponderosa pine	Larvae feed in groups and make a fairly tight tent in early summer. Old head capsules, cast skins and pellets remain in place. General appearance is similar to that of the web-spinning sawflies.
Web-spinning sawflies (*Acantholyda* spp., *Cephalcia* spp.)	Pines, spruce, plum	Silk is mixed with old skins and the pellet droppings the larvae produce. "Nests" are produced during late spring and early summer. Nests often persist.

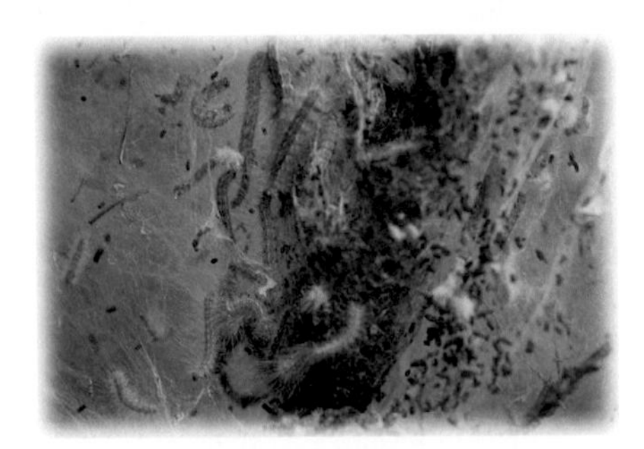

TOP: Tiger moth tent on ponderosa pine terminal.
MIDDLE: Tiger moth.
BOTTOM: Fall webworm colony.

TIGER MOTH

Lophocampa (=Halisidota) ingens (Edwards)
Lepidoptera: Arctiidae

Hosts: Pinyon, ponderosa, bristlecone and lodgepole pine. A related species (*Lophocampa argentata subalpina*) feeds on pinyon and juniper.

Damage and Diagnosis: Larvae feed on older needles, sometimes defoliating trees in early spring. The larvae feed as a group and produce a large tent enclosing the upper terminal growth.

Defoliation historically has been most common in the Black Forest region and in West Slope pinyon-juniper stands. Adults are buff-colored moths with a wingspan of about three inches, forewings reddish-brown with large silvery ovals, and hind wings off-white.

The larvae are black caterpillars covered with gold and black tufts of hairs. Mature larvae reach one and one half inches. They are the only caterpillars actively feeding on pine through winter and early spring. Adults are buff-colored moths with a wingspan of about three inches, forewings dark brown with large white ovals, and hind wings white.

Life History and Habits: Tiger moths spend the winter as caterpillars in groups within silken tents. The tents are constructed by binding needles together, usually in the top of the tree. The larvae continue to feed and develop throughout winter when warm weather permits and become full-grown in late spring. They pupate in early summer and adults emerge in July and August to mate and lay eggs. Newly hatched larvae feed for a short time, then build silken nests in which they overwinter. There is one generation produced per year.

Management: Tiger moth outbreaks are almost always restricted to native stands rather than landscape plantings. Control is rarely warranted and numerous biological controls, particularly parasitic wasps, typically cause outbreaks to collapse within one or two years. Specific controls are listed under **Leaf feeding caterpillars** in the supplement.

FALL WEBWORM

Hyphantria cunea (Drury)
Lepidoptera: Arctiidae

Hosts: Cottonwood and chokecherry are the most common hosts, but over 100 species of hardwood trees may be eaten.

Damage and Diagnosis: The larvae feed on leaves and build unsightly silken tents. Heavy infestations can defoliate trees. Wandering larvae are sometimes a serious nuisance. Fall webworm is the most common tent-making caterpillar in Colorado with outbreaks particularly common along parts of the West Slope and Platte River Valley.

Color of the caterpillars is pale but highly variable, yellow to brown, sometimes greenish. The body is covered with tufts of long, silky, gray hairs arising from black and orange tubercles. Caterpillars are about one inch long when mature. The adults are attractive satiny-white moths sometimes with brown or black spots. The wingspan is about one and one half inches.

Life History and Habits: The fall webworm spends the winter as a pupa within a light-colored cocoon on the ground or under bark. In late spring and early summer, adults emerge and egg laying takes place. Eggs are laid in masses on the leaves. The newly hatched larvae feed together and spin lightly woven webbed tents that enshroud the leaves on which they are feeding. As the caterpillars grow, they continue to build tents which may reach several feet in diameter by late summer. The larvae mature in the late summer and early fall. They wander from the plant and search for a protected location to pupate.

One generation per year is normal for this species. However, egg laying occurs over an extended period of several weeks with tents being initiated from late June through July. A second generation may occur in warmer areas of the state.

Management: Although an abundant and highly conspicuous defoliator, fall webworm damage tends to occur late enough in the season to cause little effect on plants. Only in areas where repeated, heavy defoliation occurs, or where there is a severe nuisance problem associated with wandering caterpillars, is control warranted.

Fall webworm is attacked by a great many natural enemies including tachinid flies, parasitic wasps and various hunting wasps. Sustained outbreaks may be related to excessive use of area-wide spraying for mosquitoes, which is detrimental to natural enemies of the fall webworm.

Fall webworm is highly susceptible to many insecticides, particularly pyrethroids. Specific treatments are outlined in the control supplement under the section on **Fall Webworm**. However, control is made difficult due to the tents and to the large size of the trees that typically are attacked, which produce problems with providing sufficient coverage. If trees are to be sprayed, treatments are best applied when tents are small, particularly if an insecticide is used that must be eaten, such as *Bacillus thuringiensis*. The addition of a soap or some other surface-active material which can allow better wetting of the tent silk should improve control.

Tents of the fall webworm can be pulled out and destroyed. However, burning the tents and caterpillars while on the tree can easily do more damage than is caused by the insect.

TOP: Fall webworm, *Hyphantria cunea*, adult.
CENTER: Douglas-fir tussock moth larva.
BOTTOM: Defoliation of blue spruce by Douglas-fir tussock moth.

DOUGLAS-FIR TUSSOCK MOTH

Orgyia pseudotsugata (McDunnough)
Lepidoptera: Lymantriidae

Hosts: Colorado blue spruce has been most commonly damaged, but Engelmann spruce, Douglas-fir and white (concolor) fir are less common hosts.

Damage and Diagnosis: Larvae feed on the needles, typically defoliating the tree in a top-down pattern. Defoliation is rapid and tops may be killed, sometimes after only a single season of severe injury. Following repeated attacks over several seasons whole trees may die or succumb to attack by bark beetles. Douglas-fir tussock moth is one of the most serious pests of landscape trees along much of the Front Range. Occasionally, it occurs as a forest pest in Colorado, most recently in the Pike National Forest, in an area south of Dekkers.

The larvae range from 3/4 to one inch long when mature. They are generally gray with distinctive tufts of hairs along the front of the back. The hairs of the caterpillars may cause skin irritations, particularly following repeated exposure.

The adult males are moths with rusty-colored forewings and gray-brown hind wings, with a wing-span of about one inch. Females are thick-bodied and wingless, found in association with their pupal case.

Life History and Habits: Douglas-fir tussock moth spends the winter as eggs, laid within a mass covered with the hairs of the female. Eggs hatch in the spring, often in late May. The small, hairy caterpillars migrate to the new growth. There also tends to be a migration to the top of the tree and many newly emerged larvae may subsequently be dispersed by wind. Since the adult female moths do not fly, wind-blown movement of young larvae is an important means of initiating new infestations.

The caterpillars first feed solely on the newer foliage, and partially eaten needles may wilt and turn brown. Later, the older caterpillars will move to older needles as the more tender needles are eaten. During feeding, particularly when disturbed, larvae may

TOP: Tussock moth, *Dasychira vagans*, larva on willow.
CENTER: American dagger moth caterpillar.
BOTTOM: Leaf petiole clipping typical of dagger moth.

drop from branch to branch on long silken strands. By mid-July or August, the larvae become full grown and many may migrate away from the infested tree. They pupate in brownish spindle-shaped cocoons in the vicinity of the infested trees.

The adults emerge from late July through mid-August. The males are winged but the females have only minute, non-functional wings. Mating of the females occur in the immediate vicinity of the pupal case and they then lay characteristic masses of eggs covered with grayish hairs. There is one generation per year.

Related Species: Several other species of tussock moths can be found although they are not considered economically important. *Dasychira grisefacta* (Dyar) is occasionally abundant in forests, where it feeds on a wide range of conifers including pinyon, ponderosa pine, Douglas-fir, and spruce. A related species, *Dasychira vagans* (Barnes and McDunnough), feeds on birch, aspen, poplar, and willow, reportedly preferring the latter. The whitemarked tussock moth, *Orygia leucostigma* (J.E. Smith), also is occasionally found in the region. This is an important pest species in the eastern United States and feeds on an extremely wide range of shrubs, fruit trees, shade trees and even conifers.

Management: Several natural controls affect Douglas-fir tussock moth populations. At least seven species of parasitic wasps and a tachinid fly have been identified as parasites that are locally present. Caterpillars may be killed by general predators, notably spiders. A nuclear polyhedrosis virus disease, known as the "wilt disease", also can be an important mortality agent during outbreaks. Perhaps most important are extreme winter and freezing temperatures during spring that kill eggs and newly emerged larvae.

Chemical controls can be effective but need to be applied thoroughly to the top of the tree. In addition, younger larvae are much more effectively controlled than older larvae, so treatment timing is best shortly after eggs have hatched. Pyrethroids (e.g., Astro, Tempo, Talstar, Scimitar) and pyrethrins (Pyrenone) are particularly effective against Douglas-fir tussock moth; carbaryl (Sevin, Sevimol) and Orthene are also effective. The addition of insecticidal soap or some other surfactant may assist in control of these hairy caterpillars. Young caterpillars are susceptible to *Bacillus thuringiensis* products (Thuricide, Dipel, etc.), but control during outbreaks has been erratic, perhaps due to difficulties in providing adequate coverage of the new growth on which the caterpillars primarily feed. Older caterpillars are somewhat less sensitive to Bt-products than are young larvae.

Outbreaks of Douglas-fir tussock moth are cyclical. Some estimate of outbreak potential can be made by surveying the site for the presence of egg masses in winter and early spring. Where egg masses can be easily found in the vicinity of the planting, a high risk exists for subsequent injury.

AMERICAN DAGGER MOTH

Acronicta americana (Harris)
Lepidoptera: Noctuidae

Hosts: Primarily silver maple but occasionally other maples including boxelder

Damage and Diagnosis: Larvae feed on foliage and have the unique habit of clipping off partially eaten leaves in order to avoid detection by predatory birds. Leaves on the lawn with portions of the petioles attached are characteristic and indicate dagger moth feeding overhead. The larvae somewhat resemble yellow-haired "woolly bears" and may be found wandering in the vicinity of infested trees.

Life History and Habits: The adult stage is a moderately large (wingspan of about two inches) brown moth named after a dagger-like marking on the forewing. After emerging from the overwintering pupa, adult moths are present through much of spring and lay eggs on the newly emerged foliage of maple. The eggs hatch in late spring and larvae feed throughout the summer. In early fall they crawl down the tree trunk in search of pupation sites and may be encountered dozens of yards from the host tree at this time. Finding a suitable sheltered spot they spin a dense silken cocoon and pupate within it. There is one generation per year.

Related Species: Other dagger moths can be found in the region, including the poplar dagger moth (*A. leporina* L.), that feeds on poplar, willow and birch. Habits and general appearance of the caterpillars are generally similar.

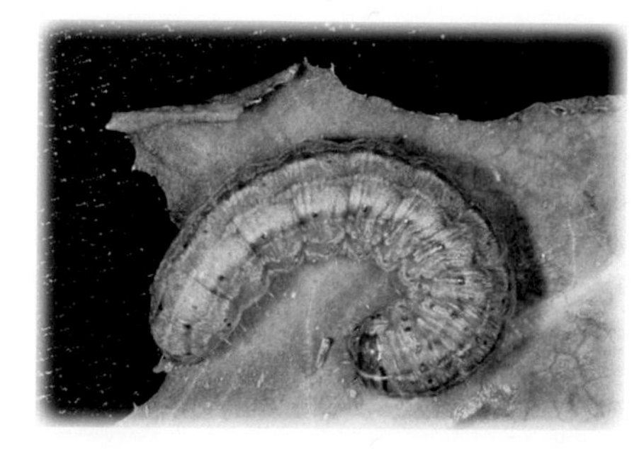

TOP: Army cutworm larva. (J. Capinera)

ARMY CUTWORM

Euxoa auxiliaris Grote

Lepidoptera: Noctuidae

Host Plants: Army cutworm feeds on a very wide range of plants. Within the region it is most damaging to alfalfa and winter wheat. It is a serious pest of many seedling plants including those grown in nursery beds; juniper has been most badly damaged in northern Colorado. Army cutworm has also been observed damaging buds of grape.

The adult stage is commonly known as the 'miller moth' and is a serious nuisance pest at times east of the Rockies. The moths feed on various flowers, concentrating on plants such as *Prunus*, lilac, spirea, cotoneaster, and Russianolive. Spruce and dense growing junipers, pines and cotoneaster are favored day time resting sites for 'miller moths' during their annual migrations.

Damage and Diagnosis: Army cutworm is the most common 'cutworm' in the High Plains Rocky Mountain region. The gray-brown larvae feed at night, most often around the soil surface. Young tender plants may be cut or girdled and killed. Larvae are also fair climbers and may destroy buds. Peak activity is during mid to late spring.

Adults of this insect are known as 'miller moths'. The miller moth can be an extremely annoying pest as it gets into buildings and vehicles during its late spring migrations to the mountains.

Life History and Habits: Eggs are laid amongst lush vegetation during September and October at lower elevation/plains areas. Eggs hatch and there is some feeding and development during fall, the amount depending on temperature. In spring, from mid-April through May, feeding accelerates and the caterpillars do plant damage at this time. Pupation occurs in the soil.

The adults that emerge have a peculiar habit among insects. Rather than laying eggs for a new generation they spend the summer in a state of semidevelopment. No eggs are matured, but instead the moths seek sources of carbohydrates and cool moist conditions for a period of increasing fat reserves. This causes them to migrate to high elevations and can involve migrations of hundreds of miles. All summer is spent at these locations where they feed on mountain flowers and rest under rocks or other cover. In early fall there is a reverse migration to lower elevations at which time eggs are laid.

Management: Sites that have lush vegetation, are irrigated, or mulched are attractive to egg laying females in early fall. Fall tillage can kill the overwintering caterpillars but if this is not possible then these sites should be monitored closely for early signs of injury in spring. High numbers of moths in fall is a risk factor for increased damage the subsequent season.

Pyrethroids (Astro, Talstar), chlorpyrifos (Dursban), and spinosad (Conserve) are among the materials effective against army cutworm that are registered for use in nurseries.

TOP: Speckled green fruitworm. (F. Peairs)
CENTER: Oak looper.
BOTTOM: Linden looper.

SPECKLED GREEN FRUITWORM

Orthosia hibisci (Guenee)
Lepidoptera: Noctuidae

Hosts: Several dozen trees and shrubs may be fed on by speckled green fruitworm. Apple, oak, and willow are among the more common hosts. Caterpillars sometimes damage rose buds and apple fruit in late spring.

Damage and Diagnosis: Speckled green fruitworm is a lime-green caterpillars, with scattered white spotting. When full grown they reach a size of about 1 and 1/2 inches. A type of "climbing cutworm" they primarily feed on leaves but are never abundant enough to be a significant defoliator. It has been most conspicuous when it occurs in outbreaks of mixed populations with other defoliators in gambel oak.

As a "fruitworm" the caterpillars also feed on flower buds and developing fruit. Damaged fruit may drop from the tree or overgrow the damaged areas and produce distorted 'catfaced' fruit.

Related and Associated Species: Caterpillars of several other species of "climbing cutworms" occasionally are associated with trees during mid to early spring. Among these are the humped green fruitworm, *Amphipyra pyramidoides* Guenee, and *Lithophane georgii* Grote. Biology of these differ somewhat from the speckled green fruitworm.

Life History and Habits: Speckled green fruitworm spends the winter as a pupa. Moths emerge very early in spring to mate and lay eggs. The developing larvae are a type of 'climbing cutworm' but remain permanently in the tree to feed. When full-grown, usually by late June, they wander from the food plant and pupate in the soil. There is one generation produced per year.

CANKERWORMS AND LOOPERS

SPRING CANKERWORM: *Paleacrita vernata* (Peck)
FALL CANKERWORM: *Alsophila pometaria* (Harris)
LINDEN LOOPER: *Erannia tiliaria* (Harris)
OAK LOOPER: *Lambdina punctata* (Hulst)
BARBERRY LOOPER: *Coryphista meadii* (Packard)
CURRANT SPANWORM: *Itame ribearia* (Fitch)
Lepidoptera: Geometridae

Hosts: Elm, honeylocust, crabapple, linden, and oak are among the more common deciduous hosts of these insects within the region. However, some species specialize in oak (oak looper), currant (currant spanworm), and mahonia (barberry looper).

Damage and Diagnosis: Cankerworms and loopers tend to be early season defoliators of a variety of deciduous trees. Within the region they are relatively rare and damage is insignificant; they can be serious pests in the midwest. The larvae are elongate caterpillars and characterized with having only two pair of prolegs, resulting in a unique "inchworm" form of locomotion.

"Measuring worms" (referring to the caterpillar behavior) and "geometers" are two other common names applied to these insects.

Life History and Habits: All of the various cankerworms and loopers associated with woody plants have a single generation and caterpillars are most active in late spring or early summer. However, life history can vary in some details:

Management: Cankerworms and loopers rarely require control as outbreaks are very sporadic. Insecticides useful for control of **Leaf feeding caterpillars** can be highly effective.

SPECIES	OVERWINTERING STAGE
Spring Cankerworm	Full grown larva in soil, emerging in spring. Adult females are wingless.
Fall Cankerworm	Usually as eggs are laid in fall. However, eastern Colorado populations often have a life cycle similar to the spring cankerworm, with spring emergence and egg laying. Adult females are wingless.
Linden Looper	Eggs laid as clusters under loose bark.
Oak Looper	Eggs laid on trunks and amongst protective debris around their Gambel oak host.
Barberry Looper	Pupae in soil, emerging as adults to mate and lay eggs in spring.
Currant Spanworm	Eggs laid around buds and on twigs in fall.

WESTERN GRAPELEAF SKELETONIZER

Harrisina brillans Browne & McDonnough
Lepidoptera: Zygaenidae

Damage and Diagnosis: A native insect originally associated with wild grapes, caterpillars of the western grapeleaf skeletonizer also feed on the leaves of cultivated grape, Virginia creeper and related ivies. Severe defoliation, though rare, can reduce subsequent yields and open canopies so that sunburning of fruit may occur. The brightly colored caterpillars attract attention and some people become sensitized to the hairs of the insect. Western grapeleaf skeletonizer is found in the southern areas of the region.

ABOVE: Western grapeleaf skeletonizer.

Life History and Habits: Winter is spent in the pupal stage, in a loose cocoon often under flaps of bark or other protective cover. Adults emerge in spring a few weeks after the first flush of growth and females lay eggs in masses on older, more shaded leaves. The newly emerge caterpillars feed gregariously, as a group, for first couple of weeks. Older larvae disperse throughout the plant. Larvae of this first generation pupate in early summer and a second generation is produced later in the season, feeding in late August and September.

Management: Several parasitic wasps and a tachinid fly are native natural enemies of the western grapeleaf skeletonizer. Also a virus disease is thought to be important in regulating populations in wild grape. Caterpillars are susceptible to *Bacillus thuringiensis.*

Common Sphinx Moths (Hornworms) of Colorado

COMMON NAME	SCIENTIFIC NAME	HOST
Elm sphinx	*Ceratomia amyntor (Geyer)*	*Elm*
Great ash sphinx	*Sphinx chersis (Hubner)*	*Ash, lilac, privet*
Wildcherry sphinx	*Sphinx drupiferarum (J.E. Smith)*	*Plum, cherry, etc.*
(no common name)	*Paonias myops (J.E. Smith)*	*Cherries, serviceberry*
Giant poplar sphinx	*Pachysphinx modesta (Harris)*	*Poplars,willow*
Columbia Basin sphinx	*Pachysphinx occidentalis (H. Edwards)*	*Poplars, willow*
Achemon sphinx	*Eumorpha achemon (Drury)*	*Grape, virginia creeper*
Common clearwing sphinx	*Hemaris thysbe (F.)*	*Honeysuckle, viburnum, hawthorn, snowberry, cherry, plum*
(no common name)	*Hyles gallaii (Rottenburg)*	*Willow weed, woodruff, bedstraw*
Whitelined sphinx	*Hyles lineata (F.)*	*Portulaca, apple, primrose, four o'clock, peony, others*
Twinspot sphinx	*Smerinthus jamaicensis (Drury)*	*Poplar, birch, elm, willow, ash, apple*
Tomato hornworm	*Manduca quinquemaculata (Haworth)*	*Potato, tomato, tobacco family plants*
Tobacco hornworm	*Manduca sexta (L.)*	*Potato, tomato, tobacco family plants*

ACHEMON SPHINX

Eumorpha achemon (Drury)
Lepidoptera: Sphingidae

Hosts: Virginia creeper and grape

Damage and Diagnosis: Larvae feed on foliage of Virginia creeper and grape but damage is late in the season and minor. However the purplish-brown larvae, a "hornless hornworm" often attract attention when they are observed in late summer and early fall as they migrate from plants.

Life History and Habits: Achemon sphinx winter as pupa, within small earthen cells constructed a few inches below ground in the general vicinity of previously infested plantings. Adults may emerge in June, but are most common during July at which time eggs are laid on host plants. The newly hatched larvae are green with very long horn; however, the horn is lost at the first molt, leaving just an "eyespot" marking. the larvae continue to grow through summer, becoming full grown in late August or September. They then leave the plants, typically wandering several yards away, to find a place where they may dig and pupate. There is one generation a year.

Related Species: the whitelined sphinx, *Hyles lineata* (F.), occasionally feeds on grape, as well as a wide range of herbaceous plants, notably primrose. The caterpillars are boldly patterened hornworms that may range from green to black with yellow markings. The adult is the most common "hummingbird moth" found in the western US.

SPINY ELM CATERPILLAR (Mourning Cloak Butterfly)

Nymphalis antiopa (L.)
Lepidoptera: Nymphalidae

Hosts: Aspen, willow, elm, hackberry, cottonwood and poplars

Damage and Diagnosis: Spiny elm caterpillars feed on foliage and can cause localized defoliation of branches. They also attract attention because they feed in groups and

TOP: Achemon sphinx.
BOTTOM: Mourning cloak/spiny elm caterpillar larva.

Tomostethus multicinctus (Rohwer)	Brownheaded ash sawfly*	Green ash, white ash
Pikonema dimmockii (Cresson)	Greenheaded spruce sawfly	Spruce
Nematus ribesii (Scopoli)	Imported currantworm*	Currants, gooseberries
Nematus ventralis Say	Willow sawfly	Willow, poplar
Nematus oligospilus Foerster		Willow
Nematus fulvicrus Provancher		Willow
Nematus vancouverensis (Marlatt)		*Populus*
Nematus tibialis Newman		Blacklocust, honeylocust
Pontania proxima (Lepeltier)	Willow redgall sawfly	Willow (leaf gall)
Pontania c-pomum (Walsh)		Willow (leaf gall)
Pristophora serrula (Wong and Ross)		Chokecherry
Pristophora rufipes Lepeltier		*Ribes*
Pristophora coactula (Ruthe)		Willow
Pristophora sycophanta Walsh		Willow
Pristophora staudingeri (Ruthe)		Willow
Pristophora lena Kincaid		Spruce
Pristophora mollis (Hartig)		*Vaccinium*
Periclista albicollis (Norton)		Oak
Blennogeneris spissipes (Cresson)		Snowberry (bud gall)
Ardis brunniventris (Hartig)		Rose
Monophadnoides geniculatus (Hartig)	Raspberry sawfly	Raspberry
Messa leucostoma (Rohwer)		*Populus*
Messa populifoliella (Townsend)	Poplar leafmining sawfly	*Populus*
Rhogogaster californica (Norton)		*Populus*
Dimorphopteryx pinguis (Norton)		Birch, alder
Phyllocolpa excavata (Marlatt)		Willow
Phyllocolpa bozemani (Cooley)	Poplar leaffolding sawfly*	Poplars
Macremphytus tarsatus (Say)	Dogwood sawfly	Dogwood
Macremphytus testaceus (Norton)	Dogwood sawfly	Dogwood
Croesus lititarsus (Norton)	Dusky birch sawfly	Birch
Fenusa pusilla (Lepeletier)	Birch leafminer*	Birch
Fenusa dohrnii (Tischbein)	European alder leafminer	Alder
Fenusa ulmi Sundevall	Elm leafminer	Elm
Euura bebbianae (Rohwer)		Willow (stem gall)
Euura brachycarpae Rowher		Willow (stem gall)
Euura s. ovulum Walsh		Willow (stem gall)
Euura s. nodum Walsh		Willow (stem gall)

TOP: Brownheaded ash sawfly, *Tomostethus multicinctus*, adult.
SECOND: Brownheaded ash sawfly, dislodged larvae massed at base of tree.
THIRD: Brownheaded ash sawfly, egg punctures in leaf.
BOTTOM: Brownheaded ash sawfly, young larvae and shothole wounds.

Euura orbitalis Norton		Willow (stem gall)
Euura perdita Rohwer		Willow (stem gall)
Family Diprionidae		
Neodiprion edulicolis Ross	Pinyon sawfly*	Pinyon, bristlecone pine
Neodiprion gillettei (Rohwer)*		Ponderosa pine
Neodiprion autumnalis (Say)*		Ponderosa pine
Neodiprion fulviceps Cresson)*		Ponderosa pine
Neodipron ventralis Ross*		Ponderosa pine
Zadiprion townsendi (Cockerell)	Bull pine sawfly*	Ponderosa pine
Zadiprion rohweri (Middleton)		Ponderosa pine
Monoctenus fulvus (Norton)	Juniper sawfly*	Juniper
Family Pamphiliidae		
Acantholyda terminalis (Cresson)*		Conifers
Acantholyda verticalis (Cresson)*		Several pines
Acantholyda depressa Middlekauff*		Ponderosa pine
Acantholyda runcinata Middlekauff*		Ponderosa pine
Cephalcia provancheri (Huard)*		Spruce
Cephalcia fascipennis (Cresson)*		Spruce
Pamphilius sitkensis (Kincaid)		*Rubus*
*Neurotoma inconspicua (Norton)**	*Plum web-spinning sawfly*	*Plum*

* *Life history descriptions included in this publication.*

BROWNHEADED ASH SAWFLY

Tomostethus multicinctus (Rohwer)
Hymenoptera: Tenthredinidae

Hosts: Ash

Damage and Diagnosis: The larvae are pale green "worms" with some light banding. They feed on ash leaves and can extensively defoliate the tree early in the season. In recent years this insect has extended its range northward from the original infestations along the Arkansas River.

Life History and Habits: The brownheaded ash sawfly spends the winter as a full-grown larva within a cocoon around the base of previously infested ash trees. Pupation occurs in early spring. Adults are small, black wasps that emerge in April and sometimes can be found in swarms around the tree. Females insert eggs into young leaves, usually around the edge, resulting in a slight distortion of the leaves.

Early stage larvae feed on the interior of the leaf, producing small pinhole feeding wounds. As they get older, larvae feed extensively on the leaf, avoiding only the main veins. Larval development and feeding occurs throughout May and by early June they are full-grown. Full grown larvae shed a papery larval skin that remains attached to the leaf and they then crawl to the ground, where they form protective cocoons.

Almost all of the insects remain dormant until the next season, producing one generation per year. However, a few sometimes emerge in early summer and produce a small, non-damaging second generation.

Management: No significant biological controls have been observed to attack brownheaded ash sawfly in Colorado. One of the greatest checks on populations are late spring frosts that kill the first flush of foliage and the early active stages of the insect. Larvae may also be readily dislodged by high winds, with few of the insects successfully reestablishing. During outbreaks brownheaded ash sawfly can become so numerous that they eat all the foliage and starve.

Larvae are easily controlled with several contact insecticides. (See **Sawflies** section in supplement on controls.) In addition, many may be dislodged with a forceful jet of water. The presence of the small pinhole wounds made by the early stage larvae is useful for detecting incipient infestations.

TOP: Larvae of the imported currantworm.
BOTTOM: Imported currantworm, newly hatched larvae and shothole foliage injury.

IMPORTED CURRANTWORM (Currant Sawfly)

Nematus ribesii (Scopoli)
Hymenoptera: Tenthredinidae

Hosts: Currants, gooseberry

Damage and Diagnosis: The larvae chew the leaves of currants and gooseberries, often extensively defoliating the plant early in the season. Foliage in the interior of the bush is first damaged but all leaves may be eaten. Yield and quality of fruit can be affected by this injury.

The larvae are generally a light green-gray with numerous black spots. (Larvae that have just molted are light green.)

Associated Insects: Larva of a moth, the currant spanworm, *Itame ribearia* (Fitch), is another insect that often feeds on the leaves of currant and gooseberry. The larvae of this incest are also spotted, but are a type of inchworm with a distinctive looping walk. These are the immature stage of a small moth and the caterpillars can be controlled with *Bacillus thuringiensis*, which is not effective against sawflies.

Life History and Habits: The imported currantworm spends the winter in a cocoon in the soil around previously infested currants and gooseberries. The adults, a dark brown wasp about one third inch long, usually emerge early in spring, although wet cool weather may delay emergence by several weeks. After mating, the female lays eggs in rows inserted into the main veins of the leaf underside. The larvae hatch about seven to ten days after eggs are laid and at first chew small shotholes in the leaf interior. Later they disperse throughout the plant and feed along the leaf margins, becoming full-grown in about three weeks. Young larvae are pale green, but develop distinctive dark spots as they grow and reach a size of about 3/4 inch.

The full-grown larvae drop to the ground and form a cocoon. Some pupate and emerge in late June and July, producing a small second generation. The majority remain dormant and emerge the following year.

Management: In small plantings, the larvae can be controlled by handpicking or shaking. Careful examination of the newly emerging leaves can also identify the eggs which may be crushed. Most eggs and larvae will be found in the interior of the shrub.

At the end of the season, rake and remove all the debris away from the base of the plants. Most of the overwintering cocoons occur in this leaf litter and can be destroyed by this practice.

Chemical control options are limited on currants and gooseberries. Some formulations of malathion and pyrethrins allow use on this crop and are effective for imported currantworm control. However, be sure to check labels carefully to insure that use on the crop is specifically allowed.

Sprays of irritants, such as soaps, dilute dishwashing detergent, and wood ashes, can be effective controls against younger larvae.

POPLAR LEAFFOLDING SAWFLY

Phyllocolpa bozemani (Cooley)
Hymenoptera: Tenthredinidae

Hosts: Poplars, cottonwood, and willow

Damage and Diagnosis: During feeding the green, dark-headed larvae secrete saliva that causes distortions of the leaf margin. The leaves fold under and the larva feeds on the lower leaf surface within the leaf roll. The rolled leaves remain folded throughout the season. This injury does not seriously affect plant health but is a curiosity that often attracts attention, particularly because other insects and spiders often hide in the leafroll.

Life History and Habits: The poplar leaffolding sawfly spends the winter as a prepupa among debris at the base of trees, pupating in early spring. Adults are active in midspring and lay eggs as leaves first emerge. Wounds made during egg laying and subsequent larval feeding cause edges of leaves to fold. Full grown larvae drop to the ground and diapause. There is normally only one generation per year, but a small second generation is possible.

Management: A great many natural controls keep populations of these sawflies under damaging levels. Aphids, which frequently colonize the leaf curls started by the sawfly, interfere with them and may indirectly kill the sawfly larvae with their excreted honeydew. A great many parasites also attack poplar leaffolding sawflies.

PEARSLUG (Pear Sawfly, Cherry Slug)

Caliroa cerasi (L.)
Hymenoptera: Tenthredinidae

Hosts: Sweet and ornamental varieties of cherry, plum, hawthorn, pear and cotoneaster

Damage and Diagnosis: Larvae feed on the upper surface of leaves, producing distinctive skeletonizing wounds. Heavily damaged leaves turn brown and drop early. Pearslug is one of the most damaging defoliators of landscape plantings, particularly in northeastern Colorado.

Life History and Habits: Adults are small (1/4 inch), black, thick waisted wasps that emerge in late June or early July. Females insert their eggs singly in circular slits on the upper surface of leaves. Eggs hatch in about two weeks and larvae chew small pits in the upper surface of leaves during early development. Later, they more extensively chew the leaves but always avoid feeding on the larger veins and lower leaf surface, producing a typical skeletonizing injury pattern.

Pearslug larvae are shiny and slug-like in general appearance but can be somewhat variable in color. Many are a dark olive-green but later instars tend to be lighter and may even have orange tones. When full grown they wander off the plant and dig a shallow cell in the soil to pupate. Pupation occurs with in small cocoon encrusted with soil particles. In about two weeks, many pearslugs emerge as adults to produce a second generation. Larvae of these typically causing peak injury in early September. Populations of this second generation are usually smaller than the first generation, since many pearslugs remain dormant until the following spring. The full-grown larvae from the September generation drop to the soil, and spin a cocoon in which they spend the winter. They pupate the following spring.

TOP: Poplar leaf folded by leaffolding sawfly.
SECOND: Pearslug/pear sawfly, larvae and leaf damage.
THIRD: Pearslug, recently hatched larva and egg.
BOTTOM: Roseslug, larva and damage.

Related Species: The roseslug, *Endelomyia aethiops* (F.), commonly feeds on rose causing a similar skeletonizing feeding pattern on that plant but generally feeds on the lower leaf surface.

Management: Pearslug is readily controlled with several insecticides. (See **Pearslug** section in supplement on controls.) Wood ashes blown on the larvae will readily dry and kill them.

"ROSESLUGS"
ROSESLUG: *Endelomyia aethiops F.*
BRISTLY ROSESLUG: *Cladius difformis* (Panzer)

Hymenoptera: Tenthredinidae

Hosts: Rose and *Rubus* spp.

Damage and Diagnosis: Several sawflies feed on the leaves of rose and raspberry. Roseslug is a smooth, pale green worm that feeds on the underside of rose leaves producing characteristic "window pane" injuries the result from leaving the thin upper leaf surface leaf intact. More elongate holes tend to be cut by bristly roseslug. Under most conditions defoliation caused by these insects is minor, but occasional serious problems occur.

Life History and Habits: Rose slug sawflies overwinter within a cocoon buried shallowly in the soil in close vicinity of previously infested plants. The adult stage is a small, thick-bodied wasp that emerges in May or June. Female wasps insert eggs into the leaf and about a week later the newly emerged larvae begin to feed. Feeding usually takes place over a period of two to three weeks and the full grown larvae then drop to the ground, dig a small chamber in the soil and prepare the cocoon for pupation. The roseslug has only a single generation per year but later season defoliation by bristly roseslug to raspberry sometimes is produced by a second generation.

JUNIPER SAWFLY

Monoctenus fulvus (Norton)
Hymenoptera: Diprionidae

Hosts: Shrub junipers

Damage and Diagnosis: The larvae feed on the tips of junipers, sometimes checking the new growth and causing a thinning of the shrub. However, the larvae do not appear to feed very heavily on shrubs, even when in high populations. Larvae are generally gray-green with an orange head.

Life History and Habits: The overwintering stage is a full-grown larva, within cocoons at the base of previously infested plants. They change to the pupal stage in early spring and adults emerge and mate in late April and early May. Females insert eggs into the tips of shoots, over the course of three to five weeks.

Larvae hatch during mid to late May. They become full-grown after four to six weeks after which the full-grown larvae drop from the plant and spin a cocoon within the leaf litter and upper soil. Peak feeding typically occurs in early June, although some larvae may be found through mid-July. There is some indication that a small number complete development and produce a second generation. However most remain dormant until the following season.

TOP: Juniper sawfly. (J. Capinera)
CENTER: Mating pair of juniper sawfly. (J. Capinera)
BOTTOM: Sawfly, *Neodiprion ventralis*, larvae feeding on needles.

CONIFER SAWFLIES

Neodiprion species
Hymenoptera: Diprionidae

Hosts: Pines (particularly ponderosa pine and pinyon), rarely other pines and Douglas-fir

Damage and Diagnosis: Most species of conifer sawfly larvae are present in spring and feed on old needles prior to budbreak. (One of the most common conifer sawflies, *N. autumnalis,* feeds during midsummer. It also preferentially feeds on older needles.) During extreme outbreaks some feeding on bark of shoots may also occur. Conifer sawfly injuries can significantly weaken trees, particularly when other pest insects also occur that feed on new growth (e.g., pine budworm). Although all sizes of pines may be infested, damage is most common to small or medium-sized trees growing in open stands on the plains or in plantations.

The larvae usually medium-brown, elongate and about 3/8 inch long when mature. They have dark brown heads and shields on the thorax and posterior. Prolegs, 8 pairs, are prominent and they typically feed in groups.

Associated Species: The "Bull pine sawfly", *Zadiprion townsendi* (Cockerell) is also found on ponderosa pine. However, its habits differ from *Neodiprion* species in that its dark olive, warty larvae are present throughout the winter months.

Life History and Habits: Several species of conifer sawflies are found in the region including, *N. edulicolus* (Ross), *N. gillettei* (Rohwer), (Say), *N. ventralis* Ross and *N. fulviceps* (Cresson). Life histories vary somewhat, but all prefer to feed on older needles.

Conifer sawflies overwinter either as eggs inserted in the needles or as fully grown larvae (prepupae) within cocoons mixed in the litter at the base of previously infested plants. For the latter, pupation occurs in late winter or early spring and the adults emerge a few weeks later. Females then lay eggs that are inserted in rows in the needles.

Pine sawfly larvae typically emerge from eggs and begin to feed on needles in April and early May; *N. autumnalis* has a midsummer peak of feeding. The larvae feed gregariously, often headed outward from the twig. They finish feeding within a few weeks, drop to the soil and spin a cocoon in which they pupate. All of the conifer sawflies have a single generation per year.

Management: "Boom or bust" populations are typical with these insects. Outbreaks can cause serious damage in one season but normally do not kill trees. Several natural controls, notably parasitic wasps, typically cause outbreaks to subside after only a few generations. Larvae are readily controlled with several contact insecticides. (See **Sawflies** section in the supplement on controls.)

BULL PINE SAWFLY

Zadiprion townsendi (Cockerell)
Hymenoptera: Diprionidae

BELOW: Bull pine sawfly, *Zadiprion townsendi*, larvae.

Host: Ponderosa pine. A related species, *Zadiprion rohweri* (Middleton), is associated with pinyon.

Damage and Diagnosis: The larvae feed on older needles, occasionally causing obvious defoliation by late winter or early spring. Damage is generally insignificant, but the presence of larvae actively feeding during the cool season may attract attention. The larvae are generally green and wormlike with some golden flecking on the body. They are typically found feeding in groups and are the only sawflies feeding on ponderosa pine during winter.

Adults are heavy-bodied wasps, approximately 1/2-inch in length. They are generally brown in color with light bands on the thorax.

Life History and Habits: The bull pine sawfly is one of the few insects that remain on trees and feed throughout the winter months, weather permitting. Like most sawflies, the larvae typically feed in groups of several dozen individuals, chewing the needles and producing large amounts of frass. The larvae have the unusual habit of curling outwards from the needle and snapping back, a behavior that they may engage in continuously during some periods. The apparent purpose of this behavior is for defense against birds or parasites that are natural enemies of the insect.

By the onset of winter, the bull pine sawfly larvae are about half grown. They remain semidormant through the cold month, clustering at the base of the needles, but occasionally feed. Feeding resumes in spring and they become full grown by May or June. At this time they drop to the ground, tunnel an inch or so into the soil, and spin a silk cocoon. Some of the larvae then transform to the pupal stage, while others may not pupate until late summer or even the following season. Adults emerge a few weeks after pupation, cutting through the cocoon to emerge. The adults do not live long and mated females insert their eggs into the pine needles, concentrating egg laying on the upper areas of the tree. Eggs hatch from July through early fall and the larvae feed throughout this period. The entire life cycle may take one to two years to complete.

WEB-SPINNING SAWFLIES

Cephalcia species, *Acantholyda species*
Family: Pamphiliidae

Appearance: Adults - Small nonstinging wasps with long, slender antennae. Larvae - Resemble free-living types of sawflies but does not have any abdominal legs except on the last segment. Larvae occur within silken tubes or mats and produce a pelleted excrement somewhat resembling rodent droppings. The larvae also produce a characteristic "devil's horn" head capsule that also is shed in the webbing/frass "nest".

Hosts: Pines, spruce, plum

Damage: The larvae feed on needles late in the season. They rarely cause serious injury but gregarious species construct unusual "nests" of webbing and frass that attract attention.

ABOVE: Web-spinning sawfly in spruce.

Life History and Habits: Winter is spent in cells formed in the soil under previously infested trees. The adults typically emerge in May or early June and females insert eggs into the needles, typically in rows. The species which attract attention are gregarious, feeding in groups and constructing silk tubes along twigs at base of needles. However, some species are solitary and produce small, fairly inconspicuous silk-tubes. All of the web-spinning sawflies feed on the needles produced the previous season, although sometimes the new growth may also die back. They become full-grown in about three to four weeks, then drop to the soil and form the overwintering chamber. There is one generation per year, but they may remain dormant and not emerge for two or three years.

Associated and Related Species: A species of pine webworm (*Tetralopha* sp.) makes nests of webbing in ponderosa pine that are similar to that of the web-spinning sawflies.

The webspinning sawfly *Neurotoma inconspicua* (Norton), is found in plum.

LEAF FEEDING BEETLES

Several leaf beetles (Family: Chrysomelidae) feed on trees and shrubs. Extensive larval feeding is usually preceded by small holes cut into leaves by adults or small areas of "skeletonized" feeding injuries made by adults or young larvae. By watching trees for these indications, the need to make an insecticide treatment can be anticipated before serious damage has occurred.

In addition to the leaf beetles, a few other types of beetles are associated leaf chewing injuries to trees and shrubs. Black vine weevil (Family: Curculionidae) is becoming increasingly important in the region due not only to leaf injuries by the adult but also root injury by larvae. Periodically, masses of blister beetles (Family: Meloidae) may consume leaves.

Foliar applications of carbaryl (Sevin, Sevimol) have been the long-time standard for leaf beetle control. Leaf beetles are also highly susceptible to pyrethroid insecticides (e.g., Talstar, Tempo, Scimitar) as well as Dursban and Orthene. Recently, new strains of *Bacillus thuringiensis* (*tenebrionis* strain) have also been developed which are effective stomach poisons for control of younger larvae of most leaf beetles. Also, larvae and egg stages of some leaf beetle species are susceptible to neem-derived insecticides.

Root weevils are considerably more difficult to control since much of their life cycle is spent underground or off the leaves. Adult controls include pyrethroids and Orthene; larvae may be controlled with drenches of Turcam/Ficam, soil incorporated pyrethroids or insect parasitic nematodes.

Specific recommendations for these insects appear in the supplement on insect control recommendations. Sections in the supplement that deal with leaf feeding beetles include **Elm Leaf Beetle** and **Black Vine Weevil**.

TOP: Cottonwood leaf beetle.
BOTTOM: Cottonwood leaf beetle larvae.

COTTONWOOD LEAF BEETLE

Chrysomela scripta (F.)
Coleoptera: Chrysomelidae

Hosts: Cottonwood primarily, but other *Populus* species and willow are occasional hosts.

Damage and Diagnosis: Larvae chew on leaves, occasionally causing severe defoliation. During early stages, cottonwood leaf beetles feed in groups and tend to produce skeletonizing injuries; later larvae feed more generally on leaves. Succulent new growth is favored. Adults feed on tender twigs and also skeletonize leaves, but do much less damage than the larvae. Damage in Colorado has primarily been along the Platte River Valley in the northeastern part of the state and the Arkansas River Valley east of I-25.

Life History and Habits: The cottonwood leaf beetle is a light tan, oval beetle marked with black spots, about 3/8 inch long. They overwinter in the adult stage, scattered around previously infested trees in protected locations, such as under leaf litter or clumps of grass. Shortly after leaves emerge, the adults begin to move back to trees and feed. After a few weeks females begin to lay eggs, which are deposited in clusters, of a couple dozen or more, on the under surface of leaves. (Cottonwood leaf beetle eggs are yellow and somewhat resemble those produced by lady beetles.)

The larvae are black and grub-like, with whitish spotting appearing as they age. Young larvae feed gregariously and skeletonize the leaf. Older larvae tend to disperse and may consume all parts of leaves excepting the largest veins. They pupate attached to the leaf, the old larval skin conspicuously present at the base of the black pupa. Adults from this first generation emerge in early summer and produce a second generation.

Related Species: Related species, *Chrysomela knabi* Brown and *C. aeneicollis* (Schaeffer), feed on willow and cause similar injuries.

Management: Cottonwood leaf beetle is generally much less common than is elm leaf beetle and rarely reaches levels that would get benefit from controls. (It is much more damaging in more eastern regions.) Cottonwood leaf beetle is susceptible to the same insecticides as is elm leaf beetle; see **Elm Leaf Beetle** in the management supplement.

ELM LEAF BEETLE

Xanthogaleruca luteola (Muller)
Coleoptera: Chrysomelidae

Hosts: Elm, particularly Siberian, Rock, and English elm, but often American elm

Damage and Diagnosis: Both the adult and larval stages chew leaves. Feeding adults are yellow-green beetles that chew holes through leaves. Larvae are generally black, with some yellow striping, and feed on the underside of leaves. Injury by larvae is characteristic, "skeletonizing" the leaf by chewing between larger veins and rarely penetrating through the upper leaf surface. Damaged areas dry out and turn brown. Heavily injured leaves will abscise. This is the most important defoliator of elm and can greatly adversely affect the appearance of trees.

In some areas an important secondary "damage" is the movement of overwintering adult beetles into homes. Overwintering beetles, in diapause, are a dark olive green and rival the boxelder bug as the most important nuisance insect invader of homes in the region.

Note: Elm leaf beetle insect is **not** involved in the transmission of the Dutch elm disease fungus, which is transmitted instead by the smaller European elm bark beetle, discussed below later.

Life History and Habits: The dark olive-green adult beetles overwinter in protected areas, including nearby buildings. In late April and early May the beetles emerge and move to elm trees to feed and mate. During this time their wing color changes to a yellow-green, signaling the end of reproductive diapause.

After a period of several weeks females begin to lay masses of eggs, typically attached to veins on the lower leaves. The bright yellow eggs are laid in masses, usually along major veins and somewhat resemble those of lady beetles. Larvae hatch after about 10 to 14 days and feed for about three weeks, undergoing three larval instars. They then crawl down the tree trunk in search of pupation sites. Most pupate at the base of the tree, but some may rest in folds of bark furrows.

Adults emerge in 10 to 15 days and most then mate and reproduce to begin a second generation. However, some may enter diapause and move to winter shelter without reproducing. The number becoming dormant after the first generation is dependent on the weather, with larger numbers continuing reproduction in warmer seasons. All beetles at the end of the summer move to winter shelter where they remain semi-dormant (in diapause) until the following season.

Management: Elm leaf beetle is under several important natural controls. Extreme winter temperatures kill many overwintering beetles, particularly if they have not found good winter shelter. Also very important are late spring frosts, which kill beetles that have already emerged and moved to the trees. Earwigs and predatory stink bugs are among the more important biological controls.

The elm leaf beetle is readily controlled with several insecticides, sprayed on the foliage. Standard applications usually currently involve use of carbaryl (Sevin), neem or some pyrethroid. Effective applications for first generation beetle larvae usually suppress populations sufficiently so that they do not be retreated later in the summer. Soil systemic applications of imidacloprid (Merit) are also highly effective controls.

The larvae can also be killed as they migrate down trunks, using a band of a contact insecticide (e.g., Sevin, Dursban, Talstar, Tempo), about one foot in width. These bands should be applied when larvae first move down the trunk. If such treatments are applied area-wide damage by second generation larvae and nuisance populations of overwintering beetles can be reduced. However, if only isolated trees are treated, migration from untreated areas will allow rapid reinfestation.

TOP: Elm leaf beetle adult, larvae, and hatched egg mass.
SECOND: Elm leaf beetle, adult feeding injury.
THIRD: Rabbitbrush beetle.
BOTTOM: *Trirhabda nitidicollis*, larva on rabbitbrush.

RABBITBRUSH BEETLE

Trirhabda nitidicollis LeConte
Coleoptera: Chrysomelidae

Hosts: Rubber rabbitbrush

Damage and Diagnosis: Adults and larvae both chew foliage, although larval feeding is much more significant. Rarely outbreaks occur that completely defoliate plants and occasionally kill rabbitbrush. Larvae also sometimes attract interest being conspicuously metallic blue-green grubs. Adults are yellow/black striped beetles.

Because the plants on which it feeds are sometimes considered rangeland weeds, the insect has been considered for biological weed control.

Life History and Habits: The rabbitbrush beetle spends the winter as eggs in small masses under ground. Eggs begin to hatch in April and appear stimulated to hatch by spring rains. The larvae crawl up the plant to feed on the new foliage.

Larvae are present primarily in May and June, with emergence often scattered over a period of several weeks resulting in a range of larval stages. Pupation occurs in the soil and adults are present in June and July. The adult beetles feed on the leaves at this time but cause little defoliation. Eggs are laid from July through September.

Related Species: Several other species of *Trirhabda* occur in the Rocky Mountain region, with sagebrush being among the more common additional host plants. *T. lewisii* Crotch also feeds on rabbitbrush and often is as common as *T. nitidicollis*.

Management: Several natural controls have been observed to feed on larvae, including lady beetles and predatory stink bugs. Applied controls include sprays of insecticides effective against other leaf beetles, such as carbaryl (Sevin) and pyrethroids.

ABOVE: Alder flea beetle larvae.

FLEA BEETLES

Coleoptera: Chrysomelidae (Alticinae)

Hosts: Several flea beetles are associated with trees and shrubs including alder, cottonwood, crabapple, poplar, sumac and willow.

Damage and Diagnosis: Adult flea beetles feed on foliage, producing small "shothole" wounds in leaves. Larvae of most species feed on roots, causing little if any significant injury. Other flea beetles develop on leaves, feeding in groups as defoliators in a manner similar to other leaf beetles.

Flea beetles tend to be small beetles, typically less than 1/4 inch long. Adults of *Altica* and *Haltica* species are usually shiny or even metallic, ranging from dark violet to green. *Disonycha* species, associated with willow, and the sumac flea beetle (*Blepharida rhois*) are considerably larger conspicuously striped. Adults of all flea beetles have an enlarged hind femur and can jump.

Larvae of species that feed on foliage are generally similar in form to other leaf beetles, such as the elm leaf beetle or cottonwood leaf beetle. They are dark brown or black, with lighter coloration on the underside.

Life History and Habits: Flea beetles spend the winter in the adult stage, under protective cover. They return to the plants in late spring and feed on leaves.

With species that develop on foliage, females lay clusters of eggs on the lower surface of leaves or stems. Larvae feed in groups when young, then disperse through the canopy as they get older. Other flea beetles lay eggs in soil cracks and larvae develop on roots. Most species appear to have two generations/year.

Some Common Species of Flea Beetles Associated with Trees and Shrubs

SCIENTIFIC NAME	COMMON NAMES	HOSTS
Blepharida rhois	Sumac flea beetle	Sumac, currant
Haltica foliaceae	Apple flea beetle	Crabapple, apple, grape, evening primrose are common hosts.
Altica ambiens	Alder flea beetle	Alder
Altica populi		*Populus*
Disonycha spp.		Willow

ABOVE: Ashgray blister beetle. (J. Capinera)

ASHGRAY BLISTER BEETLE

Epicauta fabricii LeConte
Coleoptera: Meloidae

Hosts: Honeylocust, black locust, and many other legumes, including alfalfa. Occasionally other deciduous species are attacked, such as hackberry.

Damage and Diagnosis: The adult beetles will feed on the foliage of honeylocust and other legumes. Because of an aggregation behavior, large numbers of this light, gray 1/2-inch long beetles may suddenly appear on trees. Damage is slight but the abrupt appearance - and disappearance - can attract attention and concern.

Life History and Habits: Adult beetles usually appear on honeylocust in June, typically feeding together in groups of several dozen or more, responding to aggregation pheromones. Some defoliation can occur, but the insects usually leave within a week and trees readily refoliate damaged areas.

Like many blister beetles, the immature stages develop as predators of grasshopper eggs. After feeding ash-gray blister beetle females lay eggs on soil depressions, similar to those made by ovipositing grasshoppers. The newly hatched larvae are highly active and tunnel into the soil. If successful they then feed on these eggs, transforming into a fat grub-like form. They overwinter as nearly full-grown larvae that pupate in spring. The newly emerged adults feed and mate for several weeks before egg laying begins. There is one generation produced per year.

Related Species: Many other species of blister beetles are found in the region. However, few feed on trees, with most limiting feeding to vegetables, shrubs, and alfalfa. One sporadically common species on caragana shrubs is the caragana blister beetle, *Epicauta subglabra* (Fall), which sometimes mass on this plant in late June and July.

Management: Blister beetle infestations are typically short-lived, as beetles depart suddenly. As a result, controls are rarely, if ever, needed to protect honeylocust. They can be readily controlled with most shade tree insecticides (e.g., carbaryl, chlorpyrifos, various pyrethroids).

Severity of Defoliation Injuries

A great many insects chew or tunnel into the foliage of trees and shrubs. To a great degree, plants have means to compensate for much of this injury and it is, by no means, beneficial to control all defoliating insects. Although there are no set rules as to what constitutes a damaging infestation, the plant protection manager should keep in mind a number of factors that affect plant injury by defoliators:

Percent defoliation. Obviously, the amount of leaf loss caused by insect feeding injuries can be directly related to their potential damage. However, minor leaf loss on deciduous plants, of at least 20 percent of the total leaf area (individual leaves and total leaves), is fully tolerated without causing detectable stress. Even substantially greater leaf injury can be sustained if the other factors, outlined below, minimize effects of the leaf loss.

Frequency of defoliation. Some insect pests regularly occur in high populations, and as a result these are among the more important pests of shade trees. For example, in much of the state elm leaf beetle defoliation occurs annually on Siberian elm, and Douglas-fir tussock moth caterpillars can repeatedly damage blue spruce in the metro Denver area. These repeated, sustained types of injury can produce significant stress and even mortality.

On the other hand, outbreaks of most insects that chew leaves and needles occur infrequently, often less than once in a decade. Others have life cycles which extend over more than one season (e.g., pandora moth) that serves to space out defoliation. Furthermore, natural controls are usually quite effective for these species, typically causing high populations to collapse after a season or two. Therefore, pest species which have not developed a local track record of causing regular injury rarely warrant control.

Conifers vs. hardwoods. Because conifers typically retain their needles for about three years, premature needle loss can be much more stressful than for deciduous species. Furthermore, most deciduous trees have ability to readily refoliate during the same growing season, allowing them to further compensate for leaf loss.

General health of the plant. One of the most fundamental considerations when assessing defoliation injury is the overall health of the tree or shrub. Vigorously growing deciduous plants with abundant reserves of stored food usually regrow lost leaves. Also, they can also remain healthy enough to defend against attacks by secondary pests. Defoliation is much more serious to plants that are already in poor condition.

Time of injury. As a general rule, the later in the season leaves are lost by defoliating insects, the less stressful is the damage to the tree. Insects which feed on spring growth shortly after it is produced, such as tent caterpillars or gypsy moth, remove tissues that have not had time to return to the plant the energy expended in their production. Conversely, defoliating insects which occur in late summer or fall are removing leaves that have had most of the season to return energy to the plant, and which will soon be shed due to natural processes. Almost never can a late season treatment be justified in solely in terms of protecting plant health.

With some plants, mid to late season defoliation can indirectly cause injury by inducing refoliation that does not sufficiently harden by the time that killing frosts occur. Buds may be injured as a result. There is also some reduction in starch reserves for the next season.

Part of the plant attacked. Defoliating insects all have preferred feeding sites where damage is concentrated. Those which feed on new growth, particularly current-season growth of conifers, can be extremely damaging, such as budworms. Others may concentrate on the upper parts of the plant, such as tiger moths or Douglas-fir tussock moth, which can kill out the main leader and distort the future growth pattern of the plant. On the other hand, pine sawflies and pine butterfly are examples of insects which feed on the older growth, resulting in less stressful type of injury.

Presence of secondary pests. For some defoliating insects the direct damage that they cause may be less important than their interaction with other, potentially more important pests. Several plant diseases, such as Cytospora cankers, as well as wood boring beetles and bark beetles are much more damaging to plants that are weakened and have lowered defenses. Where these secondary pests are an important local concern, defoliation is of greater consequence.

Esthetic considerations. One of the most important considerations in landscape pest management are the esthetic impacts. How plants look as a result of some leaf feeding can be extremely important to a plant owner who may demand control to limit this injury. While this is a valid personal preference, treatments for esthetics should be considered separately from those that are designed for plant protection. For many people, the threshold for injury based on esthetics is far more stringent than that required by the plant to avoid stress. As a result, need for treatments will be greatly increased if esthetics play an excessive role in treatment decision-making.

GRASSHOPPERS

REDLEGGED GRASSHOPPER: *Melanoplus femurrubrum* (DeGeer)
DIFFERENTIAL GRASSHOPPER: *M. differentialis* (Thomas)
TWOSTRIPED GRASSHOPPER: *M. bivittatus* (Say)
MIGRATORY GRASSHOPPER: *M. sanguinipes* (F.) and others

Orthoptera: Acrididae

TOP: Group of grasshoppers, resting on vegetation in late afternoon.

Damage and Diagnosis: Grasshoppers feed on the leaves and twigs of many trees and shrubs. Most damage occurs when they graze and girdle tender twigs of poplars, junipers and other plants, which cause twig die back or swellings of callous to develop around the wounds. Tree plantings surrounded by grasshopper breeding sites are highly attractive to grasshoppers.

Life History and Habits: Most of the pest species of grasshoppers in the region spend the winter as eggs. The eggs are laid underground in late summer or early fall in the form of elongate egg "pods" containing 20 to 120 eggs each. Egg laying is concentrated in dry undisturbed areas, being very heavy along roadsides, in pasturelands, and native prairie. Relatively little egg laying occurs in an irrigated yard or garden, and is largely limited to small dry areas in the yard, such as between sidewalk cracks.

Eggs hatch in late May and June. The immature nymphs take two to three months to become fully developed. Adults are present during August and remain until they are killed by a heavy frost.

Grasshoppers feed during the day, resting on shrubbery, tall plants, or man-made structures during the late afternoon and night. Movements into yards accelerate as native vegetation becomes less suitable due to summer drying or defoliation. Light frosts that may kill many host plants further concentrate them in the fall on the more cold tolerant plants (e.g., strawberries).

Management: Grasshopper outbreaks tend to occur in cycles of roughly ten to 15 years, with seriously damaging populations often lasting for two to three years. However, localized occurrence can be much more erratic. A large number of natural controls exist to bring control of outbreaks.

Cool, wet weather around the period of egg hatch is one of the most common conditions associated with declining grasshopper populations. Grasshoppers also can suffer from several diseases. The fungus disease *Entomophthora grylli* can be very destructive to populations, although spread is dependent on adequate moisture. The microsporidian *Nosema locustae* also can kill or weaken grasshoppers.

Nosema locustae is also manufactured and sold as a type of microbial insecticide under trade names such as Semaspore[R] and NoLo Bait[R]. It is most effective against young, actively growing grasshoppers; older grasshoppers are much less sensitive and rarely die, although they may be less vigorous and lay fewer eggs to survive the following season.

Insects that feed on grasshoppers include the larvae of blister beetles (predators of the eggs), robber flies, and parasitic flies. Grasshoppers are also common foods of many birds such as the horned lark, American kestrel, and Swainson's hawk.

Poultry, notably guinea hens and turkeys, feed readily on grasshoppers and can assist in controlling moderate grasshopper populations. Grasshoppers have sometimes been caught in large numbers and used as a highly nutritious poultry feed.

Since most breeding of grasshoppers occurs in dry, undisturbed sites outside of a yard or garden, control of grasshoppers in these areas is most productive. Sprays of insecticides such as carbaryl (Sevin), acephate (Orthene), permethrin or malathion can kill grasshoppers in rangeland or pastures. In addition, various baits can be used (typically carbaryl on a molasses-bran base or apple pomace base) that allow the insecticide to be applied in a more selective manner.

Related Species: Over 80 species of grasshoppers are native to the High Plains and Rocky Mountain region. Most of these do not occur in large numbers and exist as destructive pests. Many also have selective feeding habits and feed only on grasses or native shrubs (forbs) that keep them out of yards and gardens. Life cycles of the various grasshoppers can also vary, with many overwintering as nymphs or adults.

TOP: Leafcutter bee injury to foliage.
BOTTOM: Leafcutter bee cutting leaf disk from lilac.

LEAFCUTTER BEES

(*Megachile species*)
Hymenoptera: Megachilidae

Hosts: Foliage from a wide variety of plants may be utilized by leafcutter bees. Rose family plants (e.g., rose, lilac) and ash are particularly preferred.

Damage and Diagnosis: Leafcutter bees cut very characteristic semicircular notches out of leaves, using the leaf fragments to construct nesting cells. When very abundant, leafcutter bee damage can defoliate plantings.

Adult leafcutter bees somewhat resemble dark, robust honeybees. However, as they cut leaves very rapidly, within seconds, they rarely are observed on the plants. Larvae are found within the nest cells of the bees, which are constructed in decayed wood or the pith of various plants.

Life History and Habits: Leafcutter bees are solitary bees, which have each female rear young individually. When rearing young, the females bees cut leaves from rose, ash, and other plants. They work quickly and leaf cutting is very rapid, occurring in as little as ten seconds. The bees then take the leaf disks to nest sites excavated out of rotten wood, pithy plants, or other hollows. The leaves are formed into thimble-shaped rearing cells which are then packed with pollen on which the young develop.

Leafcutter bees (unlike honeybees) are native insects that are important in the pollination of several native plants, and are also extremely useful pollinators of alfalfa.

Management: Most often damage to leaves is little more than a curiosity and does not require control. In some settings, populations of this insect can be reduced by limiting breeding sites, such as exposed pith of rose and caneberries, by sealing the opening with white glue, shellac, or a thumbtack.

Many insecticides can kill the adult bees, but treatments need to be reapplied frequently. Because of the beneficial habits of leafcutter bees as pollinators, insecticides should only be used only when very serious infestations threaten.

In areas where large populations of leafcutter bees occur, such as around shelterbelts in isolated areas, damage can be severe and most controls have little effect. In these situations, the best control is to plant nonpreferred hosts and to cover susceptible plants with mesh screening.

Leaf and Needleminers

TOP: Birch leafmines with exposed larva.
CENTER: Elm leafminer tunnels.
BOTTOM: Blotch-type leaf mine produced by hawthorn leafminer. (Whitney Cranshaw)

Leafminers or needleminers are insects that have a habit of feeding internally in leaves or needles, producing tunneling injuries. Several diverse groups of insects with chewing mouthparts have developed this habit, including larvae of moths (Lepidoptera), beetles (Coleoptera), sawflies (Hymenoptera) and flies (Diptera). Many of these insects feed for their entire larval period within the leaf mine and many even pupate within the plant tissues. Other insects, such as the budworms, as leaf/needleminers only during their early growth stages, later emerging to feed externally on the plant.

Leafminers are sometimes classified by the pattern of the mine which they create. **Serpentine** leaf mines wind snake-like across the leaf. More common are various **blotch** leaf mines that are irregularly rounded. One subgroup of these are the **tentiform** leafminers, which produce bulging blotch-type mines that curve upwards, somewhat like a tent. Mines of all types are typically restricted by larger leaf veins.

Areas mined by insects die and dry out. Although injuries produced by leafmining insects can be unattractive, it is rare for them to significantly affect plant health. Also, most leafminers have important natural controls which check populations.

Injuries caused by leaf and needle mining insects can superficially resemble symptoms produced by leaf spotting fungi or other abiotic problems. They can be differentiated by pulling apart the blotchy area. If damaged by insects the leaf or needle will have a hollow area and may expose either the insect and/or its droppings (frass). Leafspotting fungi cause these areas to collapse, without any tunneling.

Insecticidal controls are best achieved if applied during or prior to egg laying and egg hatch. This timing will differ for each leafminer species. After leafminers have entered the leaves, insecticidal control becomes more difficult. Diazinon, chlorpyrifos (Dursban), and pyrethroid insecticides are contact insecticides for these preventive applications.

Insecticides with some systemic activity (acephate/Orthene, dimethoate/Cygon, Merit) can be used on some plants and provide partial control even after tunneling has begun. When using these materials, applications can be timed to coincide with observation of the first pinhead sized tunnels.

Specific recommendations for these insects appear in the supplement on insect control recommendations in the **Leafminer** sections.

BIRCH LEAFMINER

Fenusa pusilla (Lepeletier)
Hymenoptera: Tenthredinidae

Host: Birch

Damage and Diagnosis: Birch leafminer larvae develop within leaves, creating large, dark blotch mines. After the insects exit, the mined areas wrinkle and turn light brown. Defoliation occurs to heavy damaged leaves.

The mature larvae are 1/4 inch long, slightly flattened, and yellowish white in color. They, and their dark droppings, are found within the mined leaf.

Life History and Habits: Adults first appear in May when the first leaves are half grown, laying their eggs singly near the center of the new leaves. Larvae hatch and mine out the middle layers of the leaf; several mines may grow together to form one large blotch mine. After 10 to 15 days, larvae cut exit holes and drop to the soil where they burrow one to two inches below the surface and pupate within cells made of soil particles. Two to three weeks later the adults emerge and the cycle repeats. Normally there are two generations per year, although if conditions are dry many of the first generation wasps will remain dormant and not emerge until the second year.

TOP: Dark blotch leafmine of cottonwood by *Zengophora scutellaris*.
BOTTOM: Aspen leafminer, *Phyllocnistis* sp., serpentine leafmine. (H. Evans)

Related species: A closely related species, *Fenusa dohrnii* (Tischbein) commonly produces blotch leafmines on alder. The elm leafminer (following), *Fenusa ulmi* Sundevall, is locally common on elm, producing a more serpentine mine.

Management: Systemic insecticides, applied to provide coverage during the period when eggs are being laid and hatching, provide the best control. For specific recommendations see the section in the supplement on **Leafminers**.

ELM LEAFMINER

Fenusa ulmi Sundevall
Hymenoptera: Tenthredinidae

Host: Elm, particularly American elm

Damage and Diagnosis: Larvae feed between the upper and the lower leaf surfaces. Original mines are serpentine, but as larvae mature they produce large blotch mines, usually bounded by larger veins. Affected areas later turn brown, dry and drop out, producing ragged foliage.

Life History and Habits: The adult insects are small (1/8 inch) dark wasps that are active in May or early June. The females lay eggs at this time, singly in slits on the upper leaf surface. Eggs hatch within a week and larvae begin burrowing into the inner leaf layers. The larvae, flattened and whitish-green, confine the feeding to the areas between leaf veins and the mines gradually increase in width as the insects develop. In late spring, larvae emerge, drop to the ground, and burrow to a depth of about an inch. Here they spin papery cocoons and spend the duration of the summer, fall, and winter. There is one generation per year.

Management: Systemic insecticides, applied to provide coverage during the period when eggs are being laid and hatching, provide the best control. For specific recommendations see the section in the supplement on **Leafminers**.

POPLAR BLACKMINE BEETLE

Zeugophora scutellaris Suffrian
Coleoptera: Chrysomelidae

Damage and Diagnosis: The larvae feed individually in the inner tissues of the leaf creating large, black blotchy mines. Adults feed on the surface of leaves, making meandering skeletonizing injuries. Damage is most common east of the Front Range.

Hosts: Poplar, cottonwood

Life History and Habits: Adults emerge from May through July. They are about 1/4 inch long, with a dark abdomen but otherwise with primarily yellow coloration. Adults feed on the under surface of leaves and produce small skeletonizing injuries. Females also lay eggs in some of the leaf pits. The emerging larvae tunnel the leaves and feed inside the leaf until late summer. They then drop to the ground, burrow into the soil, and pupate in cells several inches below the surface. There is one generation per year.

Controls: Insecticides are most effective when applied against the adult stage before eggs are laid. Early to mid June would be most appropriate form many areas. For specific recommendations see the section in the supplement on **Leafminers**.

ASPEN LEAFMINER

Phyllocnistis populiella (Chambers)
Lepidoptera: Gracillariidae

Damage and Diagnosis: The larvae develop as leafminers, developing internally in leaves as they chew tissues between the upper and lower leaf surface. Tiny, pale-colored caterpillars may be found within mines during active infestations. External evidence are meandering (serpentine) silvery leaf mines that appear in spring and summer. Damage is cosmetic.

The adult moths occasionally becomes a minor nuisance problem as the moths move to mountain homes for winter hibernation.

Hosts: Aspen. Related species attack cottonwood, poplar and willow

Life History and Habits: The aspen leafminer overwinters as an adult moth, which is often active during much of the fall before hibernating and may wander into homes near infested trees. The moths are tiny (approximately 1/4-inch wingspan), with narrow lance-like wings mottled white and brown. In spring, they again become active and females lay eggs near the tips of young leaves. The young larvae cut through the bottom of the egg and enter the leaves, mining just under the cuticle. They grow and molt several times, with the last instar (IV) being a nonfeeding stage. They pupate within the leaves and the adults subsequently emerge in late summer. Adults feed on nectaries (sources of sugar-rich fluids) at the base of aspen leaves. There is one generation per year.

SPOTTED TENTIFORM LEAFMINER

Phyllonorycter (= Lithocolletis) blancardella (F.)
Lepidoptera: Gracillariidae

Host: Apple

Damage and Diagnosis: Larvae make small (one half inch) mines close to the lower surface of the leaf, causing a slight crimping in the leaf and making it appear tentlike (e.g., tentiform leaf mine). There may be up to 15 mines per leaf. Injury reduces fruit quality and can contribute to premature fruit drop.

Life History and Habits: Adults emerge in spring at bud break, mate, and deposit their eggs on the young leaves as they appear. The young larvae first feed on sap from the lower leaf surface. They later bore into the leaf and construct a mine just under the lower leaf surfaces, and chew small pits into the leaf which appear as small light spots. They also spin silken threads across the mined tissues, and when the silk dries and shrinks the area forms a raised ridge (tentiform leaf mine). By mid to late June the larvae pupate within the mine. Just prior to adult emergence the pupae wiggle halfway out of the mine, leaving this characteristic sign. Second generation adults appear in late June and early July. There are three generations per year, the last one spending the winter as a pupa within the fallen leaves.

Related Species: *Phyllonorycter salicifoliella* (Chambers) and *Phyllonorycter nipigon* (Freeman) commonly make similar tentiform mines in the lower leaf surface of poplar and willow. Outbreaks are fairly common at higher elevations (above 7000-ft) in the state. Infestations tend to be concentrated on lower leaves. There are probably two generations per year of these species. Related species of leafminers also attack hackberry and aspen, producing blotch mines.

Management: These leafminers are heavily parasitized by many insects. Outbreaks are typically of short duration, one to two years at most.

Control with insecticides is very difficult and rarely warranted except during sustained outbreaks. On apple, only a few Restricted Use pesticides have proved effective for control of spotted tentiform leafminer. The related species on poplars and willow are likely equally difficult to control. Furthermore, these latter plants can sometimes be injured by systemic insecticides that are used for leafminer control.

TOP: Tentiform leafminer in *Populus*, upper leaf surface.
CENTER: Tentiform leafminer mines on cottonwood leaf, bottom view.
BOTTOM: Lilac leafminer damage.

LILAC LEAFMINER

Caloptilia syringella (F.)
Lepidoptera: Gracillariidae

Hosts: Lilac, privet. Rarely ash, euonymus.

Damage and Diagnosis: The larvae first produce a blotch mine of the foliage, then tie and feed on the leaves. Injury can be extensive and unsightly. Lilac leafminers develop as very small caterpillars, about 1/3 inch when full grown and glossy green in color. The caterpillars are found within leaf mines or folded leaves.

Life History and Habits: Lilac leafminer spends the winter as a pupa or full-grown larva within the mines of dropped leaves. The adult moths emerge in spring after leaves emerge and lay eggs in small groups on the leaves. Newly hatched larvae tunnel directly into the leaf from under the egg. As they develop and grow larger they excavate a blotch-type mine and tunneling by several larvae may coalesce. As the larvae become nearly mature they leave the leaf mine and tie the edge of the leaf with silk. They continue to feed within the folded leaf and later pupate. There are probably two generations per season.

Related Species: The boxelder leafminer, *Caloptilia negundella* (Chambers), feeds on boxelder and causes similar leaf injuries. There are two generations per year and the life cycle is likely similar to that of the lilac leafminer.

Management: Raking and removal of infested leaves can be an effective cultural control.

Systemic insecticides, such as acephate (Orthene), can provide best control when applied as mines are first being produced. Alternatively, contact insects may be effective if applied when eggs are laid.

TOP: White fir needleminer injury.
BOTTOM: Spruce needleminer injury.

WHITE FIR NEEDLEMINER

Epinotia meritana (Heinrich)
Lepidoptera: Tortricidae

Host: White (concolor) fir

Damage and Diagnosis: Larvae mine and feed inside needles. Larvae prefer year-old needles, but will attack all needles during epidemics. Heavy infestations kill many needles during epidemics. Heavy infestations kill many needles and even whole trees. Most often, foliage is thinned and tree vigor reduced. Weakened trees are especially susceptible to attack by fir engraver beetles.

Adults are dusty-gray moths with a wingspan of about 3/8 inch. Forewings are mottled with patches of cream and have black bands and gray-fringed bottoms. The larvae are cream to yellow caterpillars, body sparsely covered with thin, short hairs, with a brown to black head. About 3/8 inch long when mature.

Life History and Habits: Winter is spent as immature larvae that begin feeding in spring. After mining about six needles, each larva spins a web in which it pupates in June or early July. Adults emerge 10 to 14 days later, mate within a couple days then deposit eggs. Eggs hatch in August and September; larvae subsequently bore into needles where they spend the winter. One generation is produced per year.

SPRUCE NEEDLEMINER

Endothenia albolineana (Kearfott)
Lepidoptera: Tortricidae

Host: Spruce

Damage and Diagnosis: Groups of needles are hollowed out at the base are webbed together, forming a funnel-shaped mass of dead needles. Often, several larvae may feed together producing a large mass of webbing and frass. Larvae are light greenish-brown caterpillars with a dark head, about 3/8 inch long and found associated with damaged needles.

Infestation of large trees is usually confined to lower branches, but entire crowns of small trees may be infested and defoliated. Heavy infestations occur much more commonly in landscape plantings than to forest trees.

Life History and Habits: Spruce needleminer spends the winter as a nearly full-grown larva within a cocoon constructed within mats of webbing and dead needles. They resume feeding in early spring, becoming full-grown in April or May. Adult moths emerge and lay eggs on the needles in late May and June. Eggs are laid in rows, on the underside of year-old needles.

Larvae mine the interior of needles, and damaged needles are later cut off and bound to the twigs with webbing. They feed throughout the summer, suspending feeding with the onset of cold weather. There is one generation per year.

Management: Nests can be physically removed by a jet of water in spring, before new buds open. Infested debris removed in this manner should be destroyed to kill the larvae.

Insecticide sprays are best applied during early stages of an infestation cycle, as eggs are starting to hatch. Because of the needlemining habit of this insect, systemic insecticides (e.g., Orthene, Cygon) are likely to be most effective. If larvae continue to be found in high numbers during midsummer, a second application may be of benefit.

PONDEROSA PINE NEEDLEMINER

Coleotechnites ponderosae Hodges and Stevens

Lepidoptera: Gelechiidae

Host: Ponderosa pine

Damage and Diagnosis: The tiny larvae bore into and develop within young needles, often causing the needle tips to die-back. Browning of foliage is often evidence of injury and closer inspection will reveal hollowed-out needles, typically with the base remaining green. Outbreaks have periodically occurred in several forested areas in the state.

Life History and Habits: Adults emerge in late summer to mate and lay eggs, usually inside old, previously mined needles. Eggs hatch and the tiny larvae move to green needles, bore in near the tip and begin mining. Development continues slowly through the winter and accelerates rapidly with the coming of spring. The larvae complete their development in a single needle, pupating in midsummer.

Related Species: The lodgepole needleminer, *Coleotechnites milleri* (Busck) is occasionally destructive to lodgepole pine. Pinyon needleminer, *Coleotechnites edulicola* Hodges and Stevens occasionally damages pinyon. Another species, *Coleotechnites piceaella* (Kearfott) mines the needles of spruce.

Management: A great many natural controls limit populations of ponderosa pine needle miner. Use of acephate as foliar sprays (Orthene) or injections (Acecap), applied to provide coverage during the period of egg hatch, has provided control on individual trees.

TOP: Ponderosa pine needle miner, *Coleotechnites ponderosa*, larva back lit in needle.
BOTTOM: Pine needle sheathminer.

PINE NEEDLE SHEATHMINER

Zelleria haimbachi Busck

Lepidoptera: Yponomeutidae

Hosts: Many species of two- and three- needle pines, primarily ponderosa and Jeffrey pines

Damage and Diagnosis: Larval mining in both needles and sheaths cause needles to droop, die and be shed prematurely during late spring and summer. Needles often are undercut, causing them to bend at odd angles. Outbreaks have often been coincident with pine budworm outbreaks.

The adults are silvery white moths with a wingspan of about one half inch. Forewings are light yellow with a white band lengthwise. Young larvae, found within the needles and sheaths are bright orange caterpillars. Older larvae are tan with two dull orange lines along the back.

Life History and Habits: Adults emerge in early to mid summer and lay eggs on the new growth. The eggs hatch during the summer and the young larvae tunnel the needles. Winter is spent as a first instar larva in the needle. The following spring the larvae emerge from the needles and crawl down to the base of the needle cluster. They mine inside the sheath, severing the needles and causing the primary injury. They periodically move to the sheath of new needle clusters and typically destroy six to ten clusters of needles in this fashion. There is one generation per year.

Management: While rarely necessary, spraying insecticides to prevent larval feeding in the needle sheaths can help with heavy infestations. Timing would be in spring shortly after shoots begin to elongate.

A Key to Common Insect Groups Causing Chewing Injury to Woody Plant Foliage

Note: This key is organized by couplets. Start from the beginning (1 a and b) and read the description. At the end of the line will be either the correct answer or the number of the next couplet which to move. For example, if the damage that you are seeing involves some removal of the surface leaf tissues, you are directed to move to couplet 2.

1a. Damage involves removal of surface leaf tissues 2
1a. Damage limited to tissues between leaf surfaces (leafminers) or involves swellings of leaf (galls) 15

2a. On coniferous trees 3
2b. On deciduous trees 5

3a. Webbing associated with insect .... **Tiger moth, Spruce needleminer, Web spinning sawfly**
3b. No webbing associated with insect ....

4a. Damage primarily confined to older needles .... **Conifer sawflies, Bull pine sawfly, Pine butterfly**
4b. Damage primarily confined to current season growth .... **Budworms, Douglas-fir tussock moth**

5a. Most leaf tissues are consumed, including many of the smaller veins 7
5b. Feeding is selective and avoids leaf veins, leaving veins intact (skeletonizing) 6

6a. Skeletonizing occurs on upper leaf surface .... **Slug sawfly larvae**
6b. Skeletonizing occurs on lower leaf surface .... **Leaf beetle larvae**

7a. Webbing conspicuously associated with leaf injury 8
7b. Webbing not produced, or not conspicuous 10

Note: Some webbing can be produced by spider mites, during outbreaks. Webbing produced by spiders and cottonwood fluff may also be easily confused with silk produced by insects.)

8a. Webbing ties together only a few (less than 5) leaves and a single larva present within the webbed area .... **Leafrollers, Large aspen tortrix**
8b. Webbing very prominent and covers branch terminal growth or concentrated along branch crotch 9

9a. Webbing dense, concentrated at crotch of branches .... **Tent caterpillars**
9b. Webbing covers foliage on which caterpillars are feeding .... **Fall webworm, Uglynest caterpillar**

10a. Injury is generally confined to the interior of the leaf, consisting of roughly circular holes .... **Leaf beetle adults, Sawflies** (early stage feeding)
10b. Injury originates along leaf margin (notching) 11

11b. Injury confined to leaf margin, occurring as a series of small irregular wounds (notching) that penetrate less than 1/2 inch 12
11a. Injury may penetrate more than 1/2 inch into the leaf. Wounds may be large and irregular; or very smooth and semicircular 13

12a. Notching wounds in shallow, moderately regular pattern and are limited to only a few species of plants .. **Black vine weevil**
12b. Wounds occur in an irregular manner and are scattered among a large range of plant species .... **Grasshoppers**

13a. Wounds very regular, occurring as a semicircular cut .... **Leafcutter bees**
13b. Wounds irregular. Smaller veins as well as other leaf tissues are consumed 14

14a. Petioles of leaves severed. Generally restricted to maple .... **American dagger moth**
14b. Petioles of leaves not severed .... **Caterpillars, Sawflies**

15a. Injury involves a hollowing out of the leaf or needle .... **Leafminers, Needleminers**
15b. Injury involves as swelling or growth on the foliage .... **Gall making insects and mites**

Insects and Mites that Feed with Sucking Mouthparts

APHIDS, LEAFHOPPERS AND RELATED INSECTS (HOMOPTERA)

Insects in the order Homoptera feed on plant sap, using unique "piercing-sucking" mouthparts. The majority feed on phloem, including aphids, psyllids, some leafhoppers, soft scales, and mealybugs. These excrete sticky honeydew which can be a serious nuisance and supports the growth of sooty mold fungi. A few insects, such as the spittlebugs and the "sharpshooter" leafhoppers are xylem-feeders and some leafhoppers restrict feeding to the mesophyll-parenchyma, producing white flecking wounds (stippling) at the feeding site.

Aphids are the most commonly encountered insects on trees and shrubs in the order Homoptera. Hundreds of species occur and at least one species is associated with most woody plants, generally at non-injurious levels. However, some species are significant pests, reducing plant vigor, producing leafcurl distortions of new growth, or secreting nuisance amounts of honeydew.

Aphid development can undergo several different patterns, with some species having very complex life cycles. Overwintering as an egg on a perennial plant is the norm. All aphid forms during spring and much of the summer are females, reproducing asexually and bearing live young. However, both wingless and winged females may be produced, the latter arising following overcrowding, changes in day length, or other environmental cues.

Alternation of hosts is also a common pattern, although several species (e.g., honeysuckle witches' broom aphid, giant conifer aphids) are associated with only a single host species. Alternate hosts for most aphids are an overwintering woody plant and herbaceous summer hosts. Aphids of the same species often have very different forms on these two hosts.

Egg production varies among the different aphid families. In the "true" aphids (Aphididae family) asexual reproduction predominates; males and sexual-form females are only produced during one, late summer generation. The progeny of the mating are egg-laying females that provide the overwintering eggs. On the other hand the "woolly aphids" (Eriosomatidae family) and the "pine and spruce aphids" (Adelgidae family) produce eggs laid externally during each generation.

Honeydew, excreted by aphids, and/or the presence of ants attracted to honeydew may be useful for identifying aphid population developments. Natural controls, including natural enemies (lady beetles, lacewings, syrphid flies, parasitic wasps) usually bring aphid populations under control shortly after they become noticeable. Before any insecticide treatments are contemplated, a search of the aphid colonies for these natural enemies should be made. High numbers of these beneficial insects usually indicate that aphid problems are being controlled without need for intervention.

Aphids exposed on plants can generally be well controlled by several contact insecticides. Insecticides that generally are effective for aphid control include Diazinon, Dursban, Merit, Orthene, and insecticidal soaps. Pyrethroid insecticides (Talstar, Scimitar, Tempo) tend to be a little more erratic in performance, but can control some species. Carbaryl (Sevin) is generally a poor insecticide for aphid control and may even aggravate problems. However, carbaryl is effective for control of the closely related "woolly aphids" (Adelgidae, Eriosomatidae). When applying any contact insecticide complete spray coverage is essential for good control.

Aphids that have already produced a leafcurl or similar protective structure can only be controlled with systemic insecticides. Orthene and Merit are the only widely available systemic insecticides that can be foliar applied; dimethoate (Cygon) has similar effectiveness but more restrictive labeling. In addition, soil applications or trunk injections of acephate (Orthene) or imidacloprid (Merit) are an option for high value trees in sensitive sites where sprays cannot be used.

Some aphids, particularly those that cause early season leaf curls, often overwinter as eggs on the plant. Dormant oil sprays can help provide control of these aphids.

Specific recommendations for control of aphids appear in the supplement on control recommendations. Sections dealing with aphids include: **Aphids (General)**, **Aphids (Leafcurling)**, and **Honeysuckle witches' broom aphid.**

Common Colorado Aphids that Alternate Between Woody and Herbaceous Hosts

APHID	OVERWINTERING HOST	SUMMER HOST
Green peach aphid (*Myzus persicae*)	Peach, plum, apricot, peppers, cabbage, potato,	Many garden plants
Currant aphid (*Cryptomyzus ribis*)	Currant	Motherwort, Marsh betony
Black cherry aphid (*Myzus cerasi*)	Sweet cherry, sour cherry wild mustards	Plum
Rose aphid (Macrosiphum rosae)	Rose	
Potato aphid (*Macrosiphum euphorbiae*)	Rose	Potatoes, tomatoes and many other garden plants
Rose grass aphid (*Metalophium dirhodum*)	Rose grasses, corn	
English grain aphid (*Sitobion avenae*)	Apple, corn, grains	
Rosy apple aphid (*Dysaphis plantaginea*)	Apple, pear, mountain-ash	Plantain
Bean aphid (*Aphis fabae*)	Euonymus, viburnum	Beans, beets, cucumber, carrots, lettuce, etc.
Leafcurl plum aphid (*Hyalopterus arundinis*	Plum	Various aster-family plants, clover, vinca, thistle
Sunflower aphid (*Aphis helianthi*)	Dogwood	Sunflower, pigweed, four o'clock, ragweed
Carrot-willow aphid (*Cavariella aegopodii*)	Willow	Carrot, parsley, dill
Sugarbeet root aphid (*Pemphigus populivenae*)	Narrow-leaved cottonwood	Beets, many garden plants

Aphids that Commonly Create Nuisance Problems with Honeydew

COMMON NAME (*SCIENTIFIC NAME*)	HOST PLANT	TIME OF NUISANCE PROBLEM
Linden aphid *(Myzocallis tiliae)*	Linden	Usually early fall
Western dusky-winged oak aphid (*Myzocallis alhambra*)	Oak	Usually early fall
Clear-winged aspen aphid (*Chaitophorus populifoliae*)	Aspen	Late summer - early fall
Black and green willow aphid (*Chaitophorus viminalis*)	Willow	Usually early fall
Giant willow aphid (*Tuberolachnus salignus*)	Willow	Usually early fall
Black willow aphid (*Pterocomma smithiae*)	Willow	Usually early fall
Elm leaf aphid (*Myzocallis ulmifolii*)	Elm	Summer
American walnut aphid (*Monellia caryae*)	Walnut	Late summer - early fall
Apple aphid (*Aphis pomi)*	Hawthorn, apple	Spring - early summer
Black cherry aphid (Myzus cerasi)	Cherry	Late spring
Boxelder and maple aphids (*Periphyllus* species)	*Acer* spp.	Summer - early fall
Spirea aphid (*Aphis spiraecola*)	Spirea, apple	Spring

GREEN PEACH APHID

Myzus persicae (Sulzer)
Homoptera: Aphididae

Hosts: Winter hosts include peach, apricot and, less commonly, certain cherries and plums. Summer hosts include over 200 species of herbaceous plants, including many vegetables and ornamentals. Green peach aphid is also the most commonly damaging aphid species to greenhouse crops.

Damage and Diagnosis: Green peach aphid can seriously curl and deform the emerging foliage of host plants in spring. Heavily infested terminals can sometimes be killed. Large amounts of honeydew are also produced by this insect.

Green peach aphid is also very important as a vector of several plant virus diseases that affect vegetables, particularly potatoes. In some areas of the state, notably the San Luis Valley, there are efforts to eliminate trees that support overwintering green peach aphid and/or to treat these trees annually to prevent aphids from migrating to crops in summer.

Green peach aphid is a moderate sized aphid, typically straw colored. Coloration can vary from pale green to pale pink or orange.

Related and Associated Insects: Other aphids affecting plum include mealy plum aphid, *Hyalopterus pruni* (F.), and leafcurl plum aphid, *Anuraphis helichrysi* (Kalt.). These species feed on the emergent foliage of plum and can induce leaf curls that greatly deform the new growth. Heavily infested foliage often dies by midsummer. Both species are generally green, but the mealy plum aphid is covered with a fine waxy powder whereas the pale green leafcurl plum aphids are smooth and more intensely colored. Leafcurl plum aphid may continue to be associated with plum through early July, before moving to summer hosts.

Alternate summer hosts of the leafcurl plum aphid are aster family plants, such as thistles, sunflower and aster; yarrow, shepherd's purse, and ragweed are among the other hosts. Mealy plum aphid develop on reeds and cattail in summer.

Rosy apple aphid, apple aphid, spirea aphid, and black cherry aphid are among the other species commonly associated with fruit trees in spring.

Life History and Habits: The green peach aphid spends the winter as shiny greenish black eggs laid near the buds. Eggs hatch following bud break to produce wingless females (first generation). As these develop they subsequently give live birth to a second generation. After two or three generations of wingless aphids on the winter host, winged forms are produced in late spring which abandon the plant and migrate to herbaceous summer host plants. Over 200 species of plants may serve as summer hosts for this insect, including peppers, potatoes, cabbage-family plants, and many weeds.

Management: Dormant oils are highly effective and the only available treatment for fruit trees. Systemic insecticides, such as Orthene and Merit/Marathon, can be used to control aphids within curled leaves on ornamental varieties.

BLACK CHERRY APHID

Myzus cerasi (F.)
Homoptera: Aphididae

Hosts: Sweet cherry, sour cherry, and plum (rare)

Damage and Diagnosis: These large jet-black aphids feed on the sap of new growth in spring. Leaf distortions are minor, usually a slight curling, but heavy infestations can check and even kill back some new growth. Black cherry aphid also produces large amounts of honeydew. Sweet cherry and sour cherry are particularly favored hosts but plum may also be infested.

Life History and Habits: Black cherry aphid survives winter as eggs laid near buds of cherry or plum. Eggs hatch shortly after bud break and the aphids develop on the expanding leaves. Several generations occur on this winter host with peak populations often present through much of June. As production of new growth ceases increasing numbers of winged forms are produced that leave the winter host. The alternate summer host (if it exists) is unknown, but may be a type of mustard. Late summer aphids produce overwintering eggs.

Management: Several biological controls feed on black cherry aphid. The twospotted lady beetle and larvae of various syrphid flies are most commonly associated with this insect.

Leaf curling associated with this insect is usually minor so on small plantings many can be killed by use of a forceful jet of water. Contact insecticides also can be effective, although soaps are not recommended because of potential phytotoxicity to cherry.

Dormant season applications of horticultural oil can kill overwintering eggs.

CURRANT APHID

Cryptomyzus ribis (L.)
Homoptera: Aphididae

Hosts: Currant as winter (primary) host. Motherwort and marsh betony are listed as summer hosts.

Damage and Diagnosis: Currant aphid feeds on the underside of emerging currant leaves and typically induces a leaf puckering distortion. Affected areas of the leaves are often discolored ranging from an intense red (on 'Red Lake' currant) to a more subtle orange-red color on wild currant.

Life History and Habits: Winter is spent as eggs near buds. Eggs hatch shortly after bud break and currant aphids feed for several weeks on currant before winged stages begin to disperse to summer hosts. Some aphids are associated with *Ribes* throughout the year but a large part of the population develops on motherwort or marsh betony during summer. Winged forms, including sexual stages, are produced at the end of the summer that return to Ribes, mate and produce the overwintering egg producing form.

Management: Dormant season applications of oils should effectively kill overwintering eggs. Contact aphicides, as well as systemic insecticides, can provide control, although insecticide options are very limited on edible fruit-bearing plants.

GIANT CONIFER APHIDS

Cinara spp.
Homoptera: Aphididae

Hosts: Most species of conifers, including pines, fir, Douglas-fir, and, particularly, spruce. Each *Cinara* species is specific to a particular genus of tree, some even to a particular species. Winged stages are sometimes produced but no alternate hosts are known. Approximately three dozen species occur in the region.

Damage and Diagnosis: The giant conifer aphids feed on the sap from twigs and branches, often in large groups. Heavy infestations cause a yellowing of foliage, needle drop and occasionally cause dieback of shoots. Large deposits of honeydew are often produced, promoting sooty mold growth and attracting ants and yellow jackets. Populations usually are highest in late spring.

This group includes some of the largest aphids (up to 1/4 inch), exceeded in size only by the giant willow aphid. Adults are long-legged, generally reddish-brown to brown. Eggs are shiny and black, resembling miniature jelly beans, laid conspicuously in rows on twigs and needles. The juniper aphid, *Cinara sabinae* (Gillette and Palmer), is very common on juniper and is usually the most destructive member of this genus in the state.

TOP: Currant aphid injury.
BOTTOM: Giant conifer aphids on spruce.

TOP: Rose aphids.
CENTER: Potato aphid, adult giving birth and young.
BOTTOM: Giant willow aphid colony.

Nuisance problems with giant conifer aphids are sometimes reported, as they mass on sides of buildings adjacent to infested plantings. The reason for this behavior is unknown but seems to follow periods of heavy rainfall that may dislodge many from their host plant.

Associated Species: Large aphids associated with pines can be found in the genera *Essigella* and *Eulachnus*. These feed on needles, are generally gray and can be extremely active and quick moving.

Life History and Habits: In the fall, females lay several eggs each and these overwinter on the host tree. Eggs hatch in the spring. Throughout spring and summer only females are produced, which bear live young. Males and sexual-form females are produced in late summer that mate and produce the egg laying generation.

"ROSE APHIDS"

POTATO APHID: *Macrosiphum euphorbiae* (Harris)
ROSE APHID: *Macrosiphum rosae* (L.)

Homoptera: Aphididae

Hosts: Rose is the primary host for both species. Potato aphid disperses to a wide variety of herbaceous plant hosts in summer.

Damage and Diagnosis: Aphids are very commonly associated with roses, particularly late in the season. Infestations concentrate around new shoots and flower buds. During outbreaks these insects can reduce flower size and may even kill buds. Aphids may also feed on flower petals after bud break, causing losses in flower quality.

Life History and Habits: Both rose aphid and potato aphid overwinter on rose canes as eggs laid near buds. After hatching in spring, there are several generations produced on the emergent shoots. Both winged along with the normal wingless forms are produced beginning in late spring.

Potato aphids disperse to other plants during summer and may be found on a very wide range of host plants during this period. Nightshade family hosts, such as tomatoes, are favored during this time but potato aphid can be found feeding on most vegetables, a wide variety of flowers, on common weeds such as pigweed and sowthistle, and small fruits. Winged stages of the late summer generations disperse back to rose and reproduce, contributing greatly to late season outbreaks. Both pink and green forms of the potato aphid are common and this insect may also be a serious greenhouse pest.

Rose aphid apparently restricts feeding to rose through out the year. It produces only green forms that may be winged or wingless, the latter dispersing to new plantings.

Management: General aphid predators, such as lady beetles and syrphid flies, as well as parasitic wasps commonly feed on rose aphids. These usually keep populations under fair control, although aphid outbreaks frequently occur during spring and fall when biological controls are not highly active.

Rose aphids are quite delicate, so hosing plants with a strong jet of water (syringing) can kill and dislodge many. Pruning prior to bud break can remove many of the eggs that overwinter on canes.

Most general-use insecticides, including the "rose systemics" (Disyston), are usually effective. Dilute sprays of insecticidal soaps or horticultural oils are also effective.

GIANT WILLOW APHID

Tuberolachnus salignus (Gmelin)

Homoptera: Aphididae

Host: Willow

Damage and Diagnosis: The giant willow aphid is a distinctive, purple-black aphid which can reach a size of 2/10 inch. They primarily feed on twigs and branches, often in large colonies. They can produce large amounts of honeydew, creating a serious nuisance problem which also attracts yellowjackets and honeybees. The aphids also can create dark stains if crushed. However, the feeding does not appear to substantially damage willow.

Associated Species: Another common large aphid found on twigs and branches of willow is the black willow aphid, *Pterocomma smithiae* (Monell). This species is only slightly smaller than the giant willow aphid and has distinctive orange cornicles. (Poplar and silver maple are also hosts of this insect.) Other aphid species feed on the leaves of willow, including the small black and green willow aphid, *Chaitophorus viminalis* Monell, and the carrot-willow aphid, *Cavariella aegopodii* (Scopoli).

Life History and Habits: Little is known about the giant willow aphid. Eggs are laid upon the twigs of willow in fall and hatch after new growth forms on tree in spring. There are continuous, multiple generations throughout the growing season. Populations of the giant purple willow aphids are usually most abundant late in the season at which time they can frequently be found in masses on twigs and branches. Some winged stages are also produced in late summer, which can infest new plantings. Adults can sometimes be found in early winter, walking on the snow under infested trees.

Because of their dark pigmentation, willow aphids have been used as a source of natural dye.

TOP: Winged and wingless forms of elm leaf aphid.
CENTER: Witches' broom from honeysuckle witches' broom aphid.
BOTTOM: Honeysuckle aphid injury, old witches' broom with new spring growth.

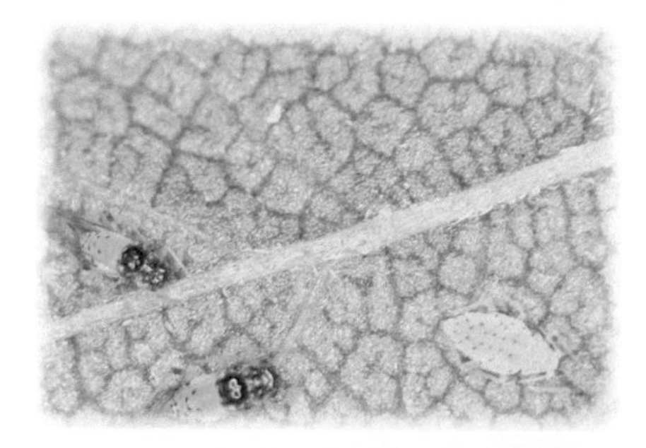

ELM LEAF APHID

Myzocallis ulmifolii (Monell)
Homoptera: Aphididae

Damage and Diagnosis: The elm leaf aphid sucks plant sap and produces abundant amounts of honeydew. Along with the European elm scale this can produce serious nuisance problems. High aphid populations cause premature leaf yellowing and drop.

Elm leaf aphids are generally pale yellow and marked with numerous dark spots on the abdomen.

Hosts: Elms, particularly American elm

Life History and Habits: Elm is the only known host. Overwintering stage are eggs laid near buds. Eggs hatch in spring and multiple, overlapping generations are produced throughout the season. By midsummer, some winged form stages are produced that may migrate to new plants. Males and sexual stage females are present in late summer and early fall, the females giving birth to the egg-laying ovipara form.

Related Species: The linden aphid, *Myzocallis tiliae* (L.), and western dusky-winged oak aphid, *Myzocallis alhambra* (Davidson), are also common on their associated hosts.

Management: Controls are similar to those for other aphids exposed on leaves. Soil injected applications of imidacloprid (Merit) have given effective control of elm leaf aphid.

HONEYSUCKLE WITCHES' BROOM APHID

Hyadaphis tartaricae (Aizenberg)
Homoptera: Aphididae

Hosts: Honeysuckle, particularly tartarian-types

Damage: Honeysuckle witches' broom aphid greatly distorts the emerging new growth. Symptoms of injury include leaf curling, small leaf size, release of dormant buds to produce spindly "witches' broom" growth. Damaged terminals fail to normally

TOP: Honeysuckle witches' broom aphids, *Hyadaphis tartaricae*, close-up.
SECOND: Snowball aphid colony, close-up.
THIRD: Leaf curl ash aphid injury.
BOTTOM: Leaf curl ash aphids.

elongate and often die during winter. Outbreaks are common in many areas of the state and can greatly affect the appearance and vigor of honeysuckle.

Associated Species: Several other aphids also occur on honeysuckle. These may produce some yellowing and leafcurling during spring and fall but the damage is of short duration and causes no significant injury to the plant.

Life History and Habits: Honeysuckle aphid is a "one host" species associated throughout its life with *Lonicera* spp. The aphid spends the winter in the egg stage, on twigs of honeysuckle. Eggs hatch in midspring, shortly after bud break. During the first few weeks populations of the aphid remain low and much of the early plant growth is relatively unaffected. However, by late May or June, aphids colonize the new growth and begin to produce leaf curling distortions.

There are several generations produced during spring and early summer. Populations usually decline as the new growth slows or ceases in late summer.

Management: Systemic insecticides sprays of acephate/Orthene can provide control for about two to three weeks. Soil injected imidacloprid (Merit) has been able to provide season-long control on irrigated sites.

Reducing fertilization and watering cuts down on the severity of deformities by slowing the production of susceptible new growth. Although eggs are laid in the old growth, pruning is not an effective control since this insect is so widespread and highly dispersive.

Syrphid fly larvae have been the most consistently important natural control, due to their ability to enter the curled leaves. Most other general aphid predators (e.g., lady beetles) restrict feeding to exposed aphids.

Some honeysuckle species and varieties are showing tolerance to insect feeding so that the intense leaf curling does not occur.

SNOWBALL APHID

Neoceruaphis viburnicola Gillette
Homoptera: Aphididae

Hosts: Certain species of *Viburnum* including *V. opulus* (particularly *V. opulus* 'Sterilis' - Snowball Viburnum), *V. acerifolium*, and *V. prunifolium*. Most other viburnums are immune to effects of this insect

Damage and Diagnosis: Snowball aphids feed on the emergent growth of certain *Viburnum* species, producing tight leaf curls. The aphids are greenish-yellow, but covered with fine wax giving them a blue-gray color. They are only found associated with plant injuries during the first month of new growth.

Life History and Habits: Snowball aphid overwinters as eggs on the twigs of the host. Eggs hatch shortly after bud break and the aphids move to the expanding leaves. Feeding induces tight leaf curling and colonies develop within these protected sites. Winged forms are produced in late spring that leave the host for an unknown summer host. Late in the season migration back to viburnum occurs followed by production of the overwintering eggs.

Managment: Dormant oil treatments should be able to kill overwintering eggs. Where foliage has already begun to be curled, sprays of a systemic insecticide such as acephate (Orthene) or imidacloprid (Merit/Marathon) can be effective.

LEAFCURL ASH APHID

Prociphilus fraxinifolii (Riley)
Homoptera: Eriosomatidae

Host: Green ash

Damage and Diagnosis: Feeding by the aphids creates very tightly rolled and thickened leaves, known as pseudo-galls, at the tips of the twigs. This injury often causes some distortion and twisting of the twig next to the damaged leaves. Expanded leaves are not susceptible to leaf curling and damage.

Leafcurl ash aphids are yellow-green with brown head but are covered with white, waxy threads. They are found within curled ash leaves in mixtures of old cast skins and droplets of wax coated honeydew. Parasitized aphids become bloated and dark-bluish black.

Life History and Habits: The life history of this insect is imperfectly understood. However the following appears to describe what has been observed in northern Colorado.

Leafcurl ash aphid overwinters in the soil, as colonies of woolly aphids on the roots of ash. Winged stages work their way through soil cracks and disperse to ash foliage shortly after bud break. Young aphids developing on the emerging leaves are capable of inducing tight leaf curling. Numerous generations are subsequently produced on ash foliage which continue to distort new growth, creating large clumps of thickened curled leaves.

Colonies begin to decline as new growth ceases. In addition, natural enemies become more abundant. Winged stages that disperse back to the roots of ash are produced by midsummer and the foliage-infesting phase of the insect passes.

Management: By midsummer, several natural controls typically curtail outbreaks, particularly after succulent new growth is no longer being produced. Earwigs, syrphid fly larvae and parasitic wasps are among the most commonly observed biological controls.

Systemic insecticides, such as Orthene and Merit, have been the only effective treatment identified for this insect. There seems to be a range in susceptibility among ash cultivars to damage by this insect with some (e.g., 'Patmore') being more commonly damaged.

WOOLLY APPLE APHID

Eriosoma lanigerum (Hausmann)
Homoptera: Eriosomatidae

Hosts: Apple, crabapple, mountain-ash, elm

Damage and Diagnosis: Feeding on elm leaves and buds in spring cause the leaves to curl into closed, stunted clusters or rosettes at the twig tips. Curled leaves enclose colonies, protecting them somewhat from predators and insecticides. The aphids usually have left elm by late June but return and may occur in high numbers in late summer and early fall on elm leaves.

The more severe injury is caused by the insect feeding on the roots of apple, crabapple, and mountain-ash, resulting in large knots (galls) on the roots. Heavily infested trees often have short, fibrous roots, and are stunted and sometimes killed. Callous tissue around aboveground wounds are also commonly infested, which inhibits healing and can produce persistent cankers.

Life History and Habits: The complete life cycle requires one year to complete, involving both a winter host of elm and summer hosts of apple, crabapple, or mountain-ash. Most *Eriosoma* species spend the winter as eggs in cracks and crevices of elm branches. However, in warmer locations or protected sites some nymphs can survive on the roots of apple. The eggs hatch in early spring and the young aphids move to the unfolding leaves and buds to feed and reproduce. After two wingless generations on elm, a winged generation follows, which migrates to the summer hosts (apple, crabapple, mountain-ash) where it feeds below ground on the roots and trunk or around wounds on the trunk. In the fall, a generation of males and females occur on elm that

TOP: Woolly apple aphid.
BOTTOM: Woolly aphid, *Eriosoma* sp., colony causing leafcurl of elm.

TOP: Woolly pine aphids.
CENTER: Pear psylla nymphs in honeydew droplets. (Gene Nelson)
BOTTOM: Sumac psyllid

mate and produce an egg laying form. Each female lays one egg and dies. Four or more generations per year.

Related Species: Related species also wintering on elm have alternative summer hosts. *Eriosoma crataegi* (Oestlund) is found on hawthorn and *E. americanum* (Riley) on amelanchier during the summer. On elm *E. americanum* produces a tight leaf curl in spring that is packed with aphids.

WOOLLY PINE ADELGIDS

Pineus species
Homoptera: Adelgidae

Hosts: Most pines host some species of *Pineus* in Colorado. Spruce is an alternate host of some *Pineus* species.

Damage and Diagnosis: This is the common group of woolly aphids found on the needles of pines during late spring. Foliage of heavily-attacked trees becomes yellowish and growth is stunted. Needle and shoot feeding can cause shoots to droop and die. Infestations also can occur on bark, contributing to decline during outbreaks. Species that have an alternate host of spruce, produce a small gall on that plant.

Life History and Habits: *Pineus* species are closely related to the Cooley spruce gall adelgid and other *Adelges* species. Similarly, they often share a complex life cycle involving alternate hosts, gall formation on the winter (primary) host, and up to six kinds of adult forms. Some species live only on one host, such as *Pineus coloradensis*. Other species have life cycles that take two years to complete.

Management: Problems on pines are usually short-lived, and the aphids disperse without causing serious damage. The woolly pine adelgids are easily controlled with contact insecticides such as insecticidal soaps or carbaryl (Sevin, Sevimol).

PEAR PSYLLA

Cacopsylla pyricola Foerster
Homoptera: Psyllidae

Hosts: Pear

Damage and Diagnosis: Pear psylla feed on the leaves of pear, excreting conspicuous droplets of honeydew. These droplets soil the leaves and fruit of the plant. Honeydew also allows sooty mold fungi to grow on the plant which discolors fruit.

High numbers of psyllids on trees can reduce plant vigor and cause it to yield poorly. Trees suffering from 'psylla shock' may take several years to recover. Pear psylla also can transmit the bacteria (phytoplasma) that produces a disease known as pear decline. Fortunately, this disease is uncommon in the region. Pear psylla occurs throughout the region but is most common west of the Rockies.

Adults are small (1/10 inch), reddish-brown winged insects, similar in overall appearance to a miniature cicada. The nymphs, found usually covered by a honeydew droplet, are flattened, green and scale-like.

Life History and Habits: Pear psylla overwinter in the adult stage. During this time they hide in protected areas (under bark, plant debris on soil, other cover) in the vicinity of previously infested trees. They become active in late winter or early spring and move to pear trees. They begin to lay yellow-orange eggs as the pear buds begin to swell and the emerging nymphs then move to feed on the tender new growth.

As the psyllid nymphs feed they are first covered with the honeydew droplet that they produce. The final stage of the developing insect has conspicuous pads where wings are developing and does not live in the honeydew. They then molt to the adult stage. Later generations lay eggs on the new leaves, often concentrating on sucker sprouts late in the season. There are normally two to three generations produced during a season. At the end of the year, dark-colored winter adult forms move to shelter.

Related Species: A very common psyllid that also creates honeydew is *Psylla negundinis* Mally, associated with boxelder. Several other *Psylla* species are associated with willow and alder. These species apparently have a life history similar to that of the pear psylla.

On sumac, the psyllid *Calophya triozomima* Schwarz occurs. This species overwinters on the plant in the nymph stage. Nymphs are generally black and may have a fringe of wax. There does not appear to be any serious injury associated with this latter insect.

Management: During very dry weather, honeydew droplets covering the nymphs may dry and crust to kill the insects. Minute pirate bugs, lady beetles and other general predators also feed on pear psylla. A chalcid wasp is an important parasite of the pear psylla in many areas.

Pear psylla thrives on tender new growth. Remove sucker shoots during midsummer to deprive the insects of favored food sources after the main leaves have hardened.

Pear psylla is very resistant to most insecticides and is difficult to control. The best controls is to treat for the insect two to three times before bloom, killing the overwintered adult stages before they have laid eggs. These treatments are most effective if other pears in the area are also treated, limiting reinfestations.

HAWTHORN MEALYBUG (Two-Circuli Mealybug)

Phenacoccus dearnessi King

Homoptera: Pseudococcidae

Hosts: Several rosaceous plants, including hawthorn, mountain-ash and Amelanchier, are reported as hosts of this insect. In Colorado, this insect has only occurred as serious pest on hawthorn.

Damage and Diagnosis: The hawthorn mealybug primarily feeds on the sap of twigs and small branches and heavy infestations weaken the plant and can cause twig dieback. The most common problems are associated with the large amounts of honeydew it excretes during feeding. Sooty mold and honeydew are often conspicuous and can greatly detract from plant appearance.

Later stages of the insects are conspicuous on the twigs, globular in form and covered with a red body that is finely covered with white wax. Small, but more elongate pale reddish-brown immature forms can be found on the bark of the trunk and larger branches during winter. Adult males are small, gnat-like winged insects present during the period of migration from branches to twigs in spring.

Life History and Habits: The hawthorn mealybug spends the winter, usually as a late stage nymph, on the trunk and larger branches, packed within cracks on the bark. In spring, females move to twigs and continue to develop, becoming full grown in May or early June. Adult males remain on the trunk until they subsequently transform to a winged adult stage that mates with the females.

After mating, the females swell greatly with hundreds of maturing eggs. The eggs hatch within the mother and crawlers emerge, although dispersal of the young nymphs is suspended during wet, cool weather. Peak production of nymphs occurs during late May and June, although it may extend into late summer. Only a single generation is thought to be produced, although the egg producing females and egg hatch has been observed in late September suggesting a small second generation.

Newly emerged nymphs feed on leaves for a brief period, but later move to protected areas on twigs where they may remain through much of the summer. Populations often are again found in high numbers on leaves during late summer, where they often are

TOP: Hawthorn mealybug, adult female and crawlers.
BOTTOM: Hawthorn mealbugs.

found aggregating in leaf folds (domatia). Migration to overwintering areas on the trunk generally occurs during September and October.

Related Species: Grape mealybug, *Phenacoccus maritimus* (Ehrhorn), has been associated with pear, grape, and catalpa in Colorado; several other plants are reported susceptible to this insect, including honeylocust and hackberry. The adult females are about 3/16-inch long, covered with a whitish wax. During outbreaks, which are rare, large amounts of the cottony wax are formed.

Life cycle is somewhat different than the hawthorn mealybug with overwintering stages being early instar nymphs within or near the cottony egg sacks. In spring, most move to twigs and leaves where they feed and develop, becoming full grown in late June and July. This first, overwintered generation are usually little observed. Eggs are laid in early summer and nymphs of this second generation develop during July and August. When full grown they move to older wood where the females again produce egg sacks. Egg production may continue until killing frosts.

Management: Hawthorn mealybug has proven fairly difficult to control, due to its waxy body covering and habit of living under loose bark and in other protected sites. Spring treatments applied to coincide with the return of overwintered stages appear most effective; the addition of a horticultural oil is useful in providing penetration and coverage of the waxy body of these insects. Alternatively, applications made in late summer or early fall directed against early instar nymphs should provide some control. Dormant season applications of oils should be applied to the trunks and under surfaces of branches where hawthorn mealybug overwinters.

Few natural enemies appear to be associated with this insect. Several species of lady beetles prey on immature stages, particularly on nymphs as they move to twigs in spring. However, after the mealybugs have settled on twigs and begun to swell with eggs they are rarely attacked.

Among hawthorn cultivars a range in susceptibility to mealybug has been observed. English hawthorn, Arnold hawthorn, and Thornless cockspur hawthorn appear particularly susceptible. Snowbird hawthorn, Russian hawthorn, and Macracantha hawthorn are less susceptible and Cordata Washington hawthorn appear resistant.

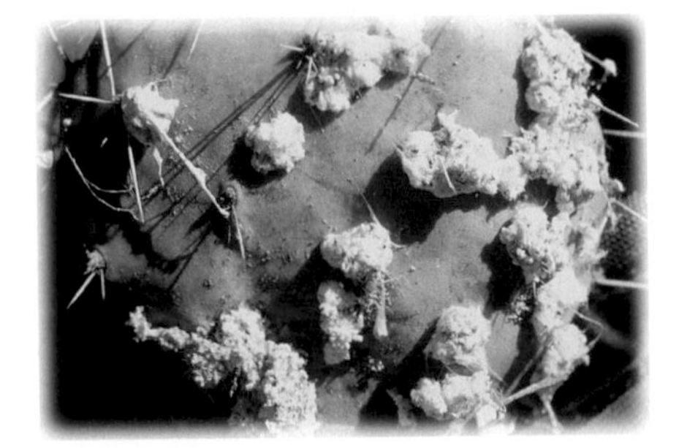

ABOVE: Cochineal.

COCHINEAL

Dactylopius confusus (Cockerell)

Homoptera: Dactylopiidae

Hosts: Primarily prickly pear but sometimes cholla types of cacti

Damage and Diagnosis: Cochineal is an unusual soft scale that produces large amounts of waxy threads over its body, similar to mealybugs. Colonies of the scale may appear as cottony masses covering large areas of the cactus pads. Heavy, sustained infestations induce wilting and sometimes death of individual pads.

Life History and Habits: Individual females can produce about 400 to 600 eggs which are extruded into a large mass of waxy threads. The newly hatched crawlers actively move about the plant and feed for several days before settling. The first stage takes about three weeks to complete after which time the females become sedentary. Males retain mobility but soon begin to produce a series of molts within small cocoons. Both the males and females are roughly synchronized in their development and mate, after which time the female swells with eggs that are laid over about a two month period. Numerous generations can be produced in a single year in the warmer areas of this insect's range, which extends well into Mexico. Conditions in much of the region likely restrict it to no more than two generations per year.

Cochineal has long been used as a natural source of a highly treasured red dye. It is still used for fabrics and is commonly found as a food coloring in some soft drinks.

This species was also introduced into Australia and South Africa as a biological control for prickly pear.

JUNIPER SPITTLEBUG

Clastoptera juniperina Bell
Homoptera: Cercopidae

Hosts: Juniper, arborvitae

Damage and Diagnosis: As the nymphs develop during late May and June they cover themselves with masses of spittle which can be unsightly. However, juniper spittlebug causes very little or no injury to established plants, although some yellow spotting may be observed.

The nymphs are dark yellow with some brown markings and can be found buried in the spittle mass that they produce. Adults are rarely seen, but are about 1/4 inch long, of oblong form and light brown with some mottled patterning.

Life History and Habits: Eggs overwinter inserted in twig tips during June and early July, and hatch the subsequent spring, often around mid May. The nymphs feed upon the xylem using piercing/sucking mouthparts. During feeding they continuously excrete fluid in the form of bubbles that completely enclose the body in the characteristic spittle mass. Development is completed in about one month after egg hatch, and nymphs in the last two weeks of development produce spittle masses that attract attention. Adults do not produce spittle masses, and are inconspicuous.

Related Species: Other spittlebugs in the genus *Clastoptera* can be found on rabbitbrush, Gambel oak, aspen and alder. Some of these, notably those associated with Gambel oak, can sometimes be extremely abundant during late spring and early summer. Life history is somewhat different in that multiple nymphs develop with the spittle mass, which is located on the leaf underside along the leaf vein. The species *Aphrophora irrorata* Ball can be found on pines.

Management: Spittle masses are most commonly observed late in the life cycle, as the masses grow large. However, since the insect shortly leaves these masses, infestations typically end within two to three weeks after becoming noticeable. Since no serious injury occurs, controls are not warranted except for aesthetic reasons. Forceful hosing with water can remove most.

TOP: Juniper spittlebug nymph exposed from spittle mass.
SECOND: Oak spittlebug nymphs.
THIRD: Mulberry whitefly.
BOTTOM: Rose leafhoppers.

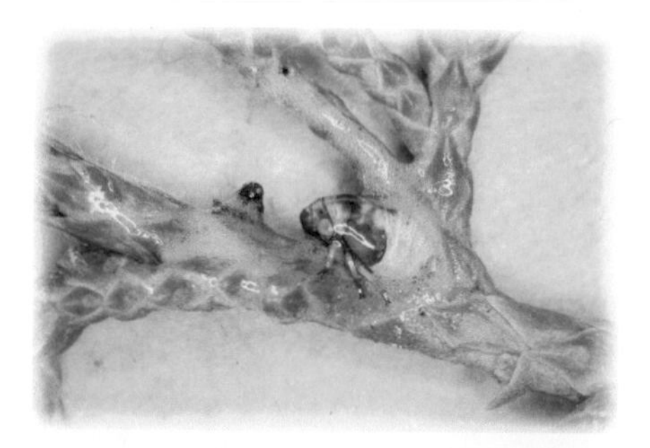

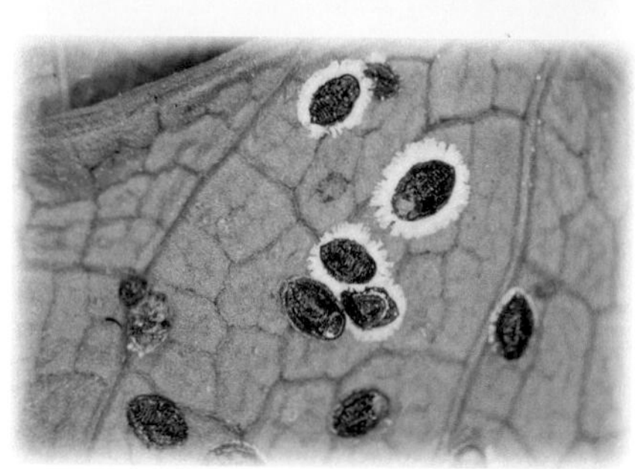

MULBERRY WHITEFLY

Tetraleurodes morei (Quaintance)
Homoptera: Aleyorodidae

Hosts: Mulberry is the only known host in Colorado. It has been reported from many other deciduous trees including hackberry, maple, dogwood and sycamore.

Damage and Diagnosis: This insect is not known to cause any significant damage to mulberry in Colorado. However the unusual appearance of the nymphs attached to the underside of leaves - a jet-black "whitefly" ringed with a white fringe, attract notice.

Life History and Habits: Life history of this insect has not been studied. Adults are active at least from June through September and this species does appear able to overwinter in Colorado.

ROSE LEAFHOPPER

Edwardsiana rosae (L.)
Homoptera: Cicadellidae

Hosts: *Rosa* spp. and *Rubus* spp. are overwintering hosts. Dogwood, oak, elm, hawthorn, apple, poplars, maple and oak are among the plants that may support this insect during summer generations.

Damage and Diagnosis: Rose leafhoppers feed on the sap of mesophyll and produce white flecking wounds on foliage. Damage occurs early in the season, after which time they disperse to summer hosts. One behavioral feature of the pale yellow nymphs is that they can not move sideways, like most leafhoppers.

Eggs are inserted into canes and occasionally these wounds have served as entry courts for pathogens.

Associated and Related Species: The white apple leafhopper, *Typhlocyba pomaria* McAtee, also feeds on rose foliage and produces similar flecking wounds. *Edwardsiana commisuralis* (Stal) is reported from dogwood from the Rocky Mountain states and west.

Life History and Habits: Eggs survive overwinter inserted into stems of rose canes. The nymphs move to foliage in late spring and feed on the underside for several weeks before becoming full grown. Nymphs are cream colored and when young have red eyes; this latter feature is lost at the last molt. The winged adults then disperse to alternate summer host plants (see above) returning only to rose in late summer for overwintering egg laying. The eggs females insert into canes may appear as small pimple-like swellings.

Management: As there is only one generation associated with rose, insecticidal controls will rarely provide any benefit. Pruning canes, particularly any that show evidence of egg laying, can remove many of the overwintering stages.

TOP: Grape leafhopper.
BOTTOM: Virginia creeper leafhopper.

VIRGINIA CREEPER (ZICZAC) LEAFHOPPER

Erythroneura ziczac Walsh

GRAPE LEAFHOPPER

Erythroneura vulnerata Fitch

Homoptera: Cicadellidae

Hosts: Virginia creeper and grape are most commonly damaged. Grape leafhopper may be found on a wide range of hosts including mints, maple, strawberry and burdock.

Damage and Diagnosis: Feeding by the nymphs produces characteristic white flecking injuries on the leaves due to destruction of the mesophyll. Small dark fecal droplets also are common around feeding sites. Heavily infested plants may appear an unsightly gray and defoliate prematurely. The leafhoppers are generally pale colored, with some colored patterning, and nymphs are highly active and will crawl quickly when disturbed.

Life History and Habits: Both leafhoppers spend the winter in the adult stage under sheltering debris in the vicinity of previously infested plants. They emerge when spring temperatures reach the mid 60s and fly to the vines to feed shortly after the new growth emerges. After several weeks, the females begin egg-laying, inserting the eggs just underneath the leaf surface.

Eggs hatch in one to two weeks and the nymphs feed on the mesophyll of cells on the lower leaf surface. They become full-grown in approximately three weeks. At least three generations typically occur in the region, with a fourth generation possible in warmer areas of the state. Feeding continues through early fall, with the adults leaving the plants for cover with cool weather.

Management: Cold and wet weather conditions in spring and fall are reported to reduce populations of grape leafhoppers. The egg parasite of the grape leafhopper, *Anagrus epos*, is common west of the Rockies and apparently is favored where grape plantings are near other fruit crops. This is because these crops host alternate species of leafhoppers also used by *Anagrus epos*, allowing its populations to increase.

Adults can be trapped on sticky surfaces, such as a sheet covered with Tanglefoot[R] or Vaseline. Hold the sticky sheet downwind of the vine shake it. Winged leafhoppers disturbed by this activity can then be trapped.

These leafhoppers are easily controlled by many insecticides. For current recommendations see the section on **Leafhoppers** in the management supplement.

HONEYLOCUST LEAFHOPPER

Macropsis fumipennis (Gillette and Baker)
Homoptera: Cicadellidae

Host: Honeylocust

Damage and Diagnosis: Honeylocust leafhopper feeds by sucking sap from the phloem of honeylocust leaves. Very little injury is produced, at most a slight flecking of the foliage. They produce honeydew but not in nuisance amounts.

Honeylocust leafhopper is commonly confused with the honeylocust plant bug. Early in the season both are present on foliage, but the latter species is far more damaging. Nymphs of the two can be separated by the leafhopper having a darker green color, more pointed abdomen, and ability to readily "crab-walk" sideways when disturbed.

Associated Insects: A treehopper, *Micrutalis calva* (Say), is also sometimes associated with honeylocust and occurs during late spring along with the leafhoppers and plant bugs. It does not appear to be damaging.

The honeylocust plant bug also occurs concurrent with the first generation of the honeylocust leafhopper. The plant bugs are much more damaging and the two species can be difficult to separate in the field. Honeylocust leafhopper tends to be darker green, a bit more elongate in body form, and can move sideways or backwards when prodded.

Life History and Habits: The life cycle of this insect is poorly understood. Overwintering stage is thought to be eggs in terminal growth. Nymphs hatch in spring and can be found feeding on new growth shortly after bud break. At least two, and probably three, generations are produced during the season as leafhoppers are present through late summer.

Management: Rarely, if ever needed. Most controls for aphids or other shade tree insects are effective, including use of insecticidal soaps.

TOP: Honeylocust leafhopper, adult.
CENTER: Honeylocust leafhopper, nymph.
BOTTOM: *Micratulis calva*, a treehopper on honeylocust.

'TRUE' BUGS (HEMIPTERA)

Members of the order Hemiptera, the 'true bugs', also have specialized piercing-sucking mouthparts designed to feed on plant fluids. However, feeding is typically more destructive than with the aphids and scales (Order: Homoptera). Most true bugs feed in a manner known as *lacerate and flush* where the mouthparts puncture and slash plant cells, followed by the production of large amounts of saliva to flush the cell contents into the mouth. As a result, areas around feeding sites usually are killed. Since most bugs also feed on tender tissues (e.g., developing leaves, fruits, flowers) these injuries may cause distortion injuries or abortion of plant parts.

Several groups of 'true bugs' are particularly important as woody plant pests. Most important in the Rocky Mountain region are the plant bugs (Family: Miridae). Most spend the winter in the adult stage in protected sites; the honeylocust plant bug (which overwinters as an egg) being an important exception to this habit, wintering in the egg stage. Lace bugs (Family: Tingidae), a group that is important in other areas of the country, are generally minor species in this region.

Note: Several other "true bugs" are notable because they feed on seeds, such as boxelder bugs and conifer seed bugs.

ASH PLANT BUG

Tropidosteptes amoenus Reuter
Hemiptera: Miridae

Host: Ash

Damage and Diagnosis: Ash plant bugs feed with piercing-sucking mouthparts in a "lacerate-and-flush" habit that kills cells around the feeding sites. White or yellow leaf speckling is a typical symptom, particularly on fully expanded leaves. Emergent leaves that ash plant bugs feed on often show necrotic spotting and distortions. Very small black "tar spots" of excrement usually appear near the damage. Injuries are usually minor.

Ash plant bug is a light brown "true bug", about 3/8 inch long, with heart-shaped making on the scutellum. The nymphs are more oval, shiny yellowish or reddish brown and lack wings.

Associated Insects: A lacebug, *Atheas* sp., also feeds on ash producing general bronzing symptoms on ash. This species is usually found near waterways on naturalized ash.

Life History and Habits: Overwintering eggs are laid under loose bark. Nymphs hatch out in late April and May and feed on the lower leaf surfaces. Most injury occurs during this period as the nymphs begin to mature by late May. This generation of adult insects inserts eggs in the midribs of the leaves and a second cycle of feeding occurs in July and August. Adults produced by this latter generation produce the overwintering eggs.

HONEYLOCUST PLANT BUG

Diaphnocoris chlorionis (Say)
Hemiptera: Miridae

Host: Honeylocust

Damage and Diagnosis: Nymphs feed on the developing buds and leaves. They use a "lacerate and flush" feeding style that produces a lot of injury to the feeding site. Young leaves and buds often are killed. Older leaves may survive but show discoloration and deformation of developing foliage due to localized necrosis around feeding points. Heavy infestations may greatly retard foliage development in spring and have been associated with twig and branch dieback.

During outbreaks many of the pale-green nymphs may be dislodged from trees following light shaking of foliage. This can result in nuisance problems when they land on people.

The honeylocust leafhopper (*Macropsis fumipennis*) typically occurs in co-infestations with honeylocust plant bug and the two species can be easily mistaken for each other. Honeylocust plant bugs are paler colored, have a slightly more blunted abdomen, somewhat giving the appearance of an aphid. The leafhoppers also may walk sideways, crablike, when disturbed a habit not shared by the honeylocust plant bug.

Life History and Habits: The overwintered eggs hatch in late April or early May. The nymphs feed on tender, emergent leaves and buds. They feed over a period of about a month with most reaching maturity in about one month. The winged adults are present for two or three weeks during which time they mate and females lay eggs into the woody tissues of twigs. There is one generation per year.

Management: During outbreaks significant damage can be done by honeylocust plant bugs and controls can significantly benefit plant growth and appearance. Vigorous hosing with water can dislodge many nymphs and may be sufficient for smaller trees. Several insecticides are effective for control. However, to get maximum benefit trees should be monitored for emergence of the insects and treated during early stages of the infestation, usually by mid May. Treatments applied during the terminal stages of the infestation, after serious symptoms have developed, will provide little benefit.

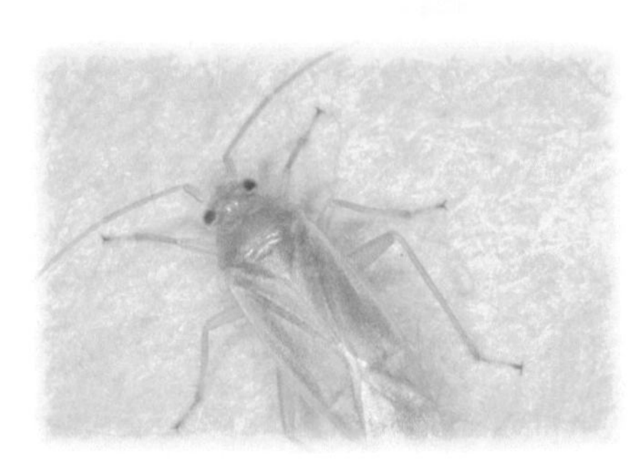

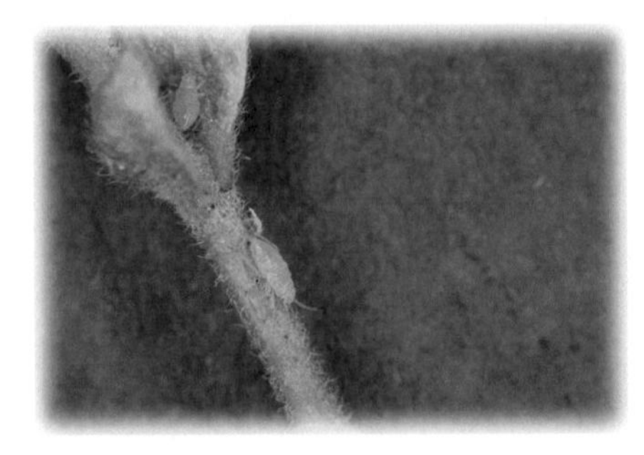

TOP: Ash plant bug.
SECOND: Ash plant bug injury.
THIRD: Honeylocust plant bug, adult.
BOTTOM: Honelylocust plant bug nymph.

TOP: Honeylocust plant bug injury.
SECOND: Lacebugs on Gambel oak.
THIRD: The opuntia bug on prickly pear.
BOTTOM: Lacebug injury to chokecherry.
(David Leatherman)

LACEBUGS

Corythucha species, *Atheas species*
Hemiptera: Tingidae

Hosts: Most regional species are found on chokecherry, oak, ash, sycamore, and net-leaf hackberry.

Damage and Diagnosis: Both nymphs and adults feed on the underside of leaves. This produces characteristic pale stippling wounds to appear on the upper leaf surface. Spots of tar-like excrement appear on the underside of leaves. In high populations, which is uncommon in the region, lacebugs may kill extensive areas of individual leaves, but it is unlikely that even this is a significant injury to plant health.

The adult bugs are of unusual appearance. They are flat, about 1/8-in long with white lacy wing covers and similar lacy extensions on sides of thorax. The nymphs are darker, covered with spines, and wingless.

Life History and Habits: Regional species have been poorly studied. Adults overwinter in bark crevices and other protective areas. They emerge in the spring when young leaves are forming and soon deposit eggs on the underside of foliage near the midrib with a brown, sticky substance. Eggs hatch in several days, the nymphs begin feeding, and mature in about a month. More than one generation may occur per year, depending on climate and species.

OPUNTIA BUG

Chelinidea vittiger Uhler
Hemiptera: Coreidae

Hosts: *Opuntia* spp. (prickly pear)

Damage and Diagnosis: Opuntia bugs are generally brown or gray with wings marked with faint yellow lines. They are often massed on plants and appear to cause some wilting. At the end of the season small numbers may move indoors, similar to boxelder bugs, and can become a minor nuisance invader.

Related and Associated Species: Another leaffooted bug that is similar in appearance is *Narnia snowi* Van Duzee, commonly associated with cholla cactus in the southern areas of the region.

Life History and Habits: Winter is spent in the adult stage, occasionally within buildings but most commonly under protective debris. They move to prickly pear very early in the year, sometimes by late winter. Eggs are laid in small masses on cactus pads and the nymphs develop over a period of a couple months. At least two generations seem to be normally produced, with populations greatest by late summer.

SPIDER MITES AND RUST MITES

Mites are not insects but instead relatives in the Class Arachnida. The spider mites (Family: Tetranychidae) are generally the most damaging group of mites. These feed on plants, whipping and breaking the upper cell layers and sucking the plant sap. This produces characteristic flecking wounds and decreases the vigor of the plants.

Spider mite development typically involves five stages. Reproduction is by eggs, which hatch into a tiny, six-legged nymph stage. There follow two eight-legged immature stages (protonymph, deutonymph) and finally the adult. Males and females occur with the males being somewhat smaller.

Spider mite populations typically develop rapidly under dry conditions, since feeding rate is partly dependent on how rapidly excess water evaporates. Plants under stress are particularly susceptible to injury. Spider mite problems can also be induced through the use of certain insecticides, such as carbaryl (Sevin), that destroy natural insect enemies of the spider mites. Among the most important predators of spider mites are minute pirate bugs, predatory plant bugs, predatory mites (Phytoseiidae family), and the "spider mite destroyer" group of lady beetles (*Stethorus* spp.).

Proper watering and fertilization of plants is a primary means of limiting spider mite injury. Forceful hosing of foliage can be particularly useful since this can crush and dislodge many mites.

Hexythiazox (Hexygon) and dicofol (Kelthane) are specific miticides producing little adverse effects on natural enemies of spider mites. Certain broad-spectrum insecticides with some mite activity include Avid, Talstar, Dursban or Orthene. (Effectiveness of these treatments can vary considerably between the different mite species.) Horticultural oils also can be effective spider mite treatments. However, during outbreaks it may be difficult to get control with any pesticide, particularly when pesticide resistant strains of twospotted spider mite are present.

The eriophyid mites (Eriophyidae) are minute mites, requiring 15X+ magnification to be observed. In addition to their small size they are characterized by having an elongate, carrot-shaped body and only two pairs of legs. Eriophyid mites are best recognized by the species that produce galls. (These are discussed under **Galls**.) However, a great many species occur on leaf surfaces and do not induce galls. These "leaf vagrants" often cause little plant injury. Exceptions are the "rust mites" that produce a rusty color on infested leaves and which may promote premature leaf drop. Interestingly, carbaryl (Sevin) is usually highly effective against eriophyid mites, in contrast to spider mites.

TOP: Twospotted spider mites and eggs.
CENTER: Twospotted spider mite webbing on ornamental pine.
BOTTOM: Twospotted spider mite injury to eggplant.

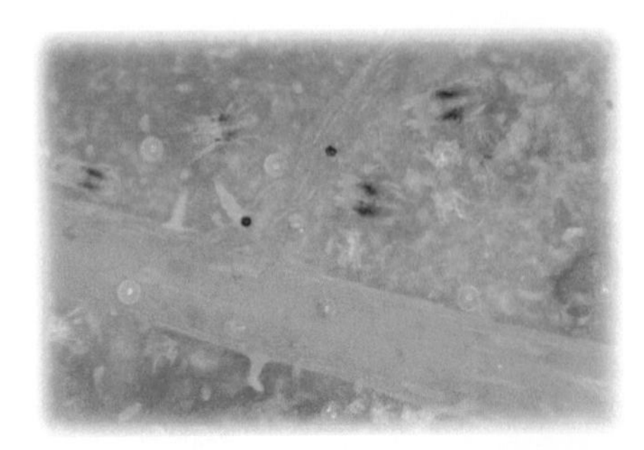

TWOSPOTTED SPIDER MITE

Tetranychus urticae Koch
Acarina: Tetranychidae

Hosts: Rose, viburnum, euonymus, dogwood, pear, raspberry, marigold and many other plants. Twospotted spider mite has the widest host range of any spider mite in the world.

Damage and Diagnosis: The twospotted spider mite pierces plant cells and feeds on the sap. Small white flecking injuries are typical around feeding sites. More generalized discoloration, typically bronzed-colored, occurs as infestations progress. Vigor of plants may be seriously reduced. Premature leaf drop occurs on heavily infested plants. Twospotted spider mite tends to produce more webbing than the other mites commonly found on shade trees and shrubs.

Twospotted spider mite is generally straw-colored and has a pair of dark blotches in the middle of the body that run along the outside. (Note: This character of having two dark spots is shared by several other mite species.) During the end of the season or at other times of colony stress orangish forms develop that indicate the mites are in a semi-dormant condition.

Many populations of twospotted spider mite show very considerable levels of pesticide resistance making management much more complex than for other spider mites.

Life History and Habits: Winter is spent as a semi-dormant adult female, usually under sheltering debris in the vicinity of previously infested host plants. A few may winter under bark cracks or other protected sites. Early spring feeding usually occurs on weeds and other herbaceous perennials. Much of the early season feeding may occur at these sites before the mites move to shrubs and trees.

During the growing season the life cycle of the twospotted spider mite is typical to that described above (i.e., egg, larva, protonymph, deutonymph, adult). Under favorable conditions of warm temperature and low humidity, the generations are completed in as little as 10 days. This can be greatly extended under cooler conditions. Adult females may lay up to five eggs per day over the course of two to three weeks.

Late in the season twospotted spider mites change to a less active non-feeding stage. At these times it changes to an orange-red color and migrates to sheltered areas until more favorable conditions indicate the recurrence of host plants. However, under protected conditions, such as are found in greenhouses, twospotted spider mite can breed continuously.

Management: Several organisms prey on spider mites in field settings; minute pirate bugs and predatory species of mites are among the most important. The bandedwinged thrips are also sometimes important predators of spider mites.

Watering and water management are very important in controlling twospotted spider mite. Overhead watering - and purposeful hosing of plants with water in a garden setting - can dislodge and kill many spider mites. Providing adequate water for plant growth needs is also important in managing spider mites. Drought and fluctuating wet/dry soil conditions can stress plants in a manner that can cause spider mite populations to increase.

The twospotted spider mite is often the most difficult mite to control with chemicals. High levels of resistance have developed in many populations. Alternating use of different classes of miticides at a site is recommended.

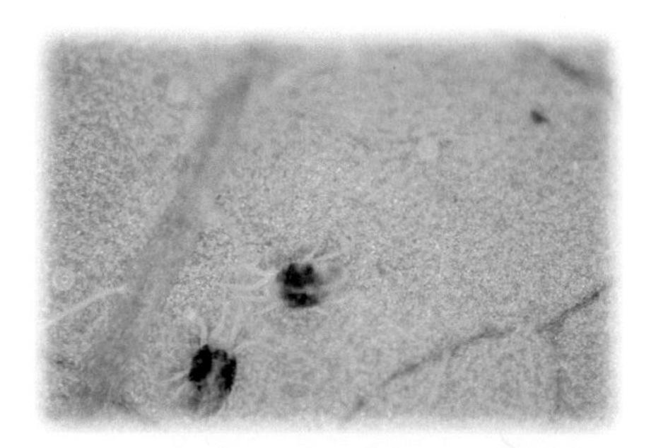

TOP: Honeylocust spider mites.
BOTTOM: Honeylocust mite bronzing of foliage.

HONEYLOCUST SPIDER MITE

Platytetranychus multidigituli (Ewing)
Acarina: Tetranychidae

Host: Honeylocust

Damage and Diagnosis: Honeylocust spider mite is a pale green mite that feeds on the underside of honeylocust leaves. As they feed their mouthparts slice the outer cell layers and they feed on the released sap, leaving small flecked wounds. Feeding is particularly concentrated along the mid-rib of the leaf underside, where eggs, old egg shells and/or the mites can usually be found. During heavy infestations leaves turn an off-yellow color and the entire crown be similarly discolored. Premature leaf drop often occurs in years when midsummer outbreaks occur. Problems with honeylocust spider mite are generally limited to trees planted along roads, in parking lots and in similar hot, dry sites.

Summer forms of the honeylocust spider mite are yellow-green and are usually found concentrated along the main leaf vein. In late summer they begin to turn more orange and overwintering forms are bright orange.

Associated Insects: An eriophyid mite is commonly associated with honeylocust spider mite on leaves of honeylocust and is thought to induce a rusty coloration of infested leaves.

Life History and Habits: Adult females, orange-red in color and in reproductive diapause are the overwintering stage. Beginning in late summer they cluster around buds or in bark cracks of branches and also may be found in masses on the lower trunk. In late spring, they resume activity, revert to their normal greenish color and begin to lay eggs on leaves. Eggs hatch and the immature nymphs feed on the leaves, molting two times before reaching the adult form. Time from egg stage to adult ranges from eight to seventeen days, depending on temperature. Multiple generations occur and in most years populations tend to increase steadily through July and decline through August. Years where there is unusual spring rainfall or cooler temperatures may see a shift in population peak of about a month.

Management: Few natural enemies have been observed attacking honeylocust spider mite in Colorado. Among the most important are the predatory plant bug *Deraeocoris nebulosus*, a predatory mite (*Amblyseius fallacis*) and a predatory thrips (*Scolothrips sexmaculatus*)

Honeylocust spider mite problems seem to be associated with dry sites, such as street tree plantings. They are rarely a significant problem when surrounded by irrigated turfgrass that apparently increase humidity.

Honeylocust spider mite is relatively easy to control with several miticides. Dormant season applications of horticultural oils, when applied to both the crown and trunk, can be effective against the overwintering females on twigs, branches and trunks.

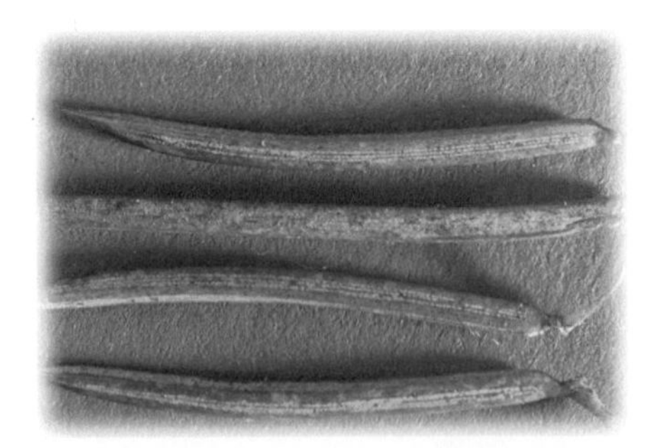

ABOVE: Spruce spider mite damage. Photograph courtesy of Oregon State University.

SPRUCE SPIDER MITE

Oligonychus ununguis (Jacobi)
Acarina: Tetranychidae

Hosts: Most conifers but injury is particularly common on spruce and junipers.

Damage and Diagnosis: Sap is removed and tiny light colored flecks are produced at feeding sites on needles. Infested trees become grayish with unthrifty appearance and may prematurely defoliate. Peak feeding injury occurs in late spring and fall, but symptoms of injury intensify during summer.

Spruce spider mites are small (1/25 inch) dark green mites or dark red, depending on the season. (Newly hatched mites are light salmon to pale red.) There is only a single dark area on the abdomen, separating them from twospotted spider mite. During heavy infestations they may produce a fine netting of silk at the base of needles.

Associated and Related Species: The ponderosa pine spider mite, *Oligonychus subnudus* (McGregor), is a common pest of pines, particularly ponderosa pine. *Platytetranychus libocedri (McGregor) is another spider mite often associated with arborvitae and juniper. It tends to become active somewhat earlier in the season than the other "cool season" mites associated with conifers.*

Eriophyid mites, tentatively identified as Nalepella halourga Keifer, have been associated with russetting injuries to spruce foliage. These have been observed to be most common in early spring.

Life History and Habits: Spruce spider mite spends the winter in the egg stage, attached near the base of needles. Eggs hatch in midspring and the newly emerged nymphs feed on on older growth. The early stage larva are salmon to pale red in color and have only six legs. As they develop they molt to eight-legged deutonymph and adult stages. All stages feed on the sap of conifer needles.

Development can be completed in about two weeks. Females lay about 30 to 40 eggs, tending to lay more eggs at cooler temperatures. When temperatures consistently exceed 80s, populations decline, many eggs go into dormancy causing populations to usually peak in late June or early July. However, some individuals mites remain throughout summer in the cooler parts of the canopy. A second, smaller population peak occurs in September and October.

Management: Among the most important natural controls are minute pirate bugs, predatory plant bugs (*Deraeocoris nebulosus*), spiders and various predatory mites. High humidity is reported to suppress development of damaging populations.

Miticides should be applied when sampling detects high mite populations. Spruce spider mite populations are best monitored by firmly tapping the branches over a sheet and checking for the presence of mites and predators. Such sampling is best done during the early stages of population development, in May and June for the first peak and again in September.

PONDEROSA PINE SPIDER MITE

Oligonychus subnudus (McGregor)
Acarina: Tetranychidae

Hosts: Pines, primarily ponderosa pine

Damage and Diagnosis: Ponderosa pine spider mites make small flecking wounds in needles to feed on the sap. Heavily infested needles become generally grayish and unthrifty in appearance.

Ponderosa pine spider mite are small (1/25 inch) green mites with some orange markings. Little webbing is produced by this mite.

Associated/Related Species: A closely related spider mite, *O. milleri* (McGregor), can also be found during spring on Scots and Austrian pine. However, it is reported not to be damaging.

Life History and Habits: Winter is spent as eggs on twigs around the base of needles. Eggs usually hatch during early spring and populations build rapidly peaking in May and June. During midsummer numbers of mites decline sharply, with most apparently remaining in temporary dormancy. Activity increases again during late summer, with a second, smaller peak in population often in September.

Management: There are several enemies of ponderosa pine spider mite. Most important appear to be predatory bugs, including minute pirate bugs and the predatory plant bug *Deraeocoris nubulosus*. In addition there are several predatory species of mites from the Bdellidae and Phytoseiidae families found associated with ponderosa pine spider mite.

Although a "cool season" mite, *O. subnudus* injury is usually most severe on plantings in dry sites, such as parking lots. Plantings made where humidity is higher should limit outbreaks. Horticultural oils and miticides such as bifenthrin (Talstar), dicofol (Kelthane), dimethoate (Cygon) and Hexygon should be effective for managing ponderosa spider mite.

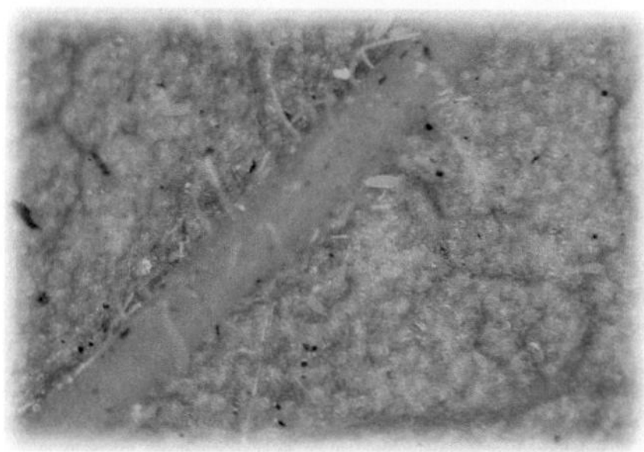

TOP: Ponderosa pine spider mite infestation.
BOTTOM: Rust mites on honeylocust.

HONEYLOCUST RUST MITE

Aculops sp.
Acarina: Eriophyidae

Hosts: Honeylocust

Damage and Diagnosis: These tiny eriophyid mites feed on the surface of the leaf undersides, bruising the epidermis and feeding on sap. They co-occur with the honeylocust spider mite and foliar discoloration common to honeylocust may result from both species. It is suspected that the more rusty brown foliar symptoms common in late summer develop from the honeylocust rust mite injuries.

Life History and Habits: Biology is unknown. Overwintering stages are presumed to be eggs laid near buds. Rust mites can be found on foliage by mid-June but peak populations typically are present in late July and early August.

Management: Several natural controls have been observed, primarily predatory thrips and predatory mites.

This species is relatively easily controlled with various insecticides and miticides. One interesting difference in control compared to the honeylocust spider mite is that carbaryl (Sevin) is quite effective and acephate (Orthene) ineffective for the honeylocust rust mite. Based on the presumed biology, dormant season sprays of horticultural oils should help control honeylocust rust mite.

TOP: Iron chlorosis on silver maple. (W. Jacobi)
SECOND: Chlorosis on rose species. (M. Schomaker)
THIRD: Salt damage to lodgepole pine. (Bill Jacobi)
BOTTOM: Salt damage to aspen. (Bill Jacobi)

Foliar Diseases

Foliar diseases that remove leaves and thus photosynthetic area do not usually kill the tree if the damage is for one year. Hardwood trees can withstand defoliation better than conifers since they normally do not retain their leaves from year to year. Conifers can be defoliated for one to several years before mortality occurs. However, all trees are stressed by defoliation and other insects and diseases may cause problems for the stressed trees.

Many agents can cause needle and leaf discoloration or death, including:

- **Fall needle drop**: Older conifer needles die normally each fall- usually a conifer will retain at least three to four year's growth of needles.
- **Insects** can remove needles - scales, leafminers, mites, aphids, defoliators such as elm leaf beetle.
- **Abiotic damages:** Frost, road salt, high temperatures, drought, winter drying, root damage from construction etc. We do not have air pollution damage to tree foliage in the region yet.
- **Fungi and bacteria**: Various fungi and bacteria can attack leaves and cause spots, complete death and defoliation. Root diseases cause leaves to yellow and die.

Diagnosis Hints: Diagnosing foliar damage is not easy if there are no distinct patterns that would suggest insect feeding or if there are no fungal fruiting bodies- small black pimples or microscopic spores.

- **Fall needle drop** occurs every year in the fall and involves older needles. Needles turn yellow to brown and fall off. If the tree retains less than three years of needles the tree is probably stressed.
- **Insects** usually cause discernable damage that looks like mining, chewing, or sucking by piercing sucking mouthparts. Aphids leave so much honeydew that leaves look wet and shinny.
- **Abiotic damages**: Abiotic damages usually do not look like an insect or fungus is moving through the tissue and thus have more uniform symptoms such as needles being all reddish brown from winter burn. The distribution of symptoms on trees and among trees in the area can give hints of the cause of damage. Abiotic damages are usually more uniformly distributed on a tree and within an area if it is weather-related etc.
- **Fungi and bacteria**: Fungi usually cause irregular damage with various color changes to hardwood leaves and spots or complete needle death on conifers. If there are not distinct raised dark colored spots (fruiting bodies) on needles there is no easy way to determine if the damage is fungal. The pattern of needle death may help- one-year's growth is usually affected and usually lower branches are affected. Bacteria usually cause a very black uniform damage to the leaf. Bacteria can also cause cankers that girdle the branch or stem causing leaf death.

ABIOTIC DISEASES

IRON CHLOROSIS

Hosts: Many plants can suffer from chlorosis, apple, aspen, cotoneaster, crabapple, stone fruits, elm, juniper, linden, pin oak, roses, silver maple, spirea and spruce are a few of the commonly affected ones in the region.

Diagnosis and Damage: Plants affected by iron chlorosis first show interveinal chlorosis, leaf veins remain green and the rest of the tissue turns pale green or yellow, the newest leaves typically being the most affected. Untreated, the condition progresses, and the interveinal tissue turns brown, similar in looks to a foliar fungal disease. Poor growth, stunting, premature leaf drop, poor or no flower and fruit production, twig and branch dieback and death of the plant may occur either as a result of the chlorosis or a secondary factor. Sometimes only part of a plant may be affected, but in severe cases, the entire plant will exhibit the symptoms.

Biology and Disease Cycle: In alkaline soils, which are prevalent in Colorado, iron and possibly other micronutrients are not readily available to plants. Plants that have a high iron requirement or those that do a poor job of taking it up often develop the problem. Among other functions, iron is needed in the synthesis of chlorophyll, the green pigment in plants. When plants are unable to absorb sufficient iron, chlorophyll manufacture is negatively affected. Because iron is relatively immobile in plants, new growth shows the symptoms of iron chlorosis first. Iron deficiency is not the only cause of leaf yellowing. Herbicide damage or other deficiencies such as nitrogen, manganese, boron or zinc may be similar to those of iron deficiency.

Conditions Favoring and Management of Disease: Iron chlorosis may also develop in plants with inadequate or stressed root systems. Any factor interfering with root function such as compacted clay soils, waterlogged soils, excessive organic matter, excessive salt levels, nutritional problems, soil sterilants can cause chlorosis symptoms to appear. Excessive amounts of phosphorus in the soil can bind iron, causing deficiency symptoms to occur.

Plant appropriate species for soil conditions in the landscape. Where possible, avoid those that are commonly susceptible to iron chlorosis, such as pin oak, silver maple and aspen. Add organic matter before planting, as it can help hold iron in a form that is readily available to plants. Determine if an environmental/cultural factor might be causing the symptoms (i.e. wet soil, dry soil, etc.) and remedy the situation if practical. Chelated iron, forms that are available to plants, can be applied to soils. Iron sulfate or other similar forms of iron are generally ineffective applied to soil since the same forces that cause iron to be unavailable in the first place act upon the applied iron. Chelating holds the element in a form that the plant can use. Iron sprays applied to the foliage are another alternative. These sprays, which can stain sidewalks and other surfaces, should be applied only during cooler weather or evenings to avoid damaging leaves. This method is often not practical or effective on large plants, but is more effective on small ones. Iron (ferric ammonium citrate) can be injected or capsules can be implanted into trunks. Injection damages the tree via the injection holes where disease organisms may enter and occasionally directly damages the tree. Holes required for capsules may attract insect or disease organisms, so these treatments are generally limited to extending the life of trees for a few more years.

LEAF SCORCH (Drought)

Hosts: All plants

Diagnosis and Damage: Leaf veins remain green and the tissue in between turns brown and brittle. Leaves may turn *brown* from the *outside edges inward*. Affected leaves may drop or remain attached. Occasionally, leaves on one side of a tree or shrub show symptoms, while the rest looks normal. Other damages may look like drought symptoms such as excess soil/road salt, root damage, herbicide damage, and some nutrient problems. Damages include poor growth, stress and death from chronic scorch or secondary insects and diseases.

Biology and Disease Cycle: Scorch is often blamed on insufficient water uptake, but other situations may play a role in its expression. Any factor that damages roots can induce scorch symptoms since poorly functioning or non-functioning roots can't absorb water well. Situations that can produce scorch symptoms include those where water is available, but plants are unable to keep up with the water loss. These situations arise during hot, dry, windy weather; from construction activities such as trenching and grading; when there is, or has been, too much water in the soil; compacted soil, improper

TOP: Leaf scorch on Ohio buckeye. (M. Schomaker)
SECOND: Leaf scorch on prunus species. (W. Jacobi)
THIRD: Symptoms of herbicide injury to bark of honeylocust trunk. (Bill Jacobi)
BOTTOM: Twig swelling symptom of herbicide injury to honeylocust. (Bill Jacobi)

planting, including too shallow or deep; over-preparation of soil; excessive fertilization; additions of mulch, plastics or pavement that restricts or suffocates roots; reflected heat and light from buildings and paved surfaces; mechanical trunk injuries; diseases that cause trunk or branch cankers which restrict water flow; advanced leaf chlorosis.

Conditions Favoring and Management of Disease: Hot, dry weather and/or periods of hot dry, wind that affect water uptake.

Management includes identification and correction of the cause or causes of the problem. Improve tree vigor through proper pruning, watering, fertilization and pest management. Check soil moisture by sampling at the drip line of the plant from about 6 to 10 inches deep. If the soil is dry when held (crumbly), water. If the soil is wet (water runs out when it is squeezed) wait to water. If soil is moist, but no water runs out, soil moisture should be adequate.

MISAPPLIED HERBICIDES

Hosts: All tree and shrub species are susceptible to damage by herbicides.

Diagnosis and Damage: Symptoms vary depending on the kind and concentration of the chemical, frequency of use, species and condition of plants, and the timing of chemical application. Some of the common symptoms of hormone-type herbicide injury on **hardwoods** are wavy or curled leaf margins, cupped leaves, twisted or drooped petioles, parallel leaf venation, chlorosis, blight and necrosis of leaves. On **conifers** symptoms can include abnormal swelling, twisting, and curling of branchlets, bleaching, browning and casting of needles. Nonselective herbicides tend to halt growth and cause chlorosis of new and old foliage. At high doses, foliar browning, foliar drop, and dieback of twigs and branches occurs. Trees and shrubs injured by these herbicides are less likely to recover than are plants injured by hormone-type herbicides. All types of herbicide damage may cause partial or total mortality of trees.

Mode of Action: Herbicides that often cause injury to non-target plants fall into two general groups, those that are used to kill established broadleaf weeds and those intended to prevent emergence of weed seedlings (preemergent) or kill all vegetation (nonselective).

Hormonal herbicides are used on turf to selectively kill broadleaf weeds. Herbicides in this group include the phenoxyacetic acids (2,4-D, 2,4-DP, MCPA, MCPP, and related compounds). These act as plant hormones that disrupt normal growth processes. Herbicides of this type move to growing points and cause development deformities. Symptoms develop several days to week after exposure.

Preemergent and nonselective herbicides, are either applied to target plants or applied to soil and taken up by roots. Those applied to soil are intended to have long-lasting residual activity, so affects can be persistent. General vegetation herbicides have various modes of action, therefore creating symptoms that vary and are not as diagnostic.

Conditions Favoring and Management of Disease: Herbicides that are mishandled or are applied under improper conditions often end up injuring non-target plants. Great care should be taken when applying herbicides to measure product carefully, and dilute properly. Avoid misapplication to non-target plants by applying sprays only at non-windy times, when drift of herbicide will be unlikely. Additionally, vapor drift can occur with hormone-type herbicides of temperatures are too high during applica-

TOP: Herbicide induced abnormal terminal growth on Austrian pine. (M. Schomaker)

SECOND: Hormonal herbicide damage - spongy bark.

THIRD: Soil sterilant effect on deciduous tree. (W. Jacobi)

BOTTOM: Dicamba herbicide causing bark splitting on silver maple. (W. Jacobi)

TOP: Banvil herbicide effect on sumac. (W. Jacobi)
SECOND: Herbicide effect on pine causing twisted leader. (W. Jacobi)
THIRD: Winter burn on ponderosa pine. (W. Jacobi)
FOURTH: Winter burn on native conifers at high elevation. (W. Jacobi)
BOTTOM: Severe winter injury on ponderosa pine. (M. Schomaker)

tion or soon after. Some plants very sensitive to hormone-type herbicides are boxelder, grape, redbud, lilac and tomato. Often they show at least mild foliar symptoms of injury from nearby use of this type herbicide. When applying granular broadleaf herbicides remember that the roots of trees extend up to five times beyond the diameter of the tree crown. Granular herbicides can leach through soil and be absorbed by tree roots present in the area of application. Glyphosate is an example of a general-purpose herbicide that is sprayed on target plants and enters only through photosynthesizing plant parts such as leaves and green stems. When it comes in contact with soil it becomes tightly bound and is soon degraded by microorganisms. This chemical is also relatively non-volatile.

WINTER DESICCATION OF EVERGREENS
(Winter Burn, Red Belt, Winter Fleck, Winter Drying)

Hosts: Evergreens

Diagnosis and Damage: Bleached, yellow or brown flecks may appear on needles. Needles may turn yellow, red or red-brown from the outside inward. Needle tips may turn brown. Entire branches may yellow and brown from the outside inward. Symptoms may be more severe on the windward side of the plant. Usually current year needles are the most severely affected. Winter fleck occurs commonly on conifers growing above 6,000 feet, mainly on the upper needle surface of two-year-old and older needles. The flecks are yellow to brown and seem to be related to snow, cold temperatures and exposure to the sky. Defoliation is not common with this damage.

In addition to the damage described above, root damage or death also occurs. Corresponding death of above ground parts may follow during the next few growing seasons, depending on the severity of damage. Because winter desiccation stresses plants, they may also succumb to secondary insect or disease organisms. While unattractive, winter injury symptoms on evergreens are eventually masked by the development of new growth in the spring and dead needle drop. Affected parts do not usually need to be removed unless buds are also dead. If buds at the tips of branches feel soft and spongy and appear green, they are alive. Dead buds are brown and brittle.

Biology and Disease Cycle: Evergreens lose moisture through needles to the surrounding environment during the winter months. In order to stay healthy, the plant must replace this lost moisture by absorbing water through the roots. Sometimes plants are unable to keep up with the loss. This is common during windy conditions and when daytime temperatures rise above freezing. Sometimes there is plenty of soil moisture, but plants are unable to absorb it because the ground is frozen. In most cases, there is inadequate soil moisture available to evergreens. Desiccation also occurs when there is fluctuating warm daytime and freezing nighttime temperatures leading to a freeze drying of tissues.

When there is inadequate soil moisture, roots die or function poorly. Water uptake is affected, resulting in brown or red, dead, above ground growth. Plant damage often continues to occur into the growing season, especially if root injury has been severe. Chronic winter drought may accumulate over time and kill or severely stress plants through the course of several years.

Conditions Favoring and Management of Disease: Typical fall/winter conditions of the region such as drying winds, lack of snow cover, highly fluctuating winter temperatures, low soil moisture and frozen soil.

TOP: Winter fleck damage. (W. Jacobi)
SECOND: Winter fleck damage. (W. Jacobi)
THIRD: Ash yellows in green ash. (W. Jacobi)
BOTTOM: Ash yellows broom in green ash. (W. Jacobi)

Management includes watering evergreens approximately monthly in the absence of adequate snowmelt or rainfall (November through March). Water during warm periods when the soil can absorb. Begin watering early in the day so that it soaks in rather than accumulates around the base of plants, which can cause rotting or bark injury. Permanently mulch around trees and shrubs, add mulch in late fall to help conserve soil moisture by reducing loss from evaporation. Plant varieties that are adapted to local winter conditions and plant so buildings protect tree from winter winds. Recent transplants can be protected with a burlap or similar windscreen located on the north and west exposure.

BACTERIAL DISEASES (PHYTOPLASMAS/SPIROPLASMAS)

ASH YELLOWS

Causal organism: phytoplasma

Hosts: Many ash (*Fraxinus*) species are hosts to ash yellows. From most to least susceptible are white ash (*F. americana*), green ash (*F. pensylvanica*), and black ash (*F. nigra*).

Diagnosis and Damage: The only reliable diagnostic field symptom is the presence of witches' brooms on the trunk and major limbs. Leaves on the brooms tend to be small, simple and **chlorotic**, - branches of brooms do not have a dominant shoot but are made up of many shoots (it is important that normal sprout growth on the trunks and large limbs called epicormic sprouts not be mistaken for witches' brooms). Small roots or leaves collected from symptomatic trees can be tested for the presence of the ash yellows phytoplasma. Over time there is a permanent reduction in tree growth. After several years of minimal growth, leaves often will only appear in tufts at the ends of branches. Leaves may fail to attain normal size and appear **chlorotic**. Branch dieback may occur following dormant periods. Infected trees do not recover. Ash yellows apparently reduces the ability of white ash to recover from other stresses, such as drought, stressful sites, etc. and often exhibit-slowed growth and dieback. However, infected green and black ashes tolerate infection in many instances without progressive deterioration. Many other factors may be involved with declining ash trees, including insect borers, over watering and other environmental stresses.

Biology and Disease Cycle: The causal organism for ash yellows is unnamed, because it has not yet been isolated and characterized. It is know that the disease is caused by a phytoplasma, a prokaryotic organism, similar to a bacteria, but simpler in structure. Insects such as leafhoppers, planthoppers and psyllids vector most phytoplasms.

Conditions Favoring and Management of Disease: Conditions that cause stress to susceptible host trees will favor disease development. Management practices that maintain tree vigor through proper cultural practices will slow disease progression and damage.

WESTERN X-DISEASE OF CHOKECHERRY

Causal organism: phytoplasma (spiroplasma)

Hosts: Hosts are chokecherry, sweet and sour cherries, several varieties of peaches, and some other *Prunus* species.

Diagnosis and Damage: Infected leaves become greenish-yellow in late June. These leaves may have a reddish tinge on their borders. In July and August the leaves turn deep red. Shoots are stunted, and rosettes result from shortened internodes at the tip. Infected fruits are somewhat pointed and are yellowish-red, not the normal deep red of healthy fruit. Both diseased and healthy fruits may be found on the same tree. Growth of infected chokecherry is reduced, internodes become shorter; the shrubs gradually decline, and ultimately die. Infected fruits are not suitable for use in jams and jellies, and their seeds do not germinate.

Biology and Disease Cycle: The western X-disease pathogen for many years was thought to be caused by a virus; however, the disease is caused by a spiroplasma. This pathogen is transmitted to *Prunus* hosts by leafhoppers when they feed on the leaves. Symptoms on leaves usually do not develop until the growing season following the year of transmission.

Conditions Favoring Disease and Management: Management of leafhopper vectors and careful choice of planting combinations in windbreak and nursery situations is recommended. Nurseries should avoid establishing beds of chokecherry near *Prunus* species that are hosts of the pathogen. Because the pathogen spreads rapidly from infected to healthy shrubs, new plantings of chokecherry should not be established near infected chokecherry. Resistant chokecherries are being identified for the northern plains and may be available in a few years.

VIRAL DISEASES

Viruses are infectious particles that consist of a nucleic acid core with a protein coat. They replicate in nature only within living cells, controlling synthetic processes of the host so that it produces more virus particles. Plant viruses as a rule are named after the first plant on which they are found.

Diagnosis and Damage: Most viral diseases of trees and shrubs cannot be diagnosed on the basis of symptoms alone. A few plant symptoms, such as mosaic patterns on leaves and chlorotic or necrotic ring spots, can be attributed to viruses with some degree of confidence. Most other symptoms caused by viruses resemble those caused by numerous other factors. Virus and virus-like symptoms are relatively well known and understood in species cultivated for orchard crops, but most viral disorders in forest, shade and ornamental trees and shrubs have not been studied. Viruses in trees and shrubs often cause no visible symptoms, or they may cause symptoms ranging from slight foliar markings and mild growth suppression to dramatic colorful patterns on leaves, stem distortions and cankers, twig and branch dieback, graft union necrosis, and slow or rapid decline of plant. Two or more viruses or viruses plus other pathogens may infect a plant simultaneously, which complicates both symptoms and diagnosis.

Biology and Disease Cycle: Plant virus particles, which are composed of a nucleic acid core with a protein coat, are very small, and only visible with an electron microscope. They are obligate parasites and cannot survive outside of their host, therefore they must be moved from one plant to another by insects, grafting, or in seed from infected parent plant. Viruses do not divide and do not produce special reproductive structures, they multiply by inducing host cells to form more virus. Plant viruses that cause significant disease become distributed systemically throughout the plant. Most plant viruses invade phloem and parenchyma tissues, but some are limited to phloem. A virus that is transmitted by one group of organisms (e.g. aphids) is not transmitted by other kinds of organisms.

Management: Viruses can not be controlled with pesticides, but instead must be controlled through other management practices. Management options include prevention of introduction of virus into plant material. Propagation facilities should, if possible,

TOP: Western X disease of chokecherry. (J. Walla)

SECOND: Western X disease of chokecherry. (J. Walla)

THIRD: Green ash with tomato ring spot/ tobacco mosaic virus foliar symptoms. (W. Jacobi)

FOURTH: Tobacco mosiac virus on green ash. (J. Costello, University of New York)

BOTTOM: Symptoms of multiple viruses on white ash. (J. Costello, University of New York)

use propagation stock and/ or seed that is certified virus free. Pruning tools should be sterilized with 10 percent bleach solution or other sterilant when used to prune hosts that may be infected with a virus disease. Protecting them against the virus insect vectors may protect plants against certain viruses. Because viruses are transported widely within infected, often-symptomless plants, known weedy hosts of viral disease in question should be eradicated from the area of host plant. Health and vigor of a plant does not help it resist infection by a virus, although plant varieties resistant to certain viral diseases do exist.

TOP: Apple union necrosis (TmRSV) symptoms on apple tree. (Larsen)
SECOND: Peach yellow bud mosaic virus symptoms on peach brances.
THIRD: Apple mosaic virus leaf symptoms on golden delicious apple. (Larsen)
FOURTH: Cherry mottle leaf virus symptoms on sweet cherry leaves. (Larsen)
BOTTOM: Vein clearing symptoms of peach mosaic virus. (Luepschen)

Virus Diagnosis Guide

HOSTS,SYMPTOMS	DISEASE
Stone fruits, rose species. Symptoms vary dependent on host and include shot holes on young leaves, chlorotic leaf mosaic and line patterns, leaf necrosis, leaf deformity, delayed fruit maturity.	Prunus necrotic ring spot virus (PNRSV)
Apple, prune plum. On apple the first symptom is pitting at the graft union that become deep invaginations in the trunk. Structurally weak tissues are laid down and trees may break off at the union. On prune symptoms include weak growth with sparse foliage. At graft union removal of bark reveals a brown line (necrotic tissue).	Tomato ringspot virus (causes apple union necrosis, prune brown line, and peach yellow bud mosaic)(TmRSV)
Grape. Variable symptoms are possible. Leaves may show a green or yellow mosaic,rings, line patterns or flecks, and exhibit a fan-like appearance.	Grapevine fanleaf virus
Apple. Expanding leaves develop pale yellow to cream-colored areas, appearing as spots, flecks blotches, vein-net, line patterns, or bands along major veins. Areas turn chrome yellow to white, become necrotic. Leaves may drop prematurely veins	Apple mosaic virus (ApMV)
Sweet cherry. Leaves show light green or whitish, diffuse discoloration between the secondary leaf veins, affected leaves often develop an irregular shape, even appearing shredded. May reduce fruit size.	Cherry mottle leaf virus
Peach, apricot, plum. Symptoms vary with host. Color-breaking of flower petals that may be smaller, crinkled and deformed. Leaves show chlorotic spots that vary from tiny flecks to streaks to vein feathering or bold blotches. These areas become necrotic and drop out.	Peach mosaic virus
Cherry. Leaf-like projections develop along the midrib on the underside of affected leaves. In severe cases, the whole leaf can be affected. Symptoms begin in lower part of tree because disease is introduced by nematode vector into the roots.	Cherry rasp leaf virus (CRLV)
Peach, plum. Peach trees produce only mild stunting, and no leaf symptoms. But peaches infected with both PDV and PNRSV showed severely retarded growth and reduced yields. Infected plum and prune develop narrow, straplike leaves that are thicker than normal and early season growth appears bunched due to shortly spaced early season internodes.	Prune dwarf virus (PDV)
Rose. Symptoms vary dependent on rose species or cultivar and conditions. Foliar symptoms range from chlorotic mottles and ringspots to light green or chlorotic line patterns to vein clearing (yellow net), vein banding, and mosaics of green and yellow or white. Leaflets may pucker. Symptom expression is	Rose mosaic complex, caused by a complex of viruses, PNRSV and ApMV appear to be the principle pathogens.

TOP: Flower color break symptoms of peach mosaic virus. (Larsen)
SECOND: Foliar symptoms of cherry rasp-leaf virus on sweet cherry and peach. (Luepschen)
THIRD: Prune dwarf virus symptoms on Italian prune. (Larsen)
BOTTOM: Apricot ring pox red blotched vein symptoms on apricot. (Larsen)

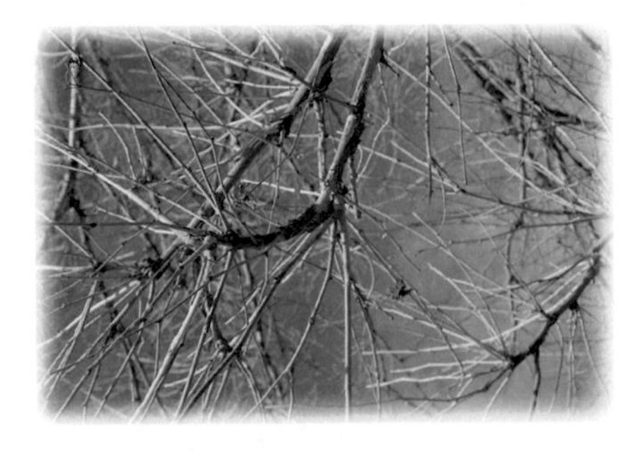

TOP: Apricot ring pox symptoms on cherry leaves. (Larsen)
SECOND: Cherry twisted leaf symptoms on cherry leaves. (Larsen)
THIRD: Sycamore anthracnose brooming symptoms. (W. Jacobi)
BOTTOM: Anthracnose symptoms on sycamore leaf. (W. Jacobi)

dependant on temperature. Diseased rose bushes are less vigorous and more susceptible to winter injury.	
Apricot, cherry. On apricot symptoms include vein banding, chlorotic spots, streaking and rings. Midsummer rings or spots may become necrotic and drop off. Affected trees may show dark purple discoloration of short section of some leaf petioles and concentric rings can develop on fruit. On sweet cherry the most consistent symptom is a twisting the midrib or petiole of leaves.	Apricot ring pox, cherry twisted leaf Unknown causal agent
Sweet cherry. Delayed budbreak, leaves are small and more erect than those are on healthy limbs or trees. Lower portions of affected trees often are bare of leaves. Diseased fruit is small and off color. Wood beneath bark may exhibit pits and grooves near the soil line.	Cherry stem pitting, caused by a graft-transmissable agent of unknown etiology that appears to be soilborne.

FUNGAL DISEASES

ANTHRACNOSE

Sexual stage: *Apiognomonia* spp., *Gnomonia* spp., *Glomerella spp.*
Asexual stage: Gloeosporium spp., *Discula* spp., *Kabatiella* spp., *Sphaceloma* spp.

Hosts: Ash, maple, rose, sycamore, and many other deciduous trees and shrubs

Diagnosis and Damage: This group of diseases is named for the foliar symptoms caused by certain groups of fungi. Symptoms vary depending on host. There are two types of foliar symptoms associated with the term anthracnose. The first and most commonly seen is necrotic irregular blotchy areas associated with veins that expand into interveinal areas. Less common, spot anthracnose, exhibits spots or round necrotic circles scattered throughout the leaf. *Dark or chlorotic margins (lines of demarcation) between healthy and diseased tissue are easily seen without the aid of a hand lens or dissecting microscope.* Twig blighting may occur in some species, such as ash, oak, and sycamore. *Symptoms may be confused with leafminer damage, leaf scorch, or frost injury.*

Fruiting structures (acervuli) appear within damaged/dead tissue or on necrotic leaf veins. Fruiting structures located within damaged tissue are not easily viewed without a microscope, however, fruiting structures produced on darkened leaf veins may be viewed with a hand lens or dissecting microscope of 30 power or more. Nondescript spores (conidia) are colorless, one-celled, ovoid to oblong in shape.

Damage is usually minor. However, on sycamore twigs leaf drop or defoliation may occur when disease is severe. Sycamores usually lose leaves in early spring, refoliate, and have the whole growing season to recover so the disease does not seem to seriously affect trees. On sycamore, leaf symptoms may occur in early spring, causing confusion with frost damage symptoms. Anthracnose on sycamore can cause twig cankers and under stressful conditions, large cankers on the main stem. If disease occurs annually, damage potential greatly increases and affected plants may decline. Trees stressed by other factors (drought, oxygen, poor soils etc) are more susceptible to the disease.

Biology and Disease Cycle: Anthracnose fungi overwinter on blighted twigs or fallen leaf debris. Initial leaf infection occurs from spring to mid-summer when moisture is frequent and temperatures cool to moderate. Spores are disseminated by wind and splashing rain to buds, shoots, and expanding leaves. Leaf symptoms, more commonly, become noticeable in mid to late summer.

Conditions Favoring Disease and Management: Periods of cool, wet weather in spring and summer favor disease development. The fungus is inhibited by temperatures above 55 F. Leaf debris should be raked up and discarded in autumn to decrease the amount of overwintering fungus. If fungus infections are on twigs, pruning may be warranted. Chemical control is usually not needed, however, if necessary, protectant fungicide applications should be applied in the spring at leaf emergence, and then again every 14 to 21 days if weather conditions remain wet and cool. Fungicide injections have given good control of twig cankers.

APPLE SCAB

Sexual stage: *Venturia inaequalis*
Asexual stage: Spilocaea pomi

Hosts: Apple and crabapple (especially the variety Radiant)

Diagnosis and Damage: *Necrotic spots* on leaves usually found along the mid rib or in association with leaf veins. Spots begin as small olive green circles *with radiate (or feathery) margins* that enlarge and darken. The fungus grows on the upper leaf surface in a branched- feather like form. Chlorosis and death of leaf tissue may result. Unlike scab lesions in high humidity areas, lesions forming under semi-arid conditions are not apt to form velvety masses of brown spores. Symptoms on fruit are also unlikely to occur in the Rocky Mountain Region unless humidity is high. Ascospores of *Venturia inaequalis* are unequally two-celled, with the upper cell shorter and wider than the lower cell. Conidia of *Spilocaea pomi* are one to two celled, and ovoid to lanceolate.

In our semiarid region damage from scab is usually minimal. Chemical control is usually not necessary. However, during spring's high humidity and on susceptible species, the disease can cause serious defoliation and become aesthetically important if left unchecked.

Biology and Disease Cycle: Scab fungi overwinter on fallen leaves and leaf debris. In the spring when overwintering ascospores become wet, they are forcibly discharged from the fruiting structure and wind disseminated. Spores land on newly emerging leaf tissue and germinate in a film of free moisture. After germination begins, however, the relative humidity must be 95 percent or greater for the process to continue. Fungal mycelium become established between the epidermis and cuticle. Lesions containing conidia are formed, which are disseminated via splashing rain or wind to begin secondary infections throughout the summer.

Conditions Favoring Disease and Management: High relative humidity and long periods of leaf wetness (at least nine hours) in the spring along with temperatures between 6 and 26 C favor disease development if a spore base is present.

Natural controls include dry weather. Preventative management of the disease involves collection of fallen leaves and leaf debris (sanitation) and the use of resistant varieties. Many fungicides are labeled for scab prevention.

BLACK SPOT OF ELM

Stegophora ulmea (syn. *Gnomonia ulmea*)

Hosts: Japanese Zelkova and many elm species including American, Chinese, and Siberian

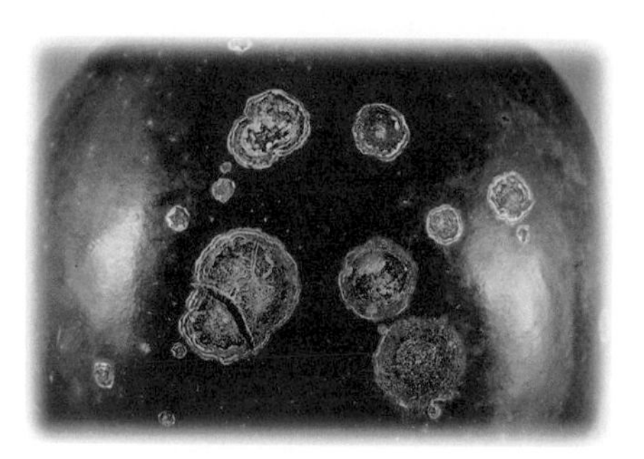

TOP: Sycamore antrhracnose symptoms on sycamore tree. (W. Jacobi)
SECOND: Apple scab on radiant crabapple. (W. Jacobi)
THIRD: Apple scab on upper leaf surface on radiant crabapple. (W. Jacobi)
BOTTOM: Apple scab fruit infection symptoms on apple tree. (Larsen)

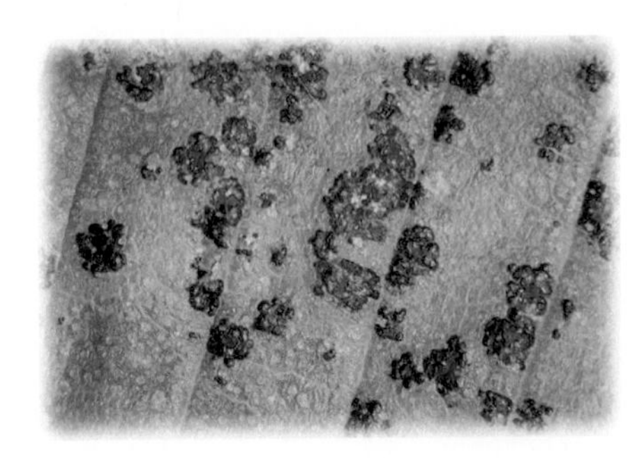

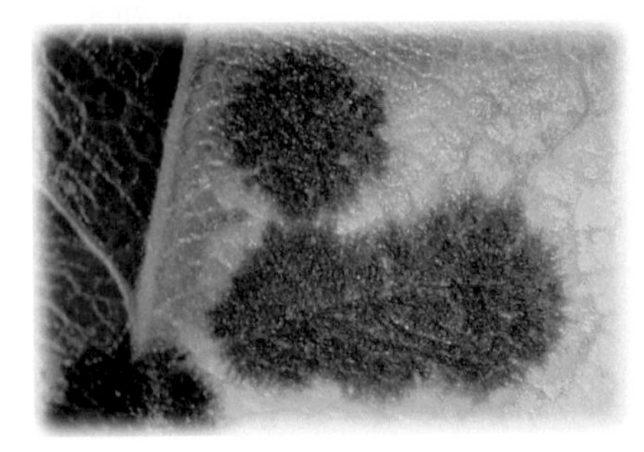

TOP: Black spot of elm.
SECOND: Close up of black spot of elm.
THIRD: Black spot on rose species. (L Pottorff)
BOTTOM: Black spot close up on rose species. (L Pottorff)

Diagnosis and Damage: This leaf spot fungus produces yellow spots on the leaves 1 mm across, in which a fruiting structure (acervulus) forms and a black stroma (specialized vegetative hyphae) forms beneath the acervulus. The stroma is visible as a black dot about 0.5 mm in diameter. Usually several acervuli and stromata develop very close together and may coalesce to irregular back masses up to 5 mm wide, surrounded by a narrow band of whitish dead tissue. Sometimes the stroma covers the entire lesion, leaving no necrotic border. From afar, symptoms include black spots and premature shedding of leaves. Severe outbreaks may impart a distinctly yellowish color and sparseness to foliage in tree crowns viewed from a distance. More conspicuous than damaging, black spot is rarely responsible for long term loss of growth or mortality.

Biology and Disease Cycle: The pathogen survives winter in fallen leaves and in buds. In spring, spores are produced that are dispersed to new foliage, starting another cycle of disease. The spores are expelled to the air in spring under conditions of alternating wetness and drying after several days with temperatures of at least 7 C. The lower-most leaves and shoots are most likely to be infected by airborne ascospores. During the early part of the growing season, lesions may form on petioles and succulent stems.

Conditions Favoring and Management of Disease: Rainy weather favors spread of the disease and high humidity favors lesion development. Management options include: collection and disposal of infected leaves in the fall to lower the amount of inoculum present to cause infection the next year; planting resistant species; and pruning host plants to encourage good air movement through the canopy and promote rapid evaporation of free water.

BLACK SPOT OF ROSES

Sexual stage: *Diplocarpon rosae*
Asexual stage: *Marssonina rosae*

Hosts: In general, hybrid teas, teas, hybrid perpetuals and polyanthas are very susceptible to black spot. Rugosa hybrids, moss roses and wichuraianas, or "old-fashioned roses", are more resistant.

Diagnosis and Damage: The disease is characterized by nearly circular black spots with fringed margins. Spots range from 2 to 12mm in diameter (1/16 - 1.5 inches) and develop on upper leaf surfaces. Small black fruiting structures may be visible inside the necrotic spots. Leaf tissue surrounding the spots may turn yellow. Leaves may drop if disease severity is high. Raised, purplish blotches may form on first year canes, if conditions are favorable. Minute black blisters (acervuli), visible with a hand lens can be found in the central parts of spots. The acervuli produce colorless, two celled spores (conidia). Fruiting structures (apothecia) of the sexual stage are uncommon. Under wet, humid growing conditions with temperatures of 65-75 F, the disease can be devastating. This is the most important disease of roses all over the world. However, there is minimal occurrence throughout the Rocky Mountain Region since the disease is less severe in semi-arid climates.

Biology and Disease Cycle: Conidia move from rose to rose by splashing water, mechanically during cultivation, or by contact with insects. They germinate on leaf tissue while the leaves are still expanding (six to fourteen days old). Conditions under which conidia will germinate are very specific, the leaf tissue must be continually wet for at least seven hours for any infection to occur. The fungus overwinters as mycelium in fallen leaf debris or in infected canes, but does not survive in the soil, or when spores may adhere to tools.

Conditions Favoring Disease and Management: No infections occur in dry air. Even with a relative humidity of 90 percent, conidia must remain immersed in water, on the leaf surface, for at least seven hours before infections will occur. Black spot is easily managed in semi-arid climates without the aid of fungicides. All leaf debris should be cleaned up and removed from the area. Avoid wetting leaves, use of drip irrigation or avoidance of overhead watering is crucial. If overhead irrigation is practiced, make sure watering is done early in the day so leaves dry quickly. Dense plantings of rose bushes should be avoided to allow for optimum air movement through the leaf canopy. Many rose cultivars have been developed with some degree of resistance to black spot.

BROWN FELT BLIGHT

Herpotrichia juniperi
Neopeckia coulteri

Hosts: These two fungi are usually found only at higher elevations (above 9,500 feet) of the region, where enough snow falls to meet their unique requirements to thrive. *H. juniperi* is found on conifers other than pine and *N. coulteri* is found on only pine species.

Diagnosis and Damage: The most conspicuous symptom of brown felt blight is the presence of felt-like growth of brown mycelium spreading over affected twigs and branches. Close examination will reveal fruiting bodies (perithecia) of the fungus immersed in the mycelium. Both *H. juniperi* and *N. coulteri* appear the same on foliage. Damage in the form of loss of foliage and branch may occur as the dense mycelial mass smothers the foliage. This disease may kill seedlings and can kill significant amounts of foliage on small trees but it has little effect on larger trees if over 50 percent of their crowns are above the snow depth.

Biology and Disease Cycle: The life cycles of *H. juniperi* and *N. coulteri* are quite similar but little is actually known about infection process or timing. Under cover of snow, the fungus envelops the branch in a gray mycelium. After snowmelt and exposure, the fungus felt stops growing and turns a dark brown. The next summer the mycelium remains inactive. The second winter, fruiting bodies (perithecia) develop on the felt under the snow.

Conditions Favoring and Management of Disease: Conditions where branches of host trees are under snow cover for extended periods of time (as often occurs at higher elevations) encourages occurrence of disease. Management options are difficult to implement in a forest situation. Reducing the duration of snow cover by means of snowbreaks that cause snow to drop in a different location and removal of infected branches to reduce inoculum are possible options.

INK SPOT

Ciborinia whetzelii (*C. bifrons, Sclerotina bifrons, S. whetzelii*)

Hosts: These fungi predominantly affect aspen and cause non-severe leaf blight on other poplars, including bigtooth aspen, eastern cottonwood, balsam poplar, and black poplar.

Diagnosis and Damage: Conspicuous numbers of dead leaves in crown; on closer examination, one to four dark *inkspot-like spots* are apparent on affected leaves. In early summer spots can be confused with leafminer insect damage. By late summer, these spots (sclerotia) drop out, leaving *shotholes* in leaves.

Typically more conspicuous than damaging, ink spot is common in young aspen stands and may cause loss of 25 to 100 percent of foliage in localized outbreaks. However it is rarely associated with long term loss of growth or mortality.

Biology and Disease Cycle: Ink spot overwinters in forest litter and in spring the sclerotia will produce stalked, cup-like fruiting bodies (appothecia). Spores are dispersed by wind and rain and infect leaf surfaces. After two to three weeks, reddish-brown blotches become visible and expand until the leaf is entirely dead. Several weeks later, one to four

TOP: Brown felt blight on spruce. (W. Jacobi)
SECOND: Brown felt blight on spruce. (M. Schomaker)
THIRD: Ink spot on aspen with black sclerotia - late season. (W. Jacobi)
BOTTOM: Early season ink spot on aspen leaf. (W. Jacobi)

TOP: Kabatina blight on juniper. (L Pottorff)
BOTTOM: Kabatina blight on juniper.

dark mycelial masses appear (inkspots). These sclerotic masses will remain on the leaf until late summer then drop to the ground, where the sclerotia will overwinter.

Conditions Favoring the Disease and Management: Incidence of ink spot is highest in dense stands of young trees and after low temperatures and high humidity in the spring during the period of spore dispersal. Management options could include removal of infected leaves on the ground which may be helpful to reduce springtime infection. Adequate spacing of trees to reduce humidity in the stand/planting will also promote conditions less favorable to the springtime dispersal of the disease. Fungicides can be applied before infection takes place on young expanding leaves, but are not effective if applied after infection.

JUNIPER TIP BLIGHTS

Kabatina juniperi
Phomopsis juniperovora

Hosts: Eastern redcedar, creeping, Rocky Mountain, and Savin juniper

Diagnosis and Damage: Both Kabatina and Phomopsis fungi can cause the *dieback of branch tips, or tip blight.* Kabatina causes tip die back in early spring and does not cause damage during the summer. Phomopsis, however, is opposite, the tip blighting symptoms appear in the summer months, not in early spring. Affected foliage initially turn dull green, then red, yellow, or brown. *Lesions or cankers containing small, black or gray fruiting structures develop on branches.* The Kabatina acervuli produce one celled, ellipsoid shaped conidia whereas the pycnidia of Phomopsis produces conidia of two shapes, short oval (alpha) spores, and long thin (beta) spores. Alpha spores have two oil droplets. Symptoms of Kabatina and Phomopsis cankers are difficult to distinguish from each other in the field. The lesions eventually girdle the small branches, leading to tip dieback. Junipers in landscape plantings sustain minimal damage. Tip blight diseases may become damaging to seedlings, in nursery settings, or to mature junipers in wind breaks.

Biology and Disease Cycle: Spores of Kabatina which infect young shoots in late summer or early fall. Symptoms develop the following spring. Spores of Phomopsis infect young shoots throughout the growing season, as long as conditions are favorable.

Conditions Favoring and Management of Disease: Warm, moist conditions favor the development of both diseases. The more moisture, the more severe the disease. Infection can occur within seven hours, with Phomopsis tip blight, on a continuously wet plant at temperatures of 65 to 70 F. Symptoms will become severe when hot temperatures of 75 to 90 F follow infection. Close proximity of one juniper to another, such as in a windbreak or nursery, favors poor air movement and prolong wetness of plant tissue. Wounding of plant tissue due to insect feeding or mechanical damage may favor Kabatina blight.

Use of a chemical is normally not needed to control tip blight diseases of juniper in the Rocky Mountain Region. The lack of moisture and humidity that occur in this area naturally limit the spread of these fungi. To manage the diseases, when they occur, prune out dead branch tips and provide adequate plant spacing to promote good air circulation. No chemical controls are labeled for Kabatina blight on juniper. In nurseries, chemical control of Phomopsis blight may be necessary during wet years. Bordeaux mixture or fixed copper compounds are labeled at seven to twenty one day intervals as needed.

LEAF RUSTS

FUNGUS	HOST	ALTERNATE HOST
Melamspora medusae	Douglas-fir and pines	Aspen and cottonwood
Puccinia montanensis	Wheat	Barberry
Melamspora occidentalis	Douglas-fir and pines	Cottonwood/ poplar
Cronartium ribicola	White pine	Currant
Phragmidium mucronatum	Rose	None
Melamspora epita	Fir, saxifraga, currant, and gooseberry.	Willow
Puccinia sparganiodes	Ash	Cordgrass
Coleosporium jonesii	Pinyon Pine	Currant

Diagnosis and Damage: The disease first appears on undersides of leaves or needles as *pustules of red, orange, or yellow spores.* Leaf yellowing, browning, curling and drop may also occur. In dry climates, rust infections tend to be most noticeable in late summer and early autumn after the host has completed most of the year's growth. As such, damage is minimal. However, yearly rust infections and accompanying leaf drop can cause a reduction in carbohydrate reserves in plant roots. This in turn may make a tree or shrub more susceptible to decline and winter injury.

Biology and Disease Cycle: Rust fungi are a group of very diverse and specialized fungi. They have complex life cycles, some with five spore stages and requiring two hosts to complete a life cycle.

Conditions Favoring and Management of Disease: Wet weather and mild temperatures are required for rust infections. Optimal temperatures range from 18 to 21 C with continual moisture on leaf surfaces for two to twenty four hours. Hot, dry weather will limit rust development.

Rust diseases are managed with sanitation, or removal of fallen leaf debris. Resistant tree, shrub and rose varieties may be available. With roses it is important to prune out old canes in the spring, to help reduce fungal carryover on canes.

Chemical control is rarely needed, and most fungicides labeled for normal leaf spots are not affective against rusts, however, several fungicides such as Bayleton are labeled for rust control on trees and ornamentals.

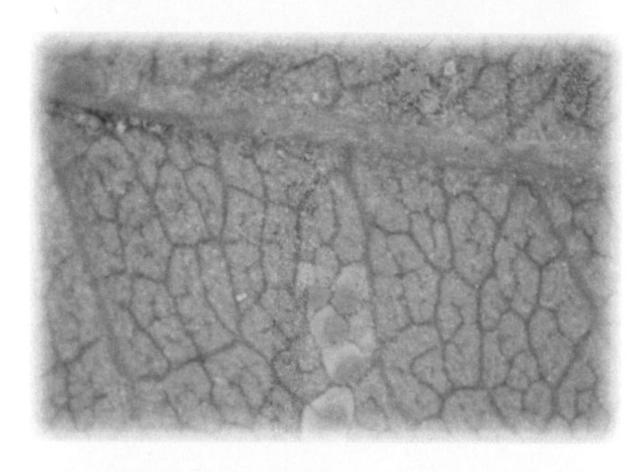

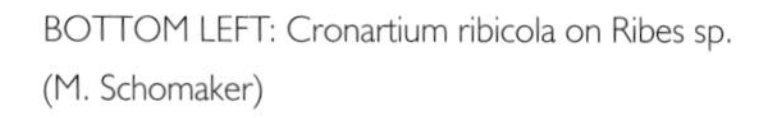

BOTTOM LEFT: Cronartium ribicola on Ribes sp. (M. Schomaker)

BOTTOM RIGHT: Melamspora epita on willow. (M. Schomaker)

TOP: Melampsora rust on aspen. (M. Schomaker)
SECOND: Needle rust on pinyon pine. (USDAFC)
THIRD: Rose rust on rose leaf. (L Pottorff)
BOTTOM: Rose rust on rose cane. (L Pottorff)

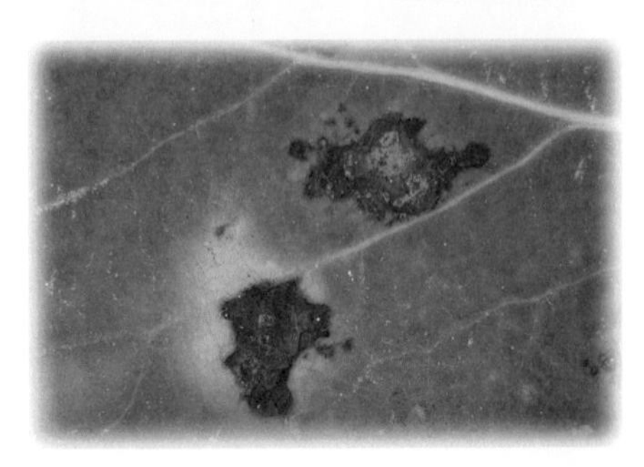

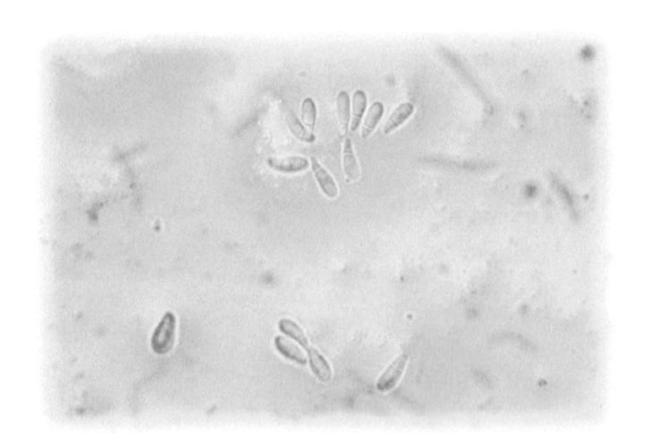

MARSSONINA BLIGHT

Marssonina populi

Hosts: Predominantly aspen; cottonwoods and poplars are also susceptible.

Diagnosis and Damage: *Dark brown flecks often with yellow margins* scattered over leaf surfaces. Spots may merge together to form blotches. Mature spots characteristically have a white center. Infected leaves are often reduced in size, turn bronze in color and shed prematurely. Elongate black lesions may develop on leaf petioles and succulent new shoots. Small lens-shaped fruiting bodies containing a white gelatinous mass of conidia form within leaf lesions. As spores are released the epidermis ruptures forming a small ring-like blister. The conidia are teardrop shaped, two celled with two or three distinct oil droplets. Conidial shape is the most diagnostic feature of this fungus.

Damage is mostly aesthetic. Tree death is rare unless there are several consecutive years of defoliation and there are other stresses such as scale insects, drought, or cytospora canker. Repeated attacks can make trees susceptible to branch dieback, other diseases, and winter injury. This leaf disease is common throughout the Rocky Mountain region.

Biology and Disease Cycle: Fungus overwinters in previously infected leaf debris on the ground. In spring, infection occurs when spores are released and carried by wind and rain to developing leaves. Secondary infection results later in the summer when fruiting bodies on the newly infected leaves produce spores and are blown to adjacent leaves.

Conditions Favoring and Management of Disease: Incidence of marssonina is highest after warm temperatures and periods of precipitation when spores are available. Management options include planting of resistant poplar clones when possible. Removal of infected leaves on the ground to help reduce primary and secondary infection. Adequate spacing of trees and avoidance of irrigating lower branches to reduce humidity in the stand/planting will promote conditions less favorable to the dispersal of the disease. Fungicides can be applied at bud break, before infection takes place on young expanding leaves, but are not effective if applied after infection.

NEEDLE CASTS (Conifer Foliar Diseases)

Hosts: Needle casts affect all species of conifers

Diagnosis and Damage: Identification of needle cast diseases is based on the appearance of discolored needles and premature death and shedding of needles. The needle cast fungi present in the region are listed in the following table. Identification is difficult without looking at spore shape and size with a compound microscope. What is most important is to determine if the tree has a needle disease, insect damage such as needle miner or abiotic damage so the correct management action can be taken. Premature needle cast should not be confused with annual needle shed. Every year, usually in the fall, conifers shed some of their oldest needles. Prior to this annual needle drop these older needles will often turn yellow or brown. Insect needle miners hollow out needles and needle scale insects can be seen as small appressed bodies on needles or twigs. Abiotic damages that may look like needle casts include salt, drought, frost- winter damage and some herbicides.

The damage caused by needle cast diseases is usually limited to loss of one year's complement of needles. Crowns of infected trees will often appear thin and chlorotic. Consecutive years of infection may reduce growth and vigor to such an extent that the

TOP: Marssonina populi on aspen leaves. (W. Jacobi)

SECOND: Marssonina populi on aspen leaf. (W. Jacobi)

THIRD: Marssonina conidia.

BOTTOM: Needlecast diseased lodgepole pine. (W. Jacobi)

host is predisposed to other diseases or insect attack; although, rarely will it result in tree mortality.

Biology and Disease Cycle: The fungi that cause needle casts are variable in their life cycles with some completing their life cycle in one year while others require two or more years. Needle cast diseases are spread by wind dispersed spores (ascospores or conidia), which are released from fruiting structures that develop on infected needles. Infection occurs in late spring or summer, and symptoms may develop as early as that fall, the next spring, or not for two or more seasons following infection. Infected needles turn *red brown to straw colored.* Fruiting bodies develop by midsummer and release spores during favorable moist and humid conditions. Diseased needles are typically cast in the late summer and fall.

Conditions Favoring and Management of Disease: Infection of conifer needles depends on moist and temperate weather conditions.

Control of needle cast fungi is typically neither practical nor necessary since the disease causes minimal damage to infected trees. Maintaining healthy and vigorous trees will encourage tree defense mechanisms. Mixed species compositions will prevent needle cast diseases from spreading to adjacent trees, as most are host specific, and may serve to reduce damage. If management is needed on high value trees or nursery stock or for visual reasons, fungicides need to be applied prior to infection periods each year.

TOP: Rhabdocline needle cast on Douglas fir. (M. Schomaker)
CENTER: Elytroderma deformans broom on ponderosa pine. (M. Schomaker)
BOTTOM: Davisomycella ponderosae on ponderosa pine. (W. Jacobi)

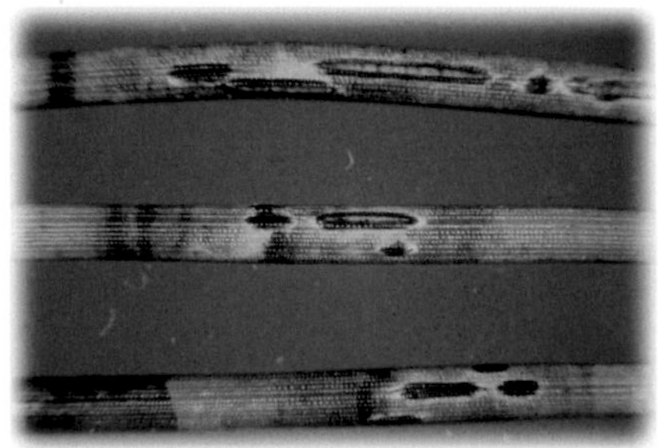

RIGHT: Davisomycella medusa on ponderosa pine. (M. Schomaker)

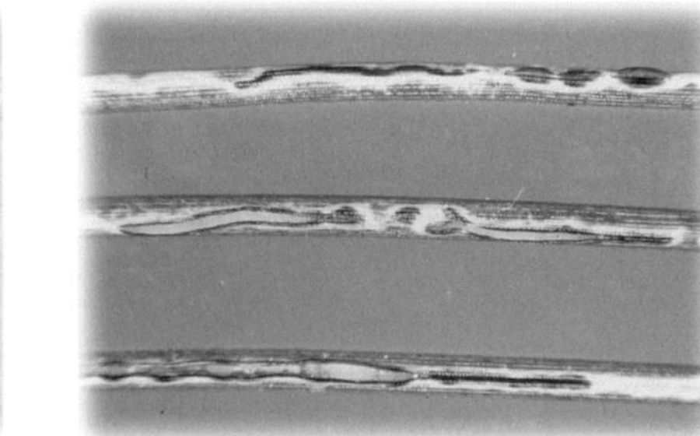

Needle Cast Fungi

FUNGAL ORGANISM	HOST	IDENTIFICATION
Bifusella saccata	Limber pine Pinyon pine	Forms large, long, shiny, black fruiting bodies on the dead tips of green needles.
Bifusella linearis	Limber pine	Forms shiny black, elongate fruiting bodies on two to three year old needles. Black crust-like fungal growths, irregular in shape and size, are frequently associated with the fruiting bodies.
Davisomycella spp.	Ponderosa pine	Form elliptic, dark brown or black, shiny, raised fruiting bodies bordered by Lodgepole pine orange-brown bands on brown, faded needles
Elytroderma deformans	Ponderosa pine Pinyon pine	Needles turn red in the spring following infection and develop dull black, elongate fruiting bodies that are scattered on all needle surfaces. Red-brown necrotic flecks in the bark of older infected twigs are diagnostic. Causes premature death of one year old needles and small, tufted brooms of twigs and branches. This is the only needle cast fungus that persists perennially in the tree.
Lirula abietis-concoloria	True firs	Forms dark brown to black, elongated fruiting bodies extending down the center of the lower surface of two year or older straw colored needles. On the upper surface of the infected needle, a single brown line is evident along the mid-rib. Infection is usually limited to 1 year's set of needles.
Lirula macrospora	Spruces	Fruiting bodies appear as long black lines that often traverse the entire length of the needle. Needles are retained on the branch after death.
Lophodermella arcuata	Limber pine	Newly developing needles are infected and remain green until the following spring when they die back from the tip and turn brown prior to bud break. Black to brown, elongate to elliptical fruiting bodies appear in early summer on all surfaces of the dead portions of infected needles.
Lophodermella cerina	Ponderosa pine	Infected needles die back from the tip turning bright red brown. Short, oval, light brown to buff fruiting bodies are formed, are often the same color as the infected needle, and develop in groups on waxy spots on green and dead needles.
Lophodermella concolor	Lodgepole pine	New needles are infected and remain green until the following spring when they turn red. Fruiting bodies appear on dead and dying needles as shallow oval depressions the same color as the needle surface.
Lophodermella montivaga	Lodgepole pine	Forms elongate, dark brown fruiting bodies with prominent central slits
Rhabdocline psuedotsugae	Douglas-fir	Infection occurs on new needles, which develop chlorotic spots in the fall following infection. In late spring on one-year-old needles, red brown spots and long brown pustules form on either side of the mid-rib on the underside of needles. Needles are cast in summer.
Virgella robusta	True firs	Forms dark brown to black, elongated fruiting bodies extending down the center of the lower surface of two year or older straw colored needles. Two distinct brown lines are formed on the upper surface of the infected needle on either side of the mid-rib. Infection is usually limited to one year's needles.

OAK LEAF BLISTER

Taphrina caerulescens

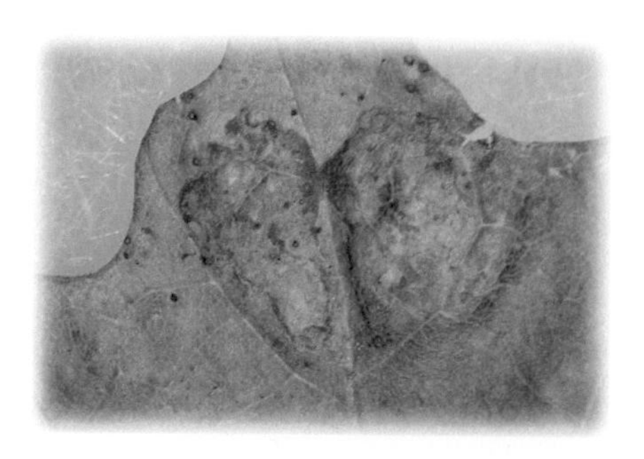

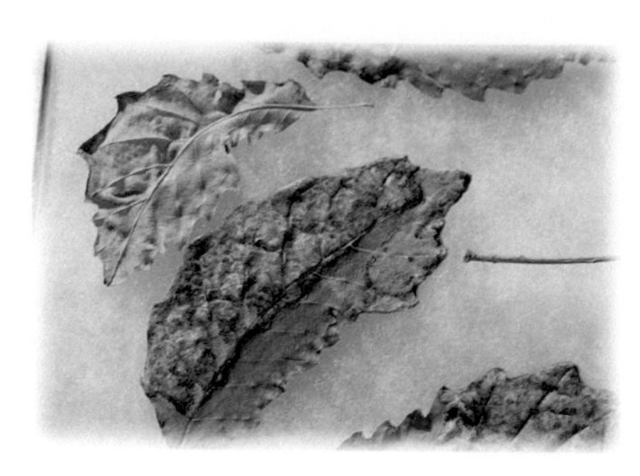

TOP: Oak leaf blister. (W. Jacobi)
BOTTOM: Oak leaf blister. (W. Jacobi)

Hosts: Red and black oaks are the most susceptible to this fungal pathogen, but all species of oak are hosts

Diagnosis and Damage: Symptoms are *blisters, bulges* or *depressions* as viewed from the reverse side, usually less than one inch in diameter on either leaf surface. These blisters form when infected cells are stimulated to enlarge, while surrounding uninfected cells remain normal size. Eventually the blistered area dies and turns brown. Symptoms may go unnoticed until a large number of leaves are severely infected or prematurely fall in late summer.

The disease may be unsightly but vigorous trees are not severely affected. However, defoliation can reduce growth and, if repeated over a period of years, may weaken a tree making it more susceptible to attack by other organisms. This disease does not seriously damage most trees.

Biology and Disease Cycle: Leaves become infected just as the buds open in the spring; expanded leaves are not susceptible. Current season leaves are infected by conidia. After infection, mycelium that stimulates blister formation is produced, and eventually produces a single layer of asci on the concave side of the blister. Asci are microscopic sacs in each of which many spores develop. The asci, at first colorless, darken after spores have been released. Ascospores then give rise to a saprophytic phase which produces conidia in the spring after overwintering on twigs or among bud scales.

Conditions Favoring and Management of Disease: Heavy infections occur following cool wet springs and may result in 50 to 85 percent defoliation of affected trees by midsummer.

Collecting and disposing of infected leaves will remove some innoculum. Maintenance of tree vigor through proper watering, pruning and pest control will reduce the effect of this disease.

PHYLLOSTICTA LEAF SPOT

Asexual stage: *Phyllosticta* spp.

Hosts: Boxelder, and other maples

Diagnosis and Damage: This leaf spot fungus produces *lesions* on the leaves that are *round* or nearly so, up to 8 mm across, yellowish brown or purple at the edges, and pale yellow in the center, where they become thin and translucent. Each lesion is bounded by a very narrow ridge. The fragile central tissue often breaks out, leaving a ragged hole. Severe outbreaks may impart a distinctly yellowish color and sparseness to foliage in tree crowns viewed from a distance.

Biology and Disease Cycle: The pathogen survives winter in fallen leaves and, in spring, produces spores that are dispersed to new foliage, starting another cycle of disease. The spores are adapted for dispersal by water rather than by wind. This explains the increase of disease during rainy seasons, and explains why this pathogen is not normally a problem in this regions semi-arid climate. After infection, the host leaf is initially defenseless but soon localizes the infection within a small area. Thus, even if infections are numerous, they usually cause no great loss of photosynthetic surface unless they trigger defoliation.

Conditions Favoring and Management of Disease; Rainy weather favors spread of the disease and high humidity favors lesion development.

Management options include collection and disposal of infected leaves in the fall to lower the amount of inoculum present to cause infection the next year. Pruning of host plants to encourage good air movement through the canopy, promoting rapid

TOP: Powdery mildew on Chinese lilac leaves. (W. Cranshaw)
BOTTOM: Powdery mildew. (W. Cranshaw)

evaporation of free water. Applying fungicides prior to infection is probably only needed in extreme cases.

POWDERY MILDEWS

Sexual stage: *Erysiphe, Sphaerotheca, Phyllactinia, Microsphaera, Podosphaera,* or *Uncinula spp.*
Asexual stage: *Oidium* spp.

Hosts: A very wide variety of plants including grasses, vegetables, ornamentals, weeds, shrubs, fruit trees, and broad-leaved trees and shrubs. Some powdery mildews are highly host specific, only affecting a single host; others may have a broader host range, affecting many different plants. Regional woody plant hosts most commonly associated with powdery mildews, and the species involved, include: aspen and cottonwood, *Erysiphe cichoracearum*; apple and crabapple, *Podosphaera leucotricha*; grape, *Uncinula necator*; honeysuckle, *Erysiphe* sp. and *Microsphaera lonicerae*; lilac, *Microsphaera penicillata*; roses, *Sphaerotheca pannosa*; snowberry, *Microsphaera diffusa*; and spirea, *Podosphaera oxyacanthae*.

Diagnosis and Damage: Early symptom expression may include a *slight reddening of leaves*. More noticeable symptoms/signs often begin in mid to late summer as *small white, talcum powder-like patches on leaf surfaces*. The disease often progresses to the point where entire leaf surfaces are covered with white growth, consisting of fungus mycelium and conidiophores (spores). *Leaves may become twisted and distorted* (as with roses) and if conditions favor disease development, affected *leaves may fall prematurely*. Powdery mildews are most commonly found on surfaces of young leaves, but may also affect the undersides of leaves, young shoots and stems, buds, flowers, and young fruit.

As environmental conditions become unfavorable, tiny, light brown-orange when young and black when mature, pinhead-sized balls form within the mass of white growth. These bodies are called cleistothecia and serve not only in sexual reproduction but also as overwintering structures.

Powdery mildews rarely cause severe damage to the plants they affect unless the damage is in early spring and young expanding leaves are attacked. Control measures seldom need to include chemical treatment as heavy infections occur late in the fall just before the leaves drop normally. Repeated (annual) infections may reduce the aesthetic and ornamental values of landscape plants. These fungi are often more abundant in semiarid regions than in areas of high rainfall, where others diseases flourish.

Biology and Disease Cycle: Powdery mildews overwinter on leaf debris as partially developed spores (ascospores) within the cleistothecia (black, pinhead-sized fruiting structures). As spring arrives, the spores mature and are released from the cleistothecia. Spores are blown or splashed onto newly emerging foliage where they germinate to form mycelium that grows only on the leaf surface. The powdery mildew fungus obtains nutrients from the plant by sending haustoria (or feeding tubes) down into the epidermal cells of the plant. Short condiophores are produced on the leaf surface containing chains of rectangular, ovoid or round spores that are carried on air currents to start secondary infections. When environmental conditions become unfavorable, the cleistothecium is formed.

Conditions Favoring and Management of Disease: Unlike most fungi, powdery mildew spores do not require free water on the leaf surface to germinate. Some species of mildews do require high relative humidity but this condition is usually met when warm days are followed by cool nights or when plants are grown in crowded, low, or shady locations without sufficient air circulation.

Powdery mildews are best controlled with preventative practices such as: use of resistant cultivars, proper plant siting (avoiding poor air circulation, shade, and over crowded conditions), and removal of dead leaves and leaf debris in the fall. Fungicides labeled for powdery mildew can be used, but will not improve already affected leaves. Alternative chemical controls include neem oil or a combination of baking soda and horticultural oil (e.g., 1 Tbsp. baking soda + 2.5 Tbsp. SunSpray horticultural oil in 1 gallon of water).

SEPTORIA LEAF SPOT

Asexual stage: *Septoria spp.*
Sexual stage: *Mycosphaerdla spp.*

Hosts: These fungal pathogens cause leaf spots on aspen, cottonwood, poplar (*S. musiva, S. populicola*), maple and dogwood (*S. cornicola*). They can also produce cankers on eastern cottonwood and some poplars. Canker occurence is rare in the region.

Diagnosis and Damage: The appearance of lesions exhibits considerable variation between and within host species. Lesions typically first appear as sunken black flecks, which expand to form circular spots 1to15 mm in diameter and may coalesce to form larger blotches of dead material. Dead tissue fades to shades of tan, brown, or white with a brown or black margin. On more resistant species, the spots may remain small (1 to 2 mm), and exhibit a silvery hue. Fruiting bodies (pycnidia) appear three to four weeks after infection, appearing as black specks on the surface of the dead tissue. Under moist conditions, the pycnidia may exude pink tendrils of conidia. Conidia are clear, elongate, slightly curved or straight with two to four cells.

Cankers appear only on trees with infected leaves, and canker severity is correlated with leaf spot severity. Most cankers originate within 1.5 meters of the ground at wounds, leticels, stipules, or leaf bases. Affected bark is initially black, but often becomes tan at the center of the affected area as pycnidia develop. Pycnidia are common on younger cankers, but are rarely found on older ones.

Varying degrees of premature defoliation occurs depending upon host susceptibility. Cankers may girdle stems of susceptible trees in two to three years, and may additionally serve as infection courts for other canker pathogens.

Biology and Disease Cycle: Septoria overwinters on dead leaves and in twigs infected the previous growing season. Initial infections are by means of ascospores released from the fruiting bodies of fallen leaves. Subsequent infections of new leaves during the summer occur by means of windblown conidia produced on leaf spots after rains.

Conditions Favoring and Management of Disease: Dispersion of ascospores is highest during periods of high moisture and temperatures around 22 to 26 C (72 to 79 F). Similarly, precipitation is the principal means by which conidia are spread to cause infections later in the growing season.

Use of resistant or tolerant species and cultivars is the best control. Spraying propagation beds, or landscape trees with a fungicide and spacing trees to provide good aeration and reduce humidity may also reduce the infection of leaves and stems. Sanitation, in the form of removing infected leaves and stems, is also helpful in reducing spring infection.

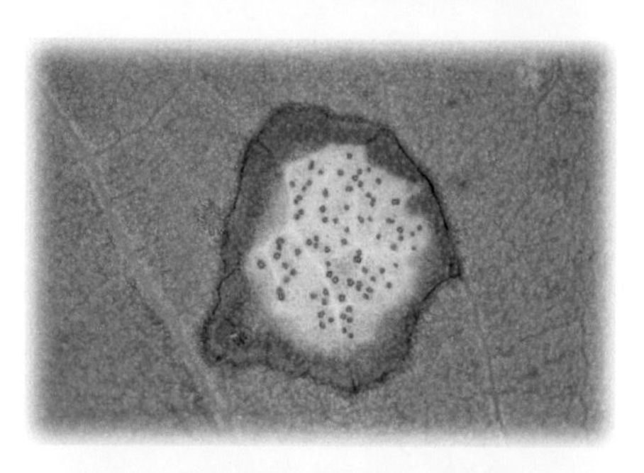

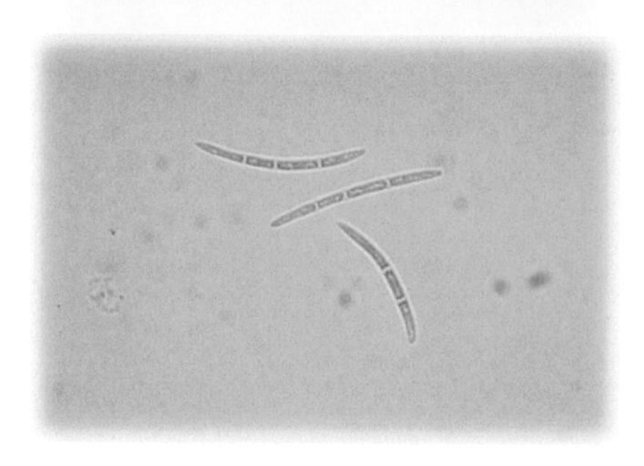

TOP: *Septoria* sp. on lanceleaf cottonwood (less leaves) vs plains cottonwood. (W. Jacobi)
SECOND: *Septoria* sp., pycnidia on lanceleaf cottonwood. (W. Jacobi)
THIRD: Leaf damaged by *Septoria* sp. (W. Jacobi)
BOTTOM: *Septoria* conidia. (W. Jacobi)

SHOT HOLE DISEASE or CORYNEUM BLIGHT

Wilsonomyces carpophilus (Coryneum carpophilum)

Hosts: Shot hole disease affects peaches, apricots and sweet cherries in this region.

Diagnosis and damage: First symptoms appear on young leaves as small reddish areas that enlarge and become purple with a white center. These spots then drop out of the leaf blade leaving a "shot hole". On fruit, lesions develop mostly on the upper surface and eventually become corky and rough. Early season infections on fruit exhibit reddish-

TOP: Coryneum blight infection on peach fruit. (Luepschen)
SECOND: Coryneum blight infection on cherry leaves. (Luepschen)
THIRD: Tar spot on willow. (M. Schomaker)
BOTTOM: Shothole symptom of coryneum infection. (David Leatherman)

purple halos surrounding tan, scab-like center spots (this damage is similar in appearance to that caused by San Jose Scale). Infections on the maturing fruit produce sunken, half inch, brownish spots rapidly, which then may coalesce, causing fruit skin to crack. Economic loss from shot hole results from fruit infection and disfigurement which renders the fruit unsalable. The fungus infects and can kill dormant buds, blossom buds and small twigs.

Biology and disease cycle: The shot hole fungus overwinters on dormant infected leaf buds, blossom buds and small twig cankers, and produces spores in early spring. Bud and twig lesions may continue to produce spores for two to three years, but the fungus does not overwinter on old infected leaves. Wind disseminated spores require free water droplets on the fruit, leaf, or twig surface in order to germinate and cause infection. Temperatures of 70 to 80 F (21-27 C) are optimum for infections, lesions can develop at 45 F (7 C) but at a slow rate. It takes from two to five days for a spore to initiate infection and cause a visible lesion.

Conditions favoring and management of disease: Shot hole is serious in years when frequent light showers occur during the summer. In this region most infections appear to take place in spring and early summer, although cool, wet periods prior to harvest can trigger infections at that time. The blight may spread rapidly within a tree, with movement from tree to tree somewhat slower. Leaf lesions produce spores that can infect the fruit whenever weather conditions are favorable. Once established in an orchard, shot hole is difficult to eradicate. A carefully planned program of chemical control and removal of dead wood is necessary to eradicate the disease.

TAR SPOTS

Rhytisma spp.

Hosts: Tar spots occur on several kinds of woody plants. The most common hosts in the region are willows, maple and boxelder.

Diagnosis and Damage: The most conspicuous symptom of tar shot is the presence of circular, raised black mats of fungal tissue scattered irregularly on the upper surface of leaves. Spots arise on maple in late spring after leaves have attained full size. Leaves with multiple spots may wither and drop prematurely but seldom so early or in such numbers as to threaten the general health of the tree. Tar spots are among the most showy and least damaging foliar diseases and do not significantly affect the health of the host tree.

Biology and Disease Cycle: New infections are initiated in the spring by ascospores produced on leaf litter on the ground. Sticky spores are ejected from fruiting bodies (apothecia) and land on young host leaves. The first symptoms of infection include areas of light green or yellowish green tissue, which eventually develop into areas of black tar-like specialized vegetative hyphae (stromata).

Conditions Favoring and Management of Disease: Moist, sheltered locations are favorable for winter survival of causal fungi and therefore following season disease development. Some tar spots are sensitive to polluted air, and are scarce in urban and industrial areas. Management options include remove of dead overwintered leaves that will provide the spores for infection the following season.

galls

Galls are abnormal plant growths induced by the action of fungi, bacteria, insects, mites and other organisms. When these occur on tree trunks and branches they are sometimes referred to as burls.

Galls resulting from activity of insects and mites typically develop around the gall maker, housing, protecting, and providing a steady food source for the organism within the gall. These galls may have many forms. Most insect galls, such as those produced by the gall midges, produce fairly simple galls that are the result of stunting and/or thickening of leaf tissues. These are sometimes referred to as indeterminate galls. Other insects and some mites produce very bizarre changes in the plant growth of radically different structure than the original tissues. These are called determinate galls and the various gall wasps that frequent oaks and rose are well known for this habit. Regardless, all gall making insects and mites create galls unique for the species and they are almost invariably restricted to a single plant species.

Galls induced by pathogens and are usually restricted to woody parts of the plant. The may similarly vary in structure with many occurring as irregular growths of irregularly organized tissues. Others may produce unusual and unique forms, such as the fruiting bodies of rust fungi.

Galls can cause problems for trees when they seriously interfere with transport of food and water. However, little if any significant injury to the plant results from most galls. On the other hand the bizarre growths often attract attention and concern. Most controls are directed at preventing these aesthetic injuries.

Since only actively growing tissues are susceptible to the gall making stimuli, most galls are produced solely during the rapid growth period of late spring. Once galls are initiated, their formation is irreversible. Although the galls are often conspicuous and may be unattractive, rarely does any real plant damage occur. Under most circumstances control of galls is not recommended.

A few gall-making insects and mites overwinter on the trees and may be controlled by dormant oil applications. However, most galls are produced by insects that move to the new growth as it develops in spring and can only be controlled by insecticides that cover the leaves during the egg laying period. Because the quality of the insecticide coverage diminishes as leaves expand, repeat applications are often necessary.

Control of gall making insects and mites are discussed in the sections Cooley spruce gall adelgid, Pinyon spindlegall midge, Honeylocust podgall midge, Poplar twiggall fly, and Eriophyid mite galls (Poplar budgall, Ash flowergall) on the management supplement.

TOP: Cooley spruce galls.
SECOND: Spruce galls produced by Cooley spruce gall (top) and *P. pinifoliae*.
THIRD: Cooley spruce gall adelgid, overwintered nymph.
BOTTOM: Cooley spruce gall adelgid, overwintered female with eggs, close-up.

GALL-MAKING INSECTS AND MITES

COOLEY SPRUCE GALL ADELGID (Cooley Spruce Gall Aphid)

Adelges cooleyi (Gillette)
Homoptera: Adelgidae

Hosts: Spruce, Douglas-fir

Damage and Diagnosis: On spruce distinctive cucumber-shaped galls develop on the new growth. These usually kill the terminal growth, although occasionally partial galls form that allow many of the upper needles to survive and the terminal continues to grow. Galls produced by this insect may be considered unsightly but rarely have significant effects on tree health. On spruce the insects are greenish-gray aphids, covered with a fine powder of wax and are found within chambers of the gall.

On Douglas-fir Cooley spruce gall adelgids are conspicuous woolly aphids that suck sap from needles. When infestations occur on developing needles symptoms often include yellowing and twisting of needles. Some secondary damage often is caused by sooty mold that grows on honeydew excreted by the insects. Large numbers also occur on cones, which may affect seed production.

Associated Species: The pine leaf adelgid, *Pineus pinifoliae* (Fitch), also makes gall in the terminal growth of spruce that resembles that of the Cooley spruce gall adelgid, but is of looser form. Galls remain green late in the season and are less conspicuous. Spruce is the primary (overwintering) host of these aphids, which infest pines during alternate stagesof the insects life cycle.

Life History and Habits: The typical life cycle involves two alternating hosts, Douglas-fir and spruce, and involves multiple forms of the insect.

On Douglas-fir, the nymphs overwinter as partially developed nymphs on the needles. They continue to develop in spring, producing a large egg mass around early May. The eggs hatch in about 20 days and the nymphs move to tips of twigs to feed primarily on the current season needles. They become full-grown in July and produce a generation of adults that are a mixture of winged and wingless forms. The wingless forms remain on Douglas-fir and have a second generation. The winged forms fly to spruce and start the cycle on this host.

On spruce, winged adults from Douglas-fir lay eggs in fall that hatch into nymphs. These nymphs overwinter at the base of needles, resume feeding in spring, and mature in May. The adults produce a very large mass of eggs on the underside of the twigs, and eggs hatch at bud break. The nymphs move to feed at the base of newly developing needles, which induce the formation of greenish-purple cone-like galls that envelop the feeding aphids. They continue to feed and develop within the galls. During late June and July, they are fully mature and emerge from cracks in the drying gall. They molt to a winged adult form and migrate to Douglas-fir. The galls continue to dry and become more conspicuously brown, but no longer contain aphids.

There is some evidence that local populations of Cooley spruce gall may be able to sustain a spruce-to-spruce life cycle in the absence of Douglas-fir.

Management: Late spring frosts and winds are very destructive to the overwintering stages of the Cooley spruce gall adelgid. Population cycles are highly variable due to these natural controls.

Spruce and Douglas-fir show a wide range in susceptibility to this insect. Many individual trees are highly resistant, apparently due to leaf waxiness.

Chemical controls directed against the overwintering stage females can be applied in fall through early spring. Treatments are best applied before the egg sack is produced, which typically begins in late April or early May. Effective treatments include carbaryl (Sevin), Dursban, permethrin (Astro) and horticultural oils. Soil injected imidacloprid

(Merit) may also control this insect, although galls may form before the insects are killed with this treatment. However, galls on Merit-treated trees may remain green.

Insecticidal soaps have been very effective for controlling Cooley spruce gall on Douglas-fir. However, control on spruce has been marginal.

Hand removal of the galls may be desirable from an aesthetic view. However, this practice has no benefit in managing the insect.

PETIOLEGALL APHIDS

Pemphigus species
Homoptera: Eriosomatidae

Hosts: Poplars and cottonwoods

Damage and Diagnosis: Depending on the species, petiolegall aphids make a variety of galls usually at the base of leaves or on petioles. Most common forms are marble-like with a transverse slit; some petiolegalls are more fluted. During spring the aphids can be found within the galls and are usually pale green with a dark thorax and covered with fine wax.

After leaving galls the aphids develop on the roots of their summer hosts. The sugarbeet root aphid, Pemphigus populivenae, can be a serious pest of sugarbeets in the region.

Life History and Habits: The various Pemphigus species found in Colorado have a one-year life cycle that involves alternation between a winter host of some Populus species, and the roots of a separate summer host.

Eggs are laid in autumn in cracks of the bark of *Populus*. The eggs hatch in spring and the nymphs (fundatrices) feed on the developing leaf petioles. Feeding induces the plants to produce a swollen overgrowth that surrounds the developing aphid. As the overwintered stage becomes full-grown, they produce young that remain in the gall until full-grown. These progeny have wings.

The galls later open along a slit, and the winged stages leave the plant during late June and July to colonize summer hosts. They develop on roots of various annual plants (e.g., sugarbeets, lambsquarters, lettuce), that they locate by following soil cracks. Several generations may be produced on these summer hosts. At the end of the summer, sexual winged stages are produced that fly back to the winter host, mate, and lay eggs.

The poplar petiolegall aphid, *Pemphigus populitransversus* Riley, forms a spherical green gall with a transverse slit on the petiole of plains cottonwood, and is also a root aphid on cabbage family crops. The sugarbeet root aphid, *Pemphigus populivenae* (Fitch), forms an elongated gall on the midvein of narrow-leafed cottonwood leaves and alternately attacks the roots of sugarbeets and other garden plants. There is some confusion regarding the taxonomy of this genus and other species may be present.

Management: Gall-forming causes little injury to the trees and controls have not been developed. Dormant oil applications should be able to control the overwintered eggs.

POPLAR VAGABOND APHID

Mordwilkoja vagabunda (Walsh)
Homoptera: Eriosomatidae

Hosts: Aspen primarily, but occasionally other *Populus*

Damage and Diagnosis: Aphids feeding on the developing leaves cause them to become highly-distorted into leathery folded leaf galls. Galled leaves tend to remain on trees and may not be visible until after normal leaf fall. Galling is concentrated on the upper third of the tree.

TOP: Cooley spruce gall adelgid on douglas-fir.
CENTER: Petiolegall.
BOTTOM: Poplar petiolegall aphids exposed in gall.

Adults are yellow-green, pear-shaped aphids, with relatively long antennae and delicate, membranous wings, about 3/16 inch long. Nymphs are greenish, turning more cream colored as they age.

Life History and Habits: Similar to other gall-making aphids, the poplar vagabond aphid uses two hosts during its life cycle. The overwintering stage are eggs laid in bark crevices or old galls on aspen or other Populus species. Eggs hatch in spring and the aphids feed on the expanding tips of the twigs. Feeding induces twig tips to form large irregularly shaped galls within which the aphids feed and reproduce. Several generations of aphids occur within the folds of the gall, and as many as 1600 individual aphids have been reported from a single gall.

In early summer, winged forms of the aphid leave the gall and fly to a summer host. Local hosts are not known from the region, but loosestrife is an important alternate host in other areas. After several generations on this host, where they may occur on leaves, stems and root, winged stages are produced that fly back to Populus in early fall, mate and produce the overwintering eggs.

Associated Species: Marginal leaf curling of aspen is sometimes associated with eriophyid mites. Also, an unidentified gall midge has been found to make reddish colored leaf folds in which it lives.

Management: Controls have not been identified. Dormant applications of horticultural oils should kill overwintering eggs.

TOP: Poplar vagabond aphid pseudo-gall on aspen.
BOTTOM: Hackberry nipplegalls, produced by hackberry nipplegall maker, *Pachypsylla celtidismamma*.

HACKBERRY NIPPLEGALL MAKER

Pachypsylla celtidismamma (Riley)
Homoptera: Psyllidae

Hosts: Hackberry species (American and net-leaf)

Damage and Diagnosis: Hackberry nipplegall psyllid produces prominent warty leaf galls, sometimes nearly covering the leaf. High levels of galling are usually restricted to only a few branches and do not produce much damage. However, the galls are sometimes considered to be unattractive. The pale-yellow nymphs can be found within the galls until they emerge in early Fall.

Life History and Habits: Hackberry nipplegall psyllids overwinter as adults in protected areas. In spring as the hackberry buds are expanding, the adults emerge and deposit eggs on the undersurfaces of the leaves. Eggs hatch and young nymphs begin to feed on the leaf. An overgrowth that appears as a raised swelling on the lower leaf surface is induced by this feeding, ultimately producing the gall that covers the insect. The nymphs develop within the gall all summer and adults emerge in late summer. There is one generation per year.

There can a great difference in the number of galls produced on different leaves within the tree or between nearby trees. This is largely due to how synchronized the leaf development is when the adult psyllids are laying eggs and eggs are hatching since leaves are suitable hosts for only a brief period during their early development.

Related Species: The hackberry blistergall psyllid, *Pachypsylla celtidivescula* Riley, produces a gall in the form of a small raised swelling, often concentrated around the base of the nipple galls on the upper leaf surface. Life cycle of this species is similar to the hackberry nipplegall maker. Blistergall psyllids are small enough to pass through most screens and sometimes enter nearby homes during fall and can be a nuisance problem in homes.

Another gall-maker, *Pachypsylla venusta* Osten Sacken, sometimes forms large galls on the petioles of net-leaf hackberry.

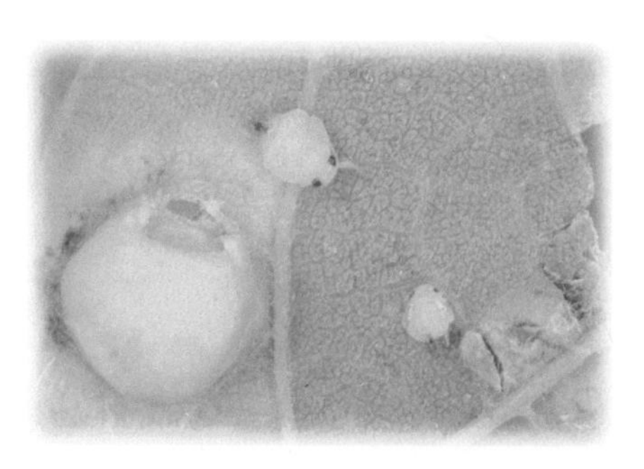

TOP: Hackberry nipple gall psyllid, eggs on bud.
CENTER: Nymphs of the hackberry nipplegall maker and hackberry blistergall psyllid, exposed from galls.
BOTTOM: Hackberry budgall psyllid, old galls.

Management: The hackberry nipplegall psyllid is commonly attacked by parasitic wasps that help reduce populations. The wasps remain in the old galls through the winter, emerging the following spring.

Hackberry psyllids are also common and important prey of many resident and migratory birds. Overwintering psyllids are favorite prey of chickadees, creepers, nuthatches, and other species. Migrating warblers, ruby-crowned kinglets, pine siskins, American goldfinches and chipping sparrows eat many adults in spring. Psyllids within galls are sometimes extracted by house finches, evening grosbeaks and fox squirrels.

Controls are rarely needed to protect tree health and most applications are made for aesthetic reasons. Insecticides should be applied to coincide with periods when eggs are being laid and systemic activity is useful for optimizing control. The eggs can be easily seen on the underside of newly expanding leaves.

HACKBERRY BUDGALL PSYLLID

Pachypsylla celtidisgemma Riley
Homoptera: Psyllidae

Hosts: Hackberry

Damage and Diagnosis: Hackberry budgall psyllids produce an enlarged, spherical swelling of the bud tissues, killing the affected bud.

Life History and Habits: Unlike the other common hackberry psyllids, the budgall psyllid spends the winter within the gall. Adults emerge in late spring, laying eggs near the developing buds. The nymphs feed on the buds, causing them to swell and form a gall. The nymphs develop within the gall throughout the year and there is only a single generation

Related Species: Another *Pachypsylla* species produces a small blistergall on the outer bark of small branches.

HONEYLOCUST PODGALL MIDGE

Dasineura gleditschiae (Osten Sacken)
Diptera: Cecidomyiidae

Hosts: Honeylocust

Damage and Diagnosis: Larvae feed on developing leaves causing the production of thickened pod-like galls. Inside currently infested leaf galls, small white maggots can be found. The galls darken, dry and drop a few weeks after adults emerge. This defoliation gives a thin appearance to the plant, particularly when most of the leaflets on a leaf are affected. Injury is usually of minor cosmetic importance but can constitute a further stress and may contribute to twig dieback. In extreme cases, attacks can be so intense that all emergent buds are killed which results in a thickened club-like swelling of the twigs.

Associated Species: Another gall midge, *Neolasioptera brevis* (Gagne) is reported to induce twig swellings and bushy growth on honeylocust. It has not been confirmed from Colorado but similar symptoms have been observed in Mesa and Morgan counties.

Life History and Habits: Honeylocust podgall midge spends the winter in the adult stage under protective cover around previously infested honeylocust plantings. The adults are tiny, delicate gray-brown flies that move to emerging honeylocust buds as they first start to break. Eggs are laid among the emerging leaves and the larvae feed on the leaflets causing them to curl and thicken into the characteristic pod gall. Larvae are cream-colored maggots that become full-grown in about three to four weeks. Pupation occurs within the gall. As the adults emerge the old pupal skin is often pulled partially out at the gall opening.

There are typically about three generations per year, with populations usually declining by early July. Later infestations may extend for additional generations where sprout growth continues to be produced or in highly fertile, irrigated sites, such as nurseries.

Management: Several natural controls are present for this species. A tiny parasitic wasp can be an important natural enemy late in the season and the predatory plant bug *Deraeocoris nebulosus* feeds on many larvae within galls. However, competition with the honeylocust plant bug has been even more important. Honeylocust plant bug competes with the honeylocust podgall midge, killing the new growth needed by the developing midges. Finally, when populations of honeylocust podgall midge get extremely high, excessive egg laying and larval feeding on the new growth can cause leaves to prematurely abscise, killing the larvae.

Pyrethroid insecticides have been the most effective chemical controls. They should be timed for periods when adults are actively laying eggs. The presence of fresh pupal skins sticking from the galls is a good indication of a recent emergence.

JUNIPER TIP MIDGE

Oligotrophus betheli (Felt)
Diptera: Cecidomyiidae

Host: Rocky Mountain juniper

Damage and Diagnosis: The developing larvae produce a reddish or yellow gall on the tips of juniper that appears as a rosette of leaves. These tips prematurely die. The pale colored maggots can be found within the gall.

Life History and Habits: Winter is spent in the larval stage within the gall. They pupate in spring, emerging from the gall as adult midges. Females mate then lay eggs on the new foliage. After egg hatch, the young larvae move to the branchlet tips and feed on the developing growth, inducing production of the characteristic gall.

WILLOW CONEGALL MIDGE

Rhabdophaga strobiloides Osten Sacken and other species
Diptera: Cecidomyiidae

Hosts: Various native willows

Damage and Diagnosis: The larvae feed on the terminal growth of willow and prevent the stems from normally elongating. This causes compact growth of the leaves somewhat resembling a cone.

Life History and Habits: The adult midges emerge in late April or early May and lays an egg on the terminal bud as it begins to swell. Feeding by the larva causes the bud to cease further develop, remaining bud-like but still capable of directing plant nutrients to the tissues. The bud swells as the small fly larva develops within a cavity in the bud. The larva spends the winter in the gall, pupates in the early spring. There is one generation produced per year.

Many other insects may later feed on and develop within the galled tissues.

TOP: Honeylocust podgalls.
SECOND: Honeylocust podgall midge adult.
THIRD: Honeylocust podgall with exposed gall midge larvae
FOURTH: Juniper tip gall midge galls.
BOTTOM: Willow cone gall caused by gall midge.

GOUTY VEINGALL MIDGE

Continaria negundinis (Gillette)
Diptera: Cecidomyiidae

Host: Boxelder

Damage and Diagnosis: The midge larvae feed on the expanding leaves, causing them to thicken greatly and curl about the midrib. Damage is generally minor, although it can be conspicuous. The larvae are creamy white maggots that may be found associated with galls.

Life History and Habits: The gouty vein gall midge has one generation per year. Winter is spent in the pupal stage around the bases of previously infested trees. The adult midges emerge in spring, about the time that new leaves are emerging on boxelder trees. Females insert their eggs into small folded leaflets.

The young midges feed on the developing leaflets, causing them to curl and thicken. They usually have completed development by early June, at which time they drop to the soil and burrow into the soil. There they produce a small cocoon in which they pupate.

Related Insects: The ash midrib gall midge, *Contarinia canadensis* Felt, can produce a thickening along the midrib of ash leaves during late spring. Life history is likely similar for this species.

Management: Controls are not recommended since only a small amount of the leaves are ever damaged. Several natural enemies are reported to attack the midge including parasitic wasps and predatory bugs.

STUBBY NEEDLEGALL MIDGE

Contarinia coloradensis (Felt)
Diptera: Cecidomyiidae

Host: Ponderosa pine

Damage and Diagnosis: Feeding by the developing larvae severely stunts the needle and results in formation of rounded, chambered gall. Galled needles die at the end of the first summer and heavy infestations can affect the appearance and vigor of the tree. However, serious outbreaks are very uncommon. The orange larvae are maggot-form and found within the chambers of the globular-form gall.

Life History and Habits: Adults emerge from soil litter from mid-April to June. Females lay eggs in small masses on developing needles. These eggs hatch into larvae that begin feeding on the needle, producing a gall in which they become enclosed. Adults emerge from galls in late summer and move to overwintering protection. There is one generation per year.

Management: Serious infestations are rare and treatments are not likely to be warranted. Control evaluations have not been conducted. However, treatments that have been effective against other gall midges (dimethoate, bifenthrin, spinosad) are likely to be effective against this species when applied during the egg laying period.

PINYON SPINDLEGALL MIDGE

Pinyonia edulicola (Gagne)
Diptera: Cecidomyiidae

Host: Pinyon. Galls can form on bristlecone pine but the insects do not complete development.

Damage and Diagnosis: Pinyon spindlegalls are swellings produced at the base of developing needles. Generally there is some discoloration (yellowing, reddening) of the affected area although this may not be evident until spring. Galls kill the infested needles

TOP: Gouty veingall midge injury on boxelder.
CENTER: Ash midrib gall midge injury.
BOTTOM: Pinyon stunt gall midge, dried galls.

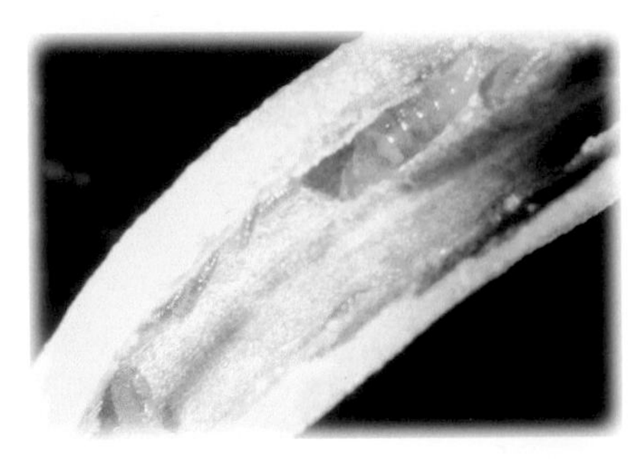

TOP: Pinyon branch heavily infested with pinyon spindle gall midge.
CENTER: Pinyon spindlegall midge larvae in gall.
BOTTOM: Stunt gall midge galls.

prematurely. By late summer such needles turn brown and drop prematurely. Following heavy infestations during which a high percentage of needles are killed, gaps in the normal pinyon foliage will appear. Established needles are not affected by the insect.

Pinyon spindlegalls uncommonly cause serious damage to pinyon and are usually only a minor aesthetic problem to trees. Rarely, outbreaks occur that affect significant numbers (greater than 25 percent) of the new needles but outbreaks are almost invariably of short duration (one to two years).

The larvae are small orange-colored maggots found within the gall. Adults are small, delicate flies, about 1/16 inch long, with an orange abdomen, present in late June or early July.

Associated Species: Another gall midge is also common on pinyon in Colorado, the pinyon stunt needlegall midge (*Janetiella* species). This insect also forms a swollen area at the base of the needles but can easily be separated from the spindlegall midge since it severely stunts the elongation of the needles. Life cycle seems to be generally similar to that of the pinyon spindlegall midge.

Life History and Habits: Adults of the pinyon spindlegall midge are small, delicate, orange flies. These are active around the pinyon from mid-June until early-July but are rarely observed. This period coincides with the development of new needles. Mating and egg laying occurs at this time. Eggs hatch in about a week.

The immature (larval) stages of the insect are orange-yellow and feed within the gall. On the average, six to fifteen midge larvae occur within each gall. The feeding of these insects causes the formation of the gall. Only the actively growing needles are capable of being affected by the insects and the gall growth is completed in about a month. The insects continue to grow inside the gall cavity and may nearly fill it when they become full-grown in early June.

After growth is completed, larvae of the pinyon spindle gall midge cease feeding and change to the pupal stage. During this period they produce a silken, frothy covering within the gall. The pupal stage lasts two to three weeks.

There is one generation per year of the pinyon spindlegall midge.

Management: Except during unusually intense outbreaks, pinyon spindle galls rarely affect tree growth and control is usually not needed to protect tree health. Natural controls exist which include rainy/windy weather during adult midge activity, various parasitic and predatory wasps, and failure of the insects to properly synchronize egg laying with susceptible periods of needle growth.

Because biological controls are so important in the natural control of pinyon spindle gall midge, it is recommended that a survey be made of parasites prior to spraying. Parasites appear different from the yellow midge larvae and can be easily detected in spring. Large numbers of parasites indicate that adequate natural control will likely be effective.

Insecticidal control is best achieved if treatments are timed during periods of egg laying and early gall growth. This generally occurs in late June and can be predicted by opening galls to watch for the silking produced during pupation that precedes adult emergence.

Dimethoate (Cygon) and spinosad (Conserve) have been among the most effective insecticide for control of pinyon spindlegall midge. Sprays should thoroughly cover newly developing needles. Because dimethoate has systemic activity, it also is capable of killing many midge larvae even after galls have started to form. However, galls remain intact even after the insects are killed. Dimethoate is also highly effective at controlling the insect after galls have started to form. Because of this, treatments can be delayed until after large numbers of new galls are observed in July or early August, before significant injury is caused. These "rescue treatments" can allow for the natural controls to establish themselves. Such treatments would only be needed if large numbers of developing galls were observed.

POPLAR TWIGGALL FLY

Hexomyza (= Melanagromyza) schineri (Giraud)
Diptera: Agromyzidae

Hosts: Primarily aspen, but poplars may also be attacked

Damage and Diagnosis: The poplar twiggall fly produces a spherical gall on the new twigs. This gall remains in place and continues to grow and enlarge as the plant develops, giving the plant a gnarled and knobby appearance. Galls do not appear to significantly weaken plants. Occasionally Cytospora canker may develop around a gall.

The galls are in the form of smooth swellings on the current season twigs. Obscured by leaves, these original galls are rarely noticed until autumn leaf fall. However, what attracts attention and particular concern is that galled tissues continue to grow and swell. Ultimately galls appear as large knots on trunks and larger branches, giving the plants a gnarled, bonsai-like appearance. Serious galling has been limited to aspen. However, it has been visible on other *Populus* species.

Areas galled by the poplar twiggall fly usually continue to grow although growth of buds adjacent to the galls is inhibited. During subsequent seasons the galled area becomes incorporated into the growing twigs and branches, ultimately appearing as large swollen bands on trunks and branches. Although these old injuries produce a permanent disfigurement of the trunks, they do not appear to threaten tree health.

Life History and Habits: The poplar twiggall fly overwinters within the gall as a full-grown, yellow-green maggot. Pupation occurs in late winter or early spring, within the gall. The majority of the pupae then drop to the ground.

At the time that the new growth is forming, the adult flies emerge from the pupae and become active. They are stout-bodied, shiny dark flies, about 1/6 inch long. During the day they may easily be found resting and sunning themselves on leaves. After mating females move to the developing twigs and insert eggs into the stems. The larvae then hatch from these eggs and produce the distinctive swelling in response to their feeding.

Areas below buds appear to be particularly favored sites for galls to be produced. As the stems continue to grow, the area where eggs were laid becomes increasingly swollen. At first, the swelling involves a fairly indistinct enlargement. However, within two months the full sized gall is usually present.

The developing gall fly is a greenish-yellow maggot that grows slowly within the gall all summer. It is very difficult to find until late summer and fall when it grows rapidly, filling a small cavity within the swollen area of the twig. Individual galls typically contain two to three larvae.

There is one primary generation per year. Some observations indicate that a few flies may have a second generation during seasons that are unusually warm and allow a period of extended development.

Apparently the fly is native to the region or at least has been present for a long time. For example, museum records of the species in Ft. Collins date to 1914. However, outbreaks began to attract attention in the mid 1980's. Originally, problems seemed limited largely to the southern Metro Denver area but recently there has been an extensive expansion of the outbreak so that it currently extends across the Front Range of Colorado and into native stands of aspen in forested areas.

Related Species: Another species of *Hexomyza* makes small swellings in the twigs of willow.

TOP: Galls of poplar twiggall fly on aspen.
SECOND: Poplar twiggall fly, developing gall on aspen.
THIRD: Poplar twiggall fly larva in gall.
BOTTOM: Poplar twiggall fly adult on aspen leaf.

Management: At least one biological control organism does occur naturally. A small chalcid wasp, *Eurytoma contractura* Bugbee, parasitizes and commonly kills large numbers of the poplar twiggall fly. Observed parasitism rates typically range from 20 to 30 percent but have exceeded 80 percent in some years. Predation of the pupae by chickadees and other birds also occurs.

The adult wasps emerge from the galls about two weeks after the adult flies first appear. Life cycle has not been developed but apparently they are larval parasites that first attack the poplar twiggall fly shortly after egg hatch.

Problems with poplar twiggall fly are most severe in succulently growing stands of aspen. It is likely that managing aspen so that growth is more moderate will result in retarded rates of gall production.

Removal of galls is commonly considered by homeowners. However, this has limited potential for control. Pruning would often require substantial branch destruction. Furthermore, it may be counterproductive if done after flies have emerged in late winter. In that case, pruning will primarily remove only those galls containing the chalcid wasp parasite of the poplar twiggall fly, decreasing the effectiveness of natural controls.

Attempts to control the poplar twiggall fly on aspen with insecticides have shown little success. Abamectin has been successful in some trials, but control is erratic and applications must be carefully timed to coincide with adult activity periods.

TOP: Poplar twiggall fly and parasitoid, *Eurytoma contractura*.
BOTTOM: Rabbitbrush galls.

RABBITBRUSH GALL MAKERS

Aciurina bigeloviae (Cockerell), other *Aciurina spp., Procecidochares spp.*
Diptera: Tephritidae

Hosts: Rabbitbrush

Damage and Diagnosis: Several different gall-forming fruit flies make galls on the stems and buds of rabbitbrush. Perhaps the most conspicuous are cottony swellings of the buds. Others produce green flowerlike growth of leaf axils or sticky globular swellings on stems. These do not appear to damage the plant but attract attention when plants are in landscape settings.

Adults are small "picture-winged" flies with patterned or generally dark wings. Larvae are pale-colored maggots found within plant galls.

Life History and Habits: The gall makers on rabbitbrush are generally similar in habits to the gall midges. Overwintering typically occurs as a partially grown larva within the gall. They complete development and adults emerge during periods when the plant is actively growing. After eggs are laid, the developing larvae stimulate gall development by their secretions and feeding activities.

Management: Controls have not been developed nor are needed. If galls are considered unattractive, they may easily be pruned out. In native stands of rabbitbrush, these gall makers are heavily parasitized by several species of wasps. Also, several gall midges may also develop within the gall, competing with the flies creating the gall.

GALL WASPS ASSOCIATED WITH OAK

Of all the insects that produce galls on woody plants, none are so numerous and diverse as the various gall wasps (Cynipidae) associated with oak. In the United States and Canada, over 550 different types of galls have been recorded from oak (76 percent of all types of galls produced by insects and mites).

Most galls on oak are produced on leaves or twigs. Common leaf galls found in Colorado take the form of small pale-colored balls, reddish cuplike swellings, bright-red raised areas along the veins, or woolly patches. The woody galls found on oak twigs include 'bullet' gall forms, that resemble a chocolate drop, or generally round balls. Galls on oak leaves are generally innocuous, although they can be a considerable curiosity. However, some of the twig galls are associated with twig dieback.

The life cycles of gall wasps can be very complex, with alternating forms and types of galls produced by some species. For example, many of the twig galling species spend the winter in the woody galls, and only females are found in these galls. They emerge in spring and lay eggs in young leaves. Galls along the leaf veins are formed by this stage, which involve both males and females. These emerge in midsummer and females insert eggs into the twigs. The developing gall wasps feed on the twigs, producing the gall. However, gall development is slow and it may take almost two years for the wasp to complete this development cycle, emerging in spring of the second year.

Intensity of attacks by gall wasps is highly variable. Much of this is related to how much susceptible new growth is present when the insects are laying eggs. Gall wasps are also heavily parasitized by other wasp species.

Effective controls have not been developed for gall wasps. Insecticide applications, if attempted, should be synchronized with periods when adult wasps emerge from galls and are laying eggs. This may occur in late fall or winter, several months before symptoms of gall formation are first observed.

TOP: Cynipid leaf galls on oak.
CENTER: Cynipid galls on oak leaf.
BOTTOM: Rough bulletgall wasp laying eggs.

ROUGH BULLETGALL WASP

Disholcaspis quercusmamma (Walsh)
Hymenoptera: Cynipidae

Hosts: Bur oak, swamp white oak

Damage and Diagnosis: Rough bulletgall wasps produce a woody, generally rounded gall on bur oak, with a slight point. Very heavy infestations can occur that largely cover twigs, reducing growth rate of the tree. Following leaf drop, the galls can be highly conspicuous. The galls also exude a honeydew-like sweet material that is attractive to bees and wasps and fosters growth of sooty molds.

Life History and Habits: Females emerge in late October and early November, after a hard frost. (A small circular hole in the gall indicates emergence.) Eggs apparently are laid in the terminal growth during the fall. Winged and wingless forms are produced; males apparently are unknown.

In late spring, the developing insect stimulates a pocket of stem tissue to produce a large rounded gall, in which the young wasp develops. Galls are pale brown and soft in early stages, later darkening and hardening. Only a single wasp develops in each gall, although sometimes other insects (inquilines) also share the gall. The larva pupates within a small cell in the center of the gall, emerging in early fall. There is one generation per season.

Management: None have been developed. Some individual trees appear to be resistant to attack, particularly those that have less fluted bark. Removal of galls before the wasps emerge may be useful controls on smaller trees.

Insecticide applications, if attempted, may give best control when applied during the period of adult emergence, to prevent subsequent egg laying. This occurs in late October or early November. Soil injections of Merit or Orthene and spring foliar applications have not been effective in CSU trials.

Natural enemies, such as parasitic wasps, are common and can be important in control. Since these wasps emerge in spring, removing galls during winter and spring, after the gall wasps have emerged, will have the adverse effect of destroying natural enemies, while not affecting the gall wasp. Emerging parasites produce a smaller emergence hole than do the gall wasps. Bird predation is also common, particularly in Fall.

ROSE GALL WASPS

(*Diplolepsis radicum* (O.S.) and other *Diplolepsis species*
Hymenoptera: Cynipidae

Hosts: Rose. Species, rugosa, and old garden roses are most commonly galled.

Damage and Diagnosis: Several species of gall wasps are associated with Rosa species, producing bizarre growths on leaves and stems known as galls. The galls may take the form of balls, spikes, or mossy growths. Little plant injury occurs from these galls, and they are primarily a curiosity. However, growth may stop beyond the point of galling.

Life History and Habits: Gall wasps overwinter in cells within the interior of the galls. They pupate in spring, within the gall. Adults emerge in midwinter, laying eggs in buds. Adults are small, inconspicuous dark wasps, and females lay eggs at the base of axillary buds. The newly hatched larvae on the stem and cause it to deform into the characteristic gall, in which they grow and develop. There is one generation per year.

Management: Other species of wasps that do not make their own galls (inquilines) often take over the galls and kill the developing gall wasps. Several parasitic wasps also attack them. Old galls can be handpicked and destroyed before adult insects emerge in midwinter. Effective controls have not been developed for gall wasps. Insecticide applications, if attempted, should be synchronized with periods when adult wasps emerge from galls and are laying eggs.

WILLOW GALL SAWFLIES

Euura species
Hymenoptera: Tenthredinidae

Hosts: Willow

Damage and Diagnosis: Several sawflies in the genus Euura form galls in the stems, twigs, or leaf petioles of willow. The swellings can become extremely numerous in willow thickets but do not appear to cause significant injury.

Life History and Habits: Winter is spent as pre-pupae in cocoons at base of plant or sometimes within the old gall. Adults emerge in spring, usually over a period of about two to three weeks. They live for only a few days, but during this time the females insert their eggs into the developing stems of willow, and the act of oviposition induces swellings to develop rapidly. Larvae, found within the gall, are pale yellow-green caterpillars with black eyes and head tinted brown. They feed for several months and sometimes several larvae will occur in a single gall. Larvae of most species emerge from the gall in late summer and move to cover on the ground. Others may cut an emergence hole in late summer, but remain in the gall to later emerge in spring. There is one generation per year.

TOP: Newly emergent rough bulletgalls.
SECOND: Rough bulletgall wasp galls
THIRD: Cynipid wasp galls on rose.
BOTTOM: Small gall on leaf of wild rose.

WILLOW REDGALL SAWFLY

Pontania proxima (Lepeletier)
Hymenoptera: Tenthredinidae

Host: Willow

Damage and Diagnosis: The willow redgall sawfly produces a pronounced, reddish bean-shaped swelling on the leaves of willow. These galls can be highly conspicuous but do not appear to seriously affect the health of the tree. High levels of galling are rarely sustained for more than a couple of seasons.

Life History and Habits: Winter is spent in the pupal stage, in soil or under protective debris. Adults emerge in spring and lay eggs in young, expanding leaves during late spring and early summer. At this time the female wasps also lay down chemicals that initiate formation of gall tissue. (Galls can be produced following oviposition even if the larvae subsequently die.) The developing larvae feed on the soft tissues within the gall, later moving into firmer tissues to feed but remaining within the gall. When full grown, they drop to the ground and spin a cocoon. They may then transform to the pupal stage or remain in the gall; soil moisture has been shown to affect this stage in related species. Ones that pupate produce a second generation that similarly confines attacks to the new growth. Others may remain dormant until the following season. Typically there are probably two generations per year under Colorado conditions.

Related Species: The related species *Pontania s-pomum* (Walsh) creates prominent, red ball-like swellings on willow leaves that develop in small clusters. Occasionally, similar fleshy leaf galls are produced on currant or gooseberry by an unknown species.

Management: Since this species does not appear to serious affect the growth of willow no applied controls have been developed.

The developing larvae are heavily parasitized by various parasitic wasps. Also many other insects may also utilize the galls in their development and incidently kill the developing willow redgall sawfly larvae.

POPLAR BUDGALL MITE

Eriophyes parapopuli (Keifer)
Acarina: Eriophyidae

Hosts: Cottonwood, particularly the variety 'Siouxland'

Damage and Diagnosis: Poplar budgall mites produce large, corky, irregularly shaped galls around the buds of poplars. Since infested buds do not produce leaves, the tree can have a thin, spindly appearance following repeated heavy attacks. The knobby galls also detract from appearance of the tree.

Life History and Habits: Biology of this mite is not well known. The mites overwinter primarily within the galls produced the previous season, as well as in protected areas around buds. In spring feeding within the old galls continues and mites begin to move to and feed on the new buds. As the buds develop, they become distorted and cover the mites. Reproduction appears to occur throughout the growing season.

There are two nymphal stages plus the adult stage. Ten to 14 days are required to complete one generation.

Management: Predator mites are commonly associated with poplar budgall mites and presumably are important natural controls. Hard winters, which can kill the galled tissues, also appear important in limiting populations. Serious infestations rarely are sustained for long due to these natural controls.

Chemical control is difficult. The mites emerge from the galls over an extended period so repeat applications would be necessary to maintain adequate control. If attempted, sprays should be initiated before bud break. Dormant oil sprays may control mites and mite eggs near bud scales, but fail to sufficiently penetrate old galls.

TOP: Sawfly, *Euura* sp., gall on willow.
SECOND: Sawfly galls on willow.
THIRD: Sawfly galls on willow.
BOTTOM: Poplar budgall.

TOP: Dried ash flowergalls.
CENTER: Cottonwood catkingall.
BOTTOM: Chokecherry fingergalls.

ASH FLOWER GALL MITE

Eriophyes fraxiniflora Felt
Acarina: Eriophyidae

Hosts: Ash

Damage and Diagnosis: The mites greatly distort the male flowers of ash. Flower growth becomes generally disorganized and remains yellow or green, producing a highly disfigured gall. These later dry and turn brown. There is no evidence that this injury significantly damages the health of ash.

Life History and Habits: The ash flower gall mites survive winters in the adult female stage under bud scales and other protected sites on the ash. As temperatures warm in spring, overwintering females commence feeding, initiate gall formation, and begin egg laying on the newly expanding buds. The tissues of the flower become disorganized and form numerous small pouches in which the mites continue to develop. In mid summer, as the galls dry, they move to the bud scales for overwinter shelter.

Related Species: *Eriophyes neoessigi* Keifer, the cottonwood catkin gall mite, infests the developing catkins of cottonwood, producing a large distorted growth.

Other eriophyid mites are common "leaf vagrants" of ash. These species have not been studied but are suspected in inducing premature leaf drop in trees that show bronzing or "brittle-leaf" conditions.

Management: Control has been erratic. There has been some success when effective pesticides have been applied shortly before bud break of flowers and again repeated. Dormant oils may kill some overwintered mites, although many are well protected under bud scales and poorly controlled.

FINGERGALL MITES

Phytoptus species
Acarina: Eriophyidae

Hosts: Chokecherry, American plum and linden are the most common plants hosting fingergall mites. Each host has its own unique fingergall mite associated with it.

Damage and Diagnosis: Fingergall mites induce small fingerlike eruptions to form on the upper surface of leaves. These galls may remain green or, as frequently occurs with chokecherry, they turn bright red. Fingergall mites are minute, not visible without magnification, and develop within the galls.

Life History and Habits: The overwintering stage are adult females that hide under bud scales and other protected crevices on twigs. At bud break they move to the developing leaves and begin to feed, inducing production of the distinctive gall. Only a single generation appears to be produced but each gall may be colonized by scores of offspring. Also other species of eriophyid mites, normally developing as "leaf vagrants" and not producing galls, may move into the galls of the fingergall mites.

BLISTER MITES

Phytoptus spp.
Acarina: Eriophyidae

Hosts: Pear, apple, crabapple

Damage and Diagnosis: Blister mites are minute eriophyid mites that feed on the surface cells of the leaf underside. These feeding wounds rupture and the mites further invade the damaged areas. The upper leaf surface opposite the original feeding injury takes a blistered appearance. Continued feeding causes injuries to coalesce producing brown scabby patches on the leaves. Injuries tend to be concentrated around the midrib.

Life History and Habits: The following is based on the biology of the Pearle blister mite, Phytoptus pyri Pagenstacher. Winter is spent as adult females under bud scales and other protected sites on the twigs. They move to the newly expanding leaves after bud break and suck the sap of cells on the lower leaf surface. Feeding sites rupture and females lay eggs within the wounds which further expands the damaged site. Individual wounds typically get about 3 mm in diameter and later dry out if they are colonized; none colonized wounds remain green or reddish. There are multiple generations per season, each taking about a month to complete, and individual from these can initiate new injuries through midsummer.

Management: Damage is rarely sufficient to warrant control. Insecticides with activity against eriophyid mites (e.g., carbaryl) can be very effective controls. Dormant season applications of horticultural oils can kill overwintering females.

Related Species: On mountain-ash *Phytoptus sorbi* Canestrini produces a simple blister-like pouch gall on leaves of mountain-ash.

MAPLE ERINEUM MITE

Eriophyes calaceris (Keifer)
Acarina: Eriophyidae

Host: Rocky Mountain maple

Damage and Diagnosis: Feeding by the mites on developing leaves induces production of a velvety growth of reddish plant hairs (erineum) on the surface of Rocky Mountain maple.

Life History and Habits: Overwintering females survive under bark crevices and bud scales. They move to the expanded buds in spring and feed on the developing leaves, causing the formation of the erineum-type galls. The mites live and reproduce within the galls through summer, returning to twigs at the end of the season.

Related Species: On boxelder *Eriophyes negundi* Hodgkiss makes an erineum of white plant hairs on the underside of leaves. The erineum is confined within a small pouch gall. A small brown felty erineum, produced by an unknown species, is also associated with a pouch gall on aspen.

ERIOPHYID MITE GALLS

Eriophyid mites (Eriophyidae family) are minute, microscopic mites that feed on plants. They are elongate in form, often somewhat carrot-shaped, and are unique among mites in having only two pairs of legs. Eriophyid mites are very commonly associated with woody plants although so poorly studied that many in Colorado are undescribed species. For most species the life cycle is fairly simple, with overwintering stages being fertile females (deutogynes) that hide under bud scales or other protected sites. Slightly different forms (protogynes) involving both sexes are present during the growing season with several, overlapping generations present.

Many eriophyid mites do not produce galls, living on the surface of leaves as "leaf vagrants". In very high populations they may cause leaf bronzing or brittleness and have been known to induce leaf abscission. Species that cause visible leaf discoloration are sometimes referred to as "rust" or "russet" mites.

Some of the simplest deformities produced by eriophyid mites are irregular leaf blisters, within which the mites develop. On other plants pouchgalls are produced, which appear as small, wart-like eruptions of the leaf surface, often with a yellowish or reddish discoloration. Where these leaf galls become very elongate they are described as fingergalls.

TOP: Blister mite damage to crabapple.
SECOND: Erioophyid mite erineum covering leaves of mountain maple.
THIRD: Eriophyid mites on ash leaf.
BOTTOM: Witches broom of hackberry caused by eriophyid mite, close-up.

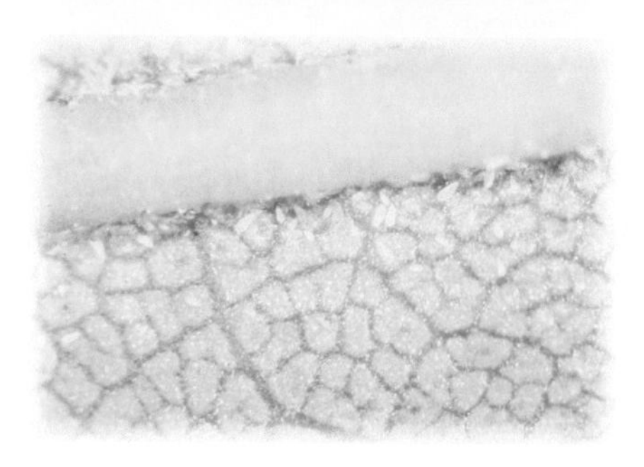

Changes in density of plant hairs are induced by the activity of some eriophyid mites. This produces dense, velvety or cottony patches, which are described as an erineum. The erineum may be associated with a pouchgall, as occurs with galls on aspen and boxelder.

Stunting and distortion of growth results when buds are infested. Eriophyid mite galls associated with flowering parts result in irregular deformities. There is also gross enlargement of the tissues, which can be extreme. For example, catkins of cottonwood may be stimulated by the activity of the cottonwood catkingall mite to grow to the size of bunch of grapes.

Common Deformities to Woody Plants Caused by Eriophyid Mites in Colorado

ERIOPHYID MITE	PLANT INJURY
Pearleaf blister mite	Causes a rusty blister on leaves of pear
Appleleaf blister mite	Causes a rusty blister on leaves of apple, crabapple
Chokecherry fingergall mites	Causes green/red gall fingergalls on the upper leaf surface of chokecherry
Plum fingergall mite	Causes a green finger gall on the lower surface of wild plum
Linden fingergall mite	Produces small finger galls on leaves of linden
Eriophyes calaceris	Produces a velvety, red growth on leaves of Rocky mountain maple
Phyllocoptes didelphis	Produces a velvety red or brown growth on leaves of aspen
Eriophyes negundi	Produces a cottony growth and pouch in leaves of boxelder
Aculops tetanothrix	Produces red or yellow pouch galls on willow
Phytoptus laevis	Produces simple pouch galls in leaves of alder
Phytoptus sorbi	Produces simple pouch galls in leaves of mountain-ash
Unknown	Produces red pouch galls on Rhus trilobita
Ash flowergall mite	Produces a distortion of male flowers on ash
Cottonwood catkingall mite	Produces a distortion of catkins on various Populus species
Unknown	Common leaf vagrant thought to be associated with leaf twisting of elm
Unknown	Produces a twisting of new leaves of boxelder
Unknown	Produces a curling of leaf edges on aspen
Unknown	Produces a small, yellow pouch gall on ash leaves
Honeylocust rust mite	Involved in producing rusty color of honeylocust leaves
Unknown	Produces leaf bronzing symptom on lilac
Pear russet mite	Causes a russetting discoloration of pear fruit
Poplar budgall mite	Distorts buds of poplars, cottonwoods
Trisetacus spp.	Symptoms commonly included rosetted growth and stunted needles of pine
Eriophyes celtis	Involved in witches' broom disorder of hackberry

PATHOGEN-ASSOCIATED GALLS

CROWN GALL

Agrobacterium tumefaciens

Hosts: More than 600 species in 93 families have been described as hosts for the crown gall bacterium. Stone and pome fruits, grapes, roses, cottonwoods, and junipers are common hosts in the region.

Diagnosis and Damage: Galls typically form on roots, as well as the root collar or root crown (where roots join the stem). Galls rarely are found on aerial portions of the tree. Galls range greatly in size, with many being less than 1 inch to approximately 1 foot in diameter. Young swellings are similar in appearance to callus, they continue to grow and eventually form galls. Gall enlargement occurs during periods of host plant growth. The gall surface generally appears similar in color to the host tissue affected. Exterior portions of the gall may partially decay during periods of host dormancy. Galls formed by A. tumefaciens lack the small holes that are characteristic of galls formed by insects and usually have a rough surface.

Galls on large, established trees seldom cause significant injury. However, galls on young trees and nursery stock can cause significant losses and mortality. Xylem function can be significantly reduced in stems of young plants. Decay fungi may invade through galls.

Biology and Disease Cycle: The crown gall bacterium overwinters in galls, plant debris in soil, and also may survive saprophytically in soil for several years. Host plant infection usually occurs through lenticels or wounds made by various cultural practices, grafting or insects. Chewing insects also can carry the bacterium from plant to plant. Once infection occurs, genetic material from the bacterium transforms host cells, which are induced to divide and enlarge in an unregulated manner. Once transformed, plant cells continue to divide. As galls weather or decay, bacteria are returned to the soil, thus completing the disease cycle. Bacteria are disseminated over long distances on diseased planting stocks or in infested soil.

Conditions Favoring and Management of Disease: Any conditions that create wounds or injuries at the root crown increase the risk of infection and subsequent disease development. Once host tissue is transformed by the infection processes, galls remain, even in the absence of the bacterium.

Prevent the introduction of the bacterium by planting only disease-free nursery stock. Remove and destroy all infected material. Also, do not grow susceptible plants in soil previously infested with the pathogen. Avoid wounding susceptible plants and control chewing insects and other pests that cause injuries. If possible, use a budding technique rather than grafting during vegetative propagation. Dip cuttings or treat fresh wounds with a suspension of the biocontrol agent A. radiobacter (strain 84). This bacterium becomes established on the plant and produces a compound that is toxic to most strains of the crown gall bacterium. Always disinfest cutting tools to reduce plant to plant transmission of the crown gall bacterium and other pathogens. Aromatic hydrocarbons applied to galls may selectively kill gall tissue. However, these products have not been extensively tested.

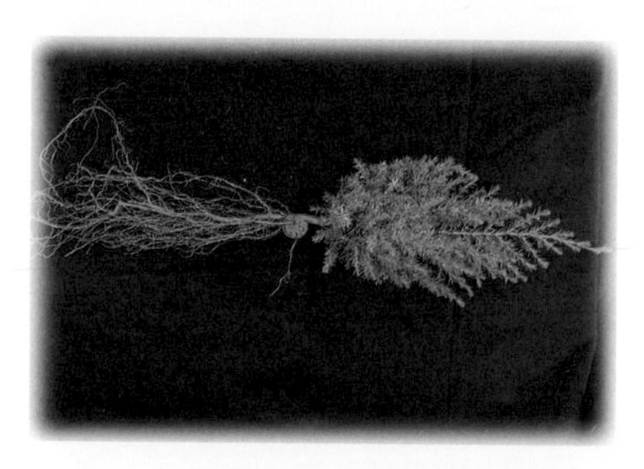

TOP: Crown gall on cottonwood. (W. Jacobi)
CENTER: Crown gall on juniper seedling. (W. Jacobi)
BOTTOM: Crown gall on cottonwood. (W. Jacobi)

Juniper Rusts
Gymnosporangium spp.

COMMON NAME	FUNGUS	SYMPTOMS ON JUNIPER	ALTERNATE HOST (most common hosts in bold type)
Cedar-apple rust Rare to nonexistent in region to date	*G. juniperi-virginianae*	Greenish brown, globose to kidney-shape annual, leaf based, galls, fleshy and soft when fresh, 10 to 30 mm in diameter. In spring golden brown to orange gelatinous horn-like projections (telia) emerge.	**Apple, crabapple**, and hawthorn
Cedar-Knot-gall rust Very common	*G. bethelii*	Reddish brown, irregular, perennial galls 3 to 15 mm in diameter, usually several galls form from a single infection.	**Hawthorn**, apple, crab apples occasionally other hosts.
Orange gall rust Common in SW Colo.	*G. speciosum*	Witches'-brooms, spindle-shaped swellings, with elongated telia, and rope like elongated swollen bark on stem and branches.	**Fendlera** and **mockorange**
Clavariiforme juniper rust Common on ground juniper	*G. clavariiforme*	Usually found on slightly swollen twigs, but can cause witches' brooms.	**Service berry**
Inconspicuous rust Common in Western Colo.	*G. incospicuum*	Telia form on unmodified, or roughened bark.	**Service berry**
Nelsonii rust	*G. nelsonii*	Galls smoothly rounded, and globose, usually single when mature.	**Service berry** and others
Juniper broom rust	*G. nidus-avis*	Witches'-brooms, or stem swellings if infection occurs a distance away from host growth points.	**Serviceberry, apple, quince**, mountain ash and hawthorn

LEFT: Orange gall rust on juniper. (W. Jacobi)
CENTER: Gymnosporangium sp. aecial stage on toba hawthorn. (W. Jacobi)
RIGHT: Gymnosporangium sp. telial stage on juniper.

Diagnosis and Damage: The most commonly observed juniper-rust fungi that occur throughout the Rocky Mountain region are listed on the preceding table. There are eleven Gymnosporangium sp reported in the Rocky Mt region. Eastern red cedar and Rocky Mountain juniper are most commonly affected by the gall or broom forming (telial) stage of the fungus, but other species of juniper may serve as hosts. Southern red cedar and common and prostrate juniper may also be affected. The leaf spot (aecial) stage of the fungus commonly affects plants in the family Rosaceae such as hawthorn, apple, crabapple, and with lesser frequency, pear, quince, mountain ash and serviceberry. Orange gall rust occurs more commonly in the southern part of the state on Utah, alligator, cherrystone and Sierra juniper.

There are two basic symptom types; **galls or brooms** form on **juniper** species and **leaf spots** (lesions) form on **apple, hawthorn or other broadleaf hosts**. Galls originate on juniper leaves and eventually become attached to twigs. Galls vary in size and shape depending on rust species (see preceding table). Galls of *G. spp* cause little or no dieback of woody twigs, but severe infections may cause some branch death. Fungi that cause witches'-brooming can deform trees and can stress trees and cause death. During warm spring rains, galls extrude chestnut-brown to orange gelatinous (telial stage) horns from dome-like swellings on the surface of galls. Occasionally, telia also may extrude from the surface of infected (swollen and discolored) leaves. After drying, telia may appear as thin dark threads. Perennial galls may produce telial horns for more than one season. Most brooms die while small, but some attain a diameter of 50 to 60 cm or more and live 15 years or longer.

Foliar lesions on broadleaf hosts initially appear as orange-yellow spots. Light-colored cylindrical projections (aecia) 3 to 4 mm long eventually form on the underside of some foliar lesions. The sides of the aecia split irregularly and release masses of reddish-brown aeciospores. The fungus may also cause deformities in green stems or fruit. Premature foliage drop can occur if foliar infection is severe.

Infection is unsightly on both hosts. Repeated defoliation of broadleaf host may result in stress and eventual decline. Green fruit may be deformed. Depending on the species, juniper rust can cause dieback of woody twigs on juniper.

Biology and Disease Cycle: The juniper rust fungi listed here have a two-year disease cycle. Following infection of the juniper by aeciospores, the fungus overwinters on juniper and cedar. Galls start to form during the following spring and approximately 20 months, from infection, is required for maturation and formation of telial horns and teliospores. Teliospores produced on the extruded horns, germinate in place and produce basidiospores. Rainsplash or wind to nearby hawthorn or other broadleaf hosts disperses basidiospores. Under conditions of high humidity, dispersal may be for several miles. Infection of broadleaf hosts results in the formation of orange-yellow spots and, later, development of cylindrical aecia on the leaf underside. Aeciospores are dispersed by wind and infect juniper during late summer or early fall, thus, completing the disease cycle.

Conditions favoring and management of disease: Moist, warm spring weather favors spore production and the infection of broadleaf hosts. The close proximity of the two hosts favors disease development by enabling completion of the disease cycle. Management options include selecting resistant broadleaf hosts that are well adapted to the local growing conditions. Because the fungus requires both plant hosts for completion of the disease cycle, elimination of one host from the vicinity will reduce new infections. Separations of at least several hundred yards, and up to two miles, are commonly recommended. Some degree of disease control may be achieved by removing and destroying galls present on the juniper and cedar hosts. In the presence of disease, protectant fungicides will provide additional control if applied at the proper time during the growing season. The aecial (broad leaf) host must be protected during the spring

TOP: *Gymnosporangium* sp. aceia on hawthorn. (W. Jacobi)

CENTER: Juniper broom rust on Rocky Mountain juniper. (W. Jacobi)

BOTTOM: Juniper broom rust fruiting bodies. (M. Schomaker)

TOP: Witches' broom on gambel oak. (M. Schomaker)
CENTER: Hackberry witches' broom. (W. Jacobi)
BOTTOM: Hackberry witches' broom. (M. Schomaker)

and early summer, until telial horns are dried and inactive. The juniper and cedar hosts must be protected during the late summer and early fall, corresponding to the time of the season when aecia have appeared on the broadleaf host.

OAK WITCHES' BROOM

Imperfect stage: *Articularia quercina*

Hosts: Occurs primarily on Gambel oak, *Quercus gambelii*, and has been observed on scrub oak, *Q. undulata*, in native settings

Diagnosis and Damage: The most obvious symptom of oak witches' broom is the brooms that may occur on the main stem or well out in the secondary branches. The brooms are generally one to two feet in diameter, and the leaves on the brooms are usually about half the size of those of non-broomed branches. Fruiting bodies (stroma) occur abundantly on the lower surfaces of leaves on brooms and impart a whitish cast to the under surface to the leaves. Powdery mildew has been associated with trees with oak witches' broom, but the mildew also occurs on non-broomed branches and trees.

The damage caused by fungal induced brooms is generally minor, except where the disease is unusually abundant. Results of heavy infection are adverse effects on the form and growth rate of the trees. Direct mortality due to the disease is rare.

Biology and Disease Cycle: Fruiting bodies of the fungus appear on oak leaves about 3 weeks after they emerge from buds in the spring and are apparent until the leaves drop in the fall. Leafing out on the brooms is a few days later than on non-broomed branches of the same trees. The fungus resides in broomed branches, both in the inner bark and outer sapwood.

Conditions Favoring and Management of Disease: Conditions that favor this disease are locations that contain susceptible hosts, especially the northern part of host range. Management options include removal of brooms by pruning. Since the fungus does enter bark and sapwood, pruning cuts would be best made some distance beyond the broom and pruning tools should be sterilized between cuts with 1:10 household bleach/water mixture, or another suitable disinfectant. No fungicidal sprays have been tested nor are labeled to control this pathogen.

WITCHES' BROOM OF HACKBERRY

A complex involving *Sphaerotheca phytophila* and *Eriophyes celtis*

Hosts: Occurs throughout the range of hackberry in the Great Plains; it is very common in eastern Kansas and Nebraska; locally common in Colorado. The principle host is hackberry.

Diagnosis and Damage: The most obvious symptom of this complex is the occurrence of multiple brooms consisting of numerous short twigs that arise close together, often at a conspicuous swelling or knot on a branch. These are not seen at branch tips, as the brooms do not occur on branches less than one year old. Buds of shoots within a broom are usually larger than normal and have more open bud scales. Witches'-brooms on hackberry are more unsightly than harmful to the tree. Brooms can cause branches to break more readily and therefore expose wood to decay fungi. Extensive brooming can reduce the vigor and stunt growth of the tree; however, trees are seldom seriously injured. Severity of attack can vary greatly among trees in close proximity to one another, but reasons for this variation are unknown.

Biology and Disease Cycle: The cause of witches'-broom of hackberry is not completely understood, but it is attributed to two agents acting in concert, a powdery mildew and an eriophyid mite. During spring and early summer, the powdery mildew fungus can sometimes be found growing on succulent stems, petioles, buds and sometimes the lower surfaces of leaves. Infection and spread may be caused by spores or mycelium within buds. Mites in all stages of development can be found throughout the year and are most numerous in late summer. They occur and overwinter beneath bud scales.

Conditions Favoring and Management of Disease: Pruning diseased twigs to sound wood can improve the tree's appearance and may help control the pathogen. Improving general growing conditions of the tree will help the tree if there is any stress from the brooming.

rusts

Rust diseases are caused by fungi that form orange-rust colored spores, thus giving this group its name. Most trees affected by rusts are conifers but several foliar rusts cause problems on hardwood trees such as cottonwoods and aspen. Most rust fungi require two unrelated plants to complete their life cycle. The other host plant is usually called an *alternate host.* Some rust fungi such as western gall rust have only one host. Rust fungi cause cankers, witches brooms, galls, and leaf spots. Thus these diseases can kill trees if a canker or cause visual problems and stress trees if the disease causes leaf spots, brooms or galls.

Rust fungi spread from host to host by wind blown spores. Infection almost always takes place through leaves. Local climate conditions and location of the alternate host affects how many infections will take place. High humidity and rain is needed for spore production and infection. The figure to the right describes a typical life cycle of a fungus that requires two hosts.

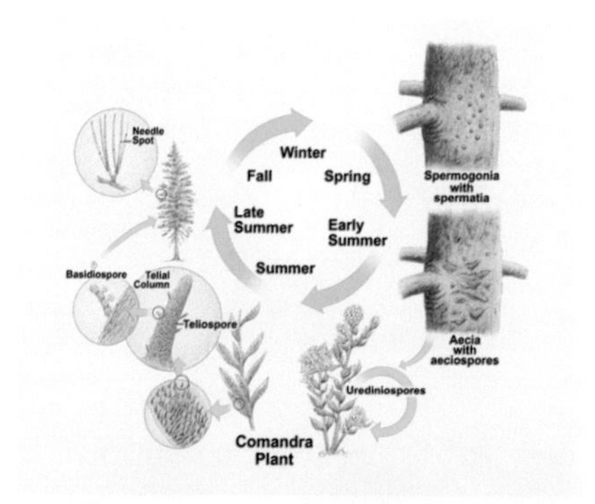

FIR BROOM RUST (Yellow Witches' Broom of Fir)

Melampsorella caryophyllacearum

Hosts: Affects many of species of true firs including subalpine fir and white fir. Requires alternate hosts in the Caryophyllaceae family including chickweed, sandwort, and starwort.

Diagnosis and Damage: This rust causes *upright, compact witches' brooms* that bear annual, yellow needles. These brooms may be mistaken for those caused by dwarf mistletoes; however, in true firs, only fir broom rust causes a marked loss of chlorophyll and annual casting of all broom needles. Infected branches and stems become swollen at the base of a broom into a spindle-shaped or nearly spherical gall. The bark on old swellings usually dies and becomes cracked, and open cankers may develop. On the leaves of the alternate host, yellow-orange spores (urediniospores and teliospores) are produced.

Damage caused as a result of this broom rust includes reduced growth, dead tops above the infection and mortality particularly in the seedling and sapling stages. Breakage may occur at the point of stem infections. Trees can survive for many years with broom rust infections.

Biology and Disease Cycle: The fungus lives systemically and perennially in both hosts and may overwinter in either host. In fir, buds and emerging twigs are infected in the spring by spores produced on the alternate host, and the fungus invades the young shoots and induces the formation of witches' brooms. The rust produces five types of spores of which two, spermatia and aeciospores, are produced on true fir; the others are produced on the alternate host.

TOP: Life cycle of a fungus that requires two hosts.
CENTER: Fir broom rust. (W. Jacobi)
BOTTOM: Fir broom rust. (W. Jacobi)

Conditions Favoring and Management of Disease: Infection of fir requires moist and temperate weather conditions and synchrony between spore (basidiospore) release and fir shoot development. Trees bearing stem cankers or brooms may be removed in precommercial and commercial thinnings and selective harvests. Brooms may be pruned out

TOP: Limb rust on ponderosa pine. (W. Jacobi)
SECOND: Midcrown branch death – limb rust on ponderosa pine. (W. Jacobi)
THIRD: White pine blister rust on Ribes sp. (M. Schomaker)
BOTTOM: White pine blister rust on limber pine. (W. Jacobi)

of trees in order to reduce risk of breakage and maintain tree vigor in areas of intensive management such as recreation areas and Christmas tree plantations.

LIMB RUST

Peridermium filamentosum

Host: Ponderosa pine. Three races of the fungus exist, one of which utilizes *Castilleja* (Indian paintbrush) species as alternate hosts; the remaining two races do not require an alternate host.

Diagnosis and Damage: Limb rust usually attacks the mid- to upper-crowns of mature trees. Progressive invasion and killing of branches is caused by the rust mycelium when it fruits and blisters on branches. The tree can have living branches above and below dead branches. This rust causes characteristic mid-crown mortality with barren branches. Both above and below the dead crown, living branches may have evidence of fruiting structures (aecia). The aecia, which are formed on twigs or small branches, are orange, long and thin, tongue shaped or narrowly conical, and often 8mm or more tall.

Successive twig and branch mortality over many years suppresses growth and leads eventually to death of the tree either through weakening defenses against secondary agents like bark beetles or by the tree's inability to provide for its own needs. Limb rust is relatively rare in the region but causes dramatic damage to old growth ponderosa pine.

Biology and Disease Cycle: Trees of all ages and sizes are infected. Infection begins on needle bearing branches where the fungus is able to grow systemically. The fungus moves up and down the stem but can not fruit and damage the cambium of the stem since the bark is thicker. This is how the infected tree can have living branches above and below dead branches. Spores (aecia) are produced in spring to mid summer on the pine host. Uredinia, telia, and basidiospores are produced on the alternate host, Indian paintbrush, in the race that requires an alternate host.

Conditions Favoring and Management of Disease: The spore production on the pine requires specific climatic requirements generally consisting of abundant rain, moist atmospheric conditions, and an optimal temperature range. Fungicides may be used to prevent infection of seedlings. Removal of infected hosts or pruning infected branches may prevent the rust from affecting other hosts in the surrounding area.

PINE BLISTER RUSTS

Cronartium spp.

Hosts:

COMMON NAME	FUNGUS	HOST	ALTERNATE HOST
White pine blister rust	*Cronartium ribicola*	Many species of five needle pine; primarily limber,bristlecone and whitebark pine	*Ribes* spp.; currants and gooseberry
Comandra blister rust	*Cronartium comandrae*	Hard pines; primarily lodgepole and rarely ponderosa	*Comandra umbellata*
Pinyon blister rust	*Cronartium occidentale*	Pinyon and singleleaf pine	*Ribes* spp.; currants and gooseberry

Diagnosis and Damage: White pine blister rust infection is first apparent as yellow spotting of needles and formation of spindle-shaped swellings at the base of infected needles.

As the infection progresses, diamond shaped cankers, orange in early summer when fruiting, are visible on young branches and stems. On older tissues, cankers have roughened, dead bark. Comandra blister rust symptoms include swelling, cracking, and elongate cankers on branches and stems. Resin is often associated with blister rust cankers. With comandra blister rust squirrels frequently gnaw off infected bark leaving a bark free canker. One to three years after infection, spermogonia appear as small drops of thick, sticky, reddish-orange liquid on the diseased bark. The following spring/summer orange aeciospores are released from pustules on the canker margin. Yellow-orange urediniospores and brown, hairlike telia are produced on the leaves of the alternate host in mid to late summer.

Seedlings and young trees are most susceptible and are usually killed within a few years by stem cankers. Mature trees are commonly infected and have dead branches in cankered area and the crown thins and eventually dies creating a spike top. Rust cankers may result in growth losses, stem deformities, reduced cone and seed production, and mortality. Main stem cankers may weaken stems and increase risk of stem failure. White pine blister rust has only recently been discovered in northern Colorado, but is common in Wyoming and other areas. The extent of damage to the regions five-needled pines is currently unknown.

Biology and Disease Cycle: The fungus develops annually on the stems and leaves of the alternate host plant, infecting pine needles in late summer and early fall where it lives perennially in the inner bark of pine host. One to four years later, spermatia are produced on the bark of the pine, and the following spring aeciospores, which can travel further than 1000 km to infect the alternate host, are produced. The rust produces five types of spores, uredial, telial, and sporidial stages take place in the alternate host. Spermatia and aeciospores, are produced on the pine host.

Conditions Favoring and Management of Disease: High humidity and moderate temperatures for about 20 hours are required for basidiospores to form, germinate, and infect the host pine. The alternate host must be within about 10 miles for many infections to occur on pine.

In small, intensively managed areas, such as around nurseries, it may be feasible to eliminate the alternate host through practices such as maintaining dense stands, herbicide applications, fire, and pulling. Pruning of infected branches is effective in preventing the fungus from moving into the main stem. Harvest heavily infested stands and replant with non-susceptible species or with more trees than normal since some will die. There has been considerable success with breeding for resistance in white pines, for white pine blister rust, although little work has been done on the five-needled pines native to the region.

SPRUCE BROOM RUST

Chrysomyxa arctostaphyli

Hosts: Primarily affects Engelmann and Colorado spruce in the region. Requires alternate host in the genus *Arctostaphylos* including kinnikinnick (bearberry) and manzanita.

Diagnosis and Damage: This rust causes conspicuous, perennial witches' brooms, which bear annual, yellow-green needles. Spermagonia appear in summer as small reddish eruptions that release a putrid odor detectable from several meters. Broom needles appear yellow-orange in midsummer when spore (aecial) pustules are formed. These brooms may be mistaken for those caused by dwarf mistletoes; however, in spruce, only spruce broom rust causes a marked loss of chlorophyll and annual casting of all broom needles. On the lower surface of the leaves of the alternate host, telia appear as localized reddish-brown spots.

TOP: *Cronartium comandrae* on comandra plant.
CENTER: *Cronartium comandrae* on lodgepole pine.
BOTTOM: *Cronartium comandrae*, squirrels exposed inner wood on lodgepole pine. (W. Jacobi)

TOP: Spruce broom rust. (W. Jacobi)
CENTER: Western gall rust canker on lodgepole pine. (W. Jacobi)
BOTTOM: Western gall rust. (W. Jacobi)

Damage caused as a result of this broom rust includes reduced growth, stem deformation, dead tops (spiketops), and tree mortality. Breakage may occur at the point of stem infections, which act as infection courts for decay fungi.

Biology and Disease Cycle: The fungus lives systemically and perennially in both hosts and may overwinter in either host. In spruce, buds and emerging twigs are infected in the spring by spores produced on the alternate host, and the fungus invades the young shoots and induces the formation of witches' brooms. The rust produces five types of spores of which two, spermatia and aeciospores, are produced on current year's needles of true fir; the others are produced on the alternate host.

Conditions Favoring and Management of Disease: Infection of fir requires moist and temperate weather conditions and synchrony between basidiospore release and spruce shoot development.

Trees bearing stem cankers or brooms may be removed in thinnings and selective harvests. Brooms may be pruned out of trees in order to reduce risk of breakage and maintain tree vigor in areas of intensive management such as recreation areas and Christmas tree plantations.

WESTERN GALL RUST OF PINE

Endocronartium harknessii

Hosts: Affected tree species include a number of native pines, including ponderosa, pinyon, lodgepole, as well as the introduced Scots and mugo pines.

Diagnosis and Damage: The disease causes the appearance of spherical galls on the branches and limbs of pines of all ages. These galls persist but are most visible in the spring when the surface ruptures to release bright orange spores. Galls are most commonly found on branches, but can be found on the main stem; this is more common in certain species, most notably jack and lodgepole pines. As galls enlarge, they cause branch dieback and often cause the host to develop witches' brooms. Trunk or 'hip' cankers are common on lodgepole pine where they do not look like a typical gall but a diamond-shaped canker.

In the Rocky Mountain region the disease is common in stands of lodgepole and ponderosa pines and can be serious in Christmas tree plantations and windbreak plantings. Tree mortality due to western gall rust occurs most frequently in plantations of seedlings, where the likelihood of a gall girdling the entire stem is highest. In mature trees, loss of growth is more likely to occur, but mortality can occur in extreme cases. Wind breakage often occurs at the gall on stems so these trees should be considered hazardous if there are targets for the tree if it should fail.

Biology and Disease Cycle: *E. harkenessii* is spread to noninfected trees by means of spores produced in the galls of infected trees. These spores, produced in the spring, are released when the gall surface ruptures. This dispersal occurs in May and June and allows the spores to infect the current year's shoots. Galls form the year after infection occurs, usually in the summer; the time of infection is thus two years in most cases. Whereas many rust fungi require an alternate host to carry out their life cycles, *E. harkenessii* does not; for this reason, it is sometimes called pine-pine gall rust. Existing galls continue to produce spores every year, making this disease increasingly capable of causing future infection as time goes on. It is this ability, coupled with the ability to directly infect other pines, that makes this disease capable of causing tremendous loss to commercial growers.

Conditions Favoring and Management of Disease: Since spores require wet plant surfaces and cool temperatures to germinate, prolonged periods of cool, wet weather promote the greatest degree of infection. Large numbers of susceptible trees in a given locale (i.e., pure stands or plantations) allow the disease to spread at a high rate. This disease is very difficult to control because of its high rate of infection and the latency period between infection and expression of the symptoms. Removing infected trees has limited effect, since latent infections from the previous year are invisible and therefore cannot be controlled. In any nursery situation, acquisition of clean stock is the best means of prevention.

conditions generally associated with *smaller branches and twigs*

Abiotic Problems

HAIL, SUNSCALD, FROST, AND WINTER EXPOSURE DAMAGE

Hosts: All tree and shrub species are susceptible to damage by weather related conditions such as hail, sunscald or frost, although severity varies among species.

HAIL DAMAGE

Hail can damage all plants by tearing leaves, defoliating branches, knocking off twigs and wounding the bark of branches. Wounds tend toward elliptical shape, vary in length from a few millimeters to 10 cm or more, and occur on the upper surface of the branch, and on the side of the plant facing the storm. This damage can provide entry points for disease pathogens, as well as causing unattractive scarring on the plant affected. Wounds may result in dieback of twigs and branches if tissues around the injuries dry out. Dieback is more likely if injury occurs during the dormant season.

TOP: Hail damage on spruce branch. (M. Schomaker)
CENTER: Hail damage apple fruit. (Luepschen)
BOTTOM: Frost injury to Douglas-fir. (David Leatherman)

WINTER SUNSCALD DAMAGE

Sunscald tends to occur late in the winter especially if there is snow on the ground. Damage is on the south or southwest side of thin barked trees that are not shaded. Intense late winter sun, direct and reflected, encourages plant cells under thin bark to become deacclimated to the cold and vulnerable to damage by freezing. This outermost tissue dies and the bark becomes pinkish or reddish and the surface slowly roughens as it dries and cracks. The damaged area eventually falls away, exposing dead sapwood. When young, many tree species are vulnerable to this type of injury, as they have yet to formed a thick layer of protective dead bark. Trees with thin, smooth bark such as maple, pear, willow, mountain ash, green and white ash, recently planted trees, and many fruit trees are most susceptible to this type of injury.

FROST AND WINTER EXPOSURE DAMAGE

Injury by spring frosts affects succulent, newly formed leaves and stems. These tissues will appear water soaked and soon become shriveled and reddish brown to dark brown or nearly black. Frost injury during the growing season is possible because plants lack any cold acclimation at that time. The likelihood that a particular species or individual will be damaged by a spring frost is related to the date when it begins spring growth. Injury by freezing during dormancy is related to the cold acclimation of a plant. Acclimation occurs gradually in response to lower mean temperatures in the fall, and deacclimation occurs with rising mean temperatures in the spring. The most common external symptoms caused by winter freezing are dieback, foliar browning, sunscald, and bark splitting near the base of the trunk.

Conditions Favoring and Management of Disease: Any of the above conditions may predispose a plant to attack by opportunistic pathogens. Injury that kills cambium,

sapwood, or newly formed tissue creates wounds, which are entry points for disease. When possible, care should be taken to prevent these type of wounds. Protect vulnerable tree trunks in late winter by shading, wrapping with insulated wrap, or painting a light reflective color. These practices that keep the trunk cooler will discourage sunscald injury. Plants that are not winter hardy in the zone that they are planted often will be damaged in a late spring frost. It is best to grow plants from similar zone seed sources when possible.

DROOPY ASPEN

Cause unknown

Hosts: Occurs only on aspen, droopy aspen have been observed in Colorado, Wyoming, New Mexico, and Utah.

Diagnosis and Damage: "Droopy aspen" is a fairly descriptive term for the symptoms of this disorder. Flexuous-rubbery, pendulous branches throughout the crown characterize affected small trees; in larger trees the lower branches show more symptoms. The affected branches have shortened internodes and enlarged nodes, a lack of lateral twig growth and foliage produces a small bunch of leaves at the branch tip. After five to ten years, the droopy branches die, as does the tree.

Biology and Disease Cycle: Preliminary studies have failed to reveal any viral particles or phytoplasma-like bodies associated with the symptoms. The cause or causal agents remain unknown. Reduced production of lignin causes the excessively flexible branches.

Conditions Favoring and Management of Disease: These abnormal trees usually are seen along roadsides, in campgrounds and as transplants in urban areas and mountain communities. These sites suggest that droopy aspen disease is associated with potential root damage from human activities. Keep root-growing space in good condition, reducing soil compaction, drought and using wood chip mulch around trees may discourage the incidence of this phenomena.

PATHOGEN-ASSOCIATED TWIG INJURIES

FIREBLIGHT

Erwinia amylovora.

Hosts: Fireblight is an especially destructive bacterial disease on apple, pear, quince and crabapple. The disease also can occur on cotoneaster, hawthorn, mountain ash, pyracantha, serviceberry, and other species in the rose family.

Diagnosis and Damage: Fireblight symptoms may be observed on blossoms, fruit, twigs, leaves, limbs and the trunk of infected trees. Blossom blight is typically found in early spring about the time of petal fall. Blossoms first appear water-soaked, and turn brown to black as they shrivel and die. The bacterium often progresses from the blighted blossoms into twigs. Blighted twigs often curl to form a characteristic *"shepherd's crook"*

TOP: Winter sunscald injury on mountain ash. (W. Jacobi)
SECOND: Frost damage on pear causing bark splitting. (W. Jacobi)
THIRD: Frost damage on new growth of white fir. (W. Jacobi)
BOTTOM LEFT: Droopy aspen symptoms. (W. Jacobi)
BOTTOM RIGHT: Droopy aspen symptoms. (W. Jacobi)

at their tips. Leaves quickly die turn brown to black and often remain attached. Infected fruit usually appears immature, shriveled and dried. The disease can progress downward in the bark to infect larger limbs and, eventually, the trunk of the tree. Canker margins may range in appearance from being indefinite and difficult to distinguish from the surrounding healthy bark, to discolored, shrunken, and delimited by a definite raised narrow margin. Occasionally the collar and roots of trees become infected through wounds or basal sprouts. Infection at these sites usually results in the immediate death of the tree. Under conditions of high humidity, bacteria may appear as a cream-colored fluid or as oozing strands or tendrils exuded from the surface of infected tissue. As the ooze dries, the exudate will appear as a silver-colored glaze.

Fireblight frequently results in dead twigs, limbs, or may kill the entire tree if not treated promptly. Decay fungi and insects can invade portions of the tree damaged by fireblight. These secondary pests may contribute to wood decay and decline of the tree.

Biology and Disease Cycle: The bacterium overwinters at the margin of cankers. In spring, when the temperature and humidity are suitable, bacteria multiply rapidly, oozing through cracks and bark pores. Insects attracted to the ooze are contaminated and inadvertently infect blossoms during visits. Bacteria move from bloosoms into twigs, limbs and the trunk. Fireblight symptoms usually appear one to three weeks following inoculation. Bacterial ooze also may be splashed by rain to infect open blossoms or wounds in bark or fruit. Wounds created by pruning, insects or hail are frequent infection sites.

Conditions Favoring and Management of Disease: Fireblight is favored by warm weather accompanied by a relative humidity above 60 percent. Blossom infections may be expected if temperatures remain between 65 F and 86 F for a day or more during flower bloom and there is at least a trace of rainfall, or the relative humidity remains above 65 percent for 24 hours. Severe epidemics may follow hailstorms because bacteria directly enter the numerous wounds created by the hail.

Management includes, planting resistant varieties, avoiding over fertilization with nitrogen or any other practice that encourages excessive growth. Pruning to remove infected tissue should be done when weather conditions and plant surfaces are dry. Remove all blighted twigs and branches by cutting 6 to 12 inches below the margin of visible infection. For cankers on large limbs or the trunk, it may be more suitable to remove the infected bark. Make a smooth continuous cut through the bark and down to the wood, 4 inches outside the canker margin, and scrape away all the bark. Immediately apply 10 percent household bleach, or another suitable disinfectant, to the bare wood. Always disinfect cutting tools between each cut by dipping in 10 percent household bleach (one part bleach to nine parts water) or by spraying the tools with a household disinfectant. All infected branches, bark, and severely infected trees should be removed from the site and destroyed.

Chemicals for fire blight management must be in place prior to or during the infection process, because the compounds available are not eradicants. Most sprays contain streptomycin or copper as the active ingredient. Copper sprays are often phytotoxic and, consequently, may russet fruits and leaves. Copper sprays are often used during dormant and blossoming periods, when russetting is not a major concern. Streptomycin is often used for fire blight management. However, frequent use of streptomycin can result in the selection of bacterial strains that are resistant to the antibiotic.

TOP: Fireblight killed branch.
SECOND: Fireblight canker with ooze on pear tree. (Luepschen)
THIRD: Fireblight canker surrounded by narrow callus ridge.
BOTTOM: Aspen stand following frost injury event. (David Leatherman)

TOP: Bacterial blight.
BOTTOM: Bacterial blight on lilac. (W. Jacobi)

BACTERIAL BLIGHT

Pseudomonas syringae pv. *syringae*.

Hosts: Bacterial blight occurs on a wide range of host plant species. It commonly affects lilac and cotoneaster.

Diagnosis and Damage: Symptoms vary with the host plant. Early symptoms of disease on foliage often appear as small brown/black spots with yellow halos. These spots may be restricted by veins and have an angular shape. Spots can enlarge and merge, resulting in dead, shriveled leaves. Infected flower clusters turn brown and die, and infected lilac often have blackened shoots. Plant parts invaded by the bacterium often turn black and lesions on woody green stems may appear as black streaks. If young stems are girdled, all distal portions will die. Infection is usually limited to current growth, and, stems one year or more in age, seldom have lesions. Lesions usually do not exude drops of fluid, in contrast to lesions associated with fire blight. Bacterial blight may result in dead flowers, leaves, and young twigs.

Biology and Disease Cycle: The bacterial blight organism commonly inhabits the surface of a wide range of plants, in the absence of disease. Bacteria can be spread from plant to plant by splashing water. Penetration of the host takes place through natural openings and wounds and canker development is optimum at temperatures of 22- 26 C (72-79 F). The onset of disease is related to the bacterium's ability to function as an ice-nucleus, which encourages freeze damage, and also to produce the toxin syringomycin of slightly-below-freezing temperatures in the presence of the bacterium result in tissue damage more severe than damage caused by either the temperature or the bacterium alone. Bacterial populations and infection of tissue, especially flowers, is more pronounced following periods of low temperature. Also, the ability to produce syringomycin contributes to the development of symptoms. The bacteria overwinter on plant surfaces, in association with buds.

Conditions Favoring Disease and Management: Splashing water aids in the dissemination of bacteria from colonized surfaces. Water soaking during rain favors penetration and invasion of plant tissues by the bacteria. Penetration of the host takes place through natural openings and wounds. Bacterial blight appears to be more severe on white-flowering varieties of lilac, and is more severe following periods of low temperature or chilling injury, presumably because ice-nucleation activity increases host tissue damage and sites for bacterial penetration.

Select varieties that are less susceptible to bacterial blight. Plant varieties adapted to the growing area in order to minimize the stresses that predispose plants to infection. Keep moisture off leaves in spring and keep plants warm since the bacterium is most active at cooler temperatures. Nurseries have found small plastic "hoop" structures placed over plants in the spring prevents bacterial blight.

BLACK KNOT OF CHERRY, CHOKECHERRY AND PLUM

Sexual: *Apiosporina morbosa* (syn. *Dibotryon morbosum*)
Asexual: *Fusicladium* sp.

Hosts: Black knot affects plums, cherries, and other trees and shrubs in the genus *Prunus* in the region.

Diagnosis and Damage: The fungus forms rough black galls along branches and twigs and rough black sunken cankers on stems which give the disease its name. At maturity these are covered with small black, globose sexual fruiting bodies (pseudothecia), which produce unequal two-celled, club-shaped, olive green ascospores. In early summer, olive green asexual fruiting structures (stroma) on galls become covered with conidiophores bearing light brown, single-celled, or infrequently, two-celled conidia. Gall formation leads to canker in the area of the gall and subsequent dieback of the twig or branch distal to the location of the gall.

Damage occurs in native areas in mountain canyons and in urban plantings. In the short term, dieback caused by the disease can substantially reduce yield of fruit trees, and in turn contributes to an overall loss of vigor, which may eventually kill the tree. Large cankers and galls may form on the trunks of trees and cause death or structural weakness leading to wind breakage and thus form hazard trees.

Biology and Disease Cycle: Infection occurs by means of spores (ascospores from perithecia and conidia from pycnidia), disseminated by wind and rain in the spring, produced in the black fruiting bodies that cover galls. The majority of infections are initiated when these spores enter through green shoots. After germinating on the surface of the shoot, the fungus penetrates into the xylem, but causes no symptoms until mid- to late summer, when swelling caused by fungal hormones becomes noticeable. Generally, it is not until the second spring after infection that the fungus produces fruiting bodies, but in areas with longer growing seasons the cycle may be completed in a single year. Once established, cankers and galls are perennial and thus will elongate along limbs, eventually spreading onto other limbs or twigs. Additionally, the fungus may spread systemically through the xylem and phloem to produce cankers on other limbs and the trunk. This cycle continues, interrupted only by winter dormancy.

Conditions Favoring Disease and Management: The disease is favored by wet, windy springs, which allow wide dissemination of ascospores, and by high densities of susceptible hosts, as in orchards and groups of wild cherries. Pruning out diseased limbs at least 10 cm beyond the gall prior to spore dissemination in the spring is fairly effective for control of the disease. Spores will develop on pruned material so all affected twigs and limbs must be taken from the site or destroyed. In some orchards, this sanitation is accompanied by application of a fungicide. Additionally, several biocontrols exist for the management of *A. morbosa*, including predaceous insects and mycoparasitic fungi, all of which feed on the black knot fruiting bodies. Generally, however, control of the disease by means of these measures in insufficient for landscape use.

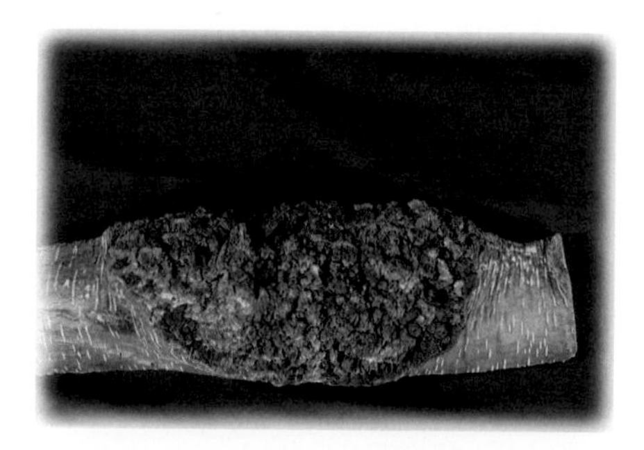

TOP: Black knot on black cherry. (W. Jacobi)
CENTER: Black knot on chokecherry. (M. Schomaker)
BOTTOM: Shoot blight on aspen shoot. (W. Jacobi)

SHOOT BLIGHT (SHEPHERD'S CROOK)

Sexual stage: *Venturia populina* or *V. tremulae*
Asexual stage: *Pollaccia spp.*

Hosts: Predominantly affects aspens, cottonwoods and poplars.

Diagnosis and Damage: Infected leaves develop irregular brown to black areas and become distorted and curled. Leaf stems may become constricted at base. Typically, the fungus spreads down through the succulent new shoot, which blackens and curls to resemble a shepherd's crook. Death of new shoots causes distorted new growth. Repeated annual infection of the current years' growth may distort tree form and cause shrubby growth. Regenerating aspen plants may be kept small bushes by this disease for years until meteorological conditions change to prohibit infection. Infections cause dieback of parts of the tree. The disease is most common in the mountainous areas throughout the Rocky Mountain region. This disease looks like spring frost damage. Distribution of frost damage may be more uniform on the tree and the area, but it is not easy to tell these apart.

Biology and Disease Cycle: Spores of this fungus overwinter in fallen leaves as well as previously diseased stems and twigs. Spores are wind-blown early in the season and infect newly expanding leaves and shoots. During extended wet periods, secondary infection may result when fungus spores are rain-splashed to other parts of the tree growth.

Conditions Favoring and Management of Disease: Warm temperatures and long-term precipitation cycles. Dense groups of trees are especially susceptible. This disease is most common on young aspen and poplars because the fungus needs high humidity to infect. Humidity is higher near the ground so the disease is common on leaves near the ground up to 6 to 8 feet. As the season progresses, uninfected tissue may become resistant to the disease. Winter pruning of the diseased shoots may help. Raking and destroying fallen leaves in landscape plantings during the growing season reduces primary and secondary infections.

Scale Insects

Scale insects suck sap from plants in a manner similar to other members of the order Homoptera, such as aphids. However, scales are characterized by their production of a protective waxy covering and many produce a waxy egg sack or covering, known as a *test*. This cover also can protect scales from effects of most insecticides.

There are several families of scale insects in the region, which have a range in form from the true 'hard' or 'armored scales', such as oystershell scale, to those that very much resemble aphids and mealybugs. Some of the differences between the 'armored scales' and 'soft scales' are indicated in the table below.

Comparison of General Characteristics of Armored Scales (Diaspididae) Versus the 'Soft Scales' (Coccidae, Asterolecaniidae, Eriococcidae).

	ARMORED (HARD) SCALE	SOFT SCALES
Scale covering attached to the insect	No	Yes
Eggs laid inside a sack-like test	Yes	No
Usual overwintering stage in Colorado	Eggs	Mated female or 2nd instar
Typical occurrence of crawlers	Late April-early June	Early-June to early August
Ability to move after crawler stage	No	Limited summer movement typical from leaves to twigs
Production of honeydew	No	Often produce abundant amounts of honeydew
Feeding site in plant	Internal contents of palisade/ parenchyma cells underneath the settled scale	Phloem

Proper tree care, including pruning of heavily scale-infested branches, is a primary management strategy for scales. However, with certain host-plant/scale combinations supplemental controls are needed.

The armored or hard scales primarily overwinter on the plants as eggs within a membranous *test* underneath the old mother scale covering. Other species overwinter as partially grown nymphs. These overwintering scales can be killed by use of oil sprays applied as dormant treatments in spring before bud break. However, control of hard scales is more erratic than soft scale control due to the protective covering.

Horticultural oils are also useful for scale control on many plants after new growth emerges. Hard scales are very susceptible to oils during the first few weeks after egg hatch, when the covering is relatively thin. However, on some plants these summer uses of oils are limited by phytotoxicity concerns.

Use of insecticide sprays for scale insect control require applications that coincide with scale-egg hatch. The brief period after egg hatch is the *crawler* stage when the insect is mobile and has not yet produced a protective waxy coating. Occurrence of the crawler stage differs by scale species and weather. After the insect settles and secretes wax it is largely impervious to insecticides. Insecticidal soaps, summer horticultural oils, diazinon, Dursban, carbaryl (Sevin), Orthene, Tempo, Talstar, and Scimitar have labeling for control of at least some species of scale insects when applied during the crawler stage (*crawler sprays*). Repeat applications may be needed to maintain coverage through egg hatch with species that have an extended period of egg hatch.

Scale crawler activity can be monitored in many ways. Branches known to be infested can be shaken over a collecting surface (e.g., paper, trays) and examined for the presence of the tiny crawlers. Often it helps to use a surface that is smooth and contrasts with the color of the crawler, i.e., light-colored surfaces for detection of dark crawlers such as pine needle scale; darker surfaces for detection of light crawlers such as oystershell scale. Alternatively, crawlers can be trapped on double-sided sticky tape, which, if examined regularly, can also be used as a method for sampling relative scale activity.

Specific recommendations for control of scale insects appear in the supplement on control recommendations. Sections dealing with scales include: **Oystershell scale**, **Pine needle scale**, and **Soft scales (General)**.

TOP: Oystershell scales close-up.
BOTTOM: Oystershell scale colony on trunk.

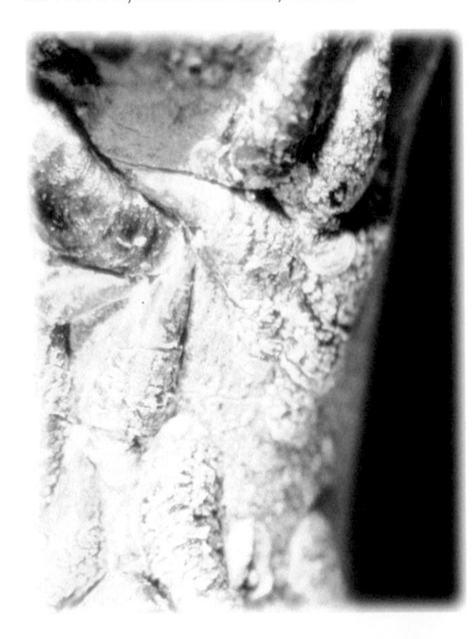

General Timing of Scale Crawler (First Instar) Appearance For Common Colorado Scale Insects

SCALE	FIRST GENERATION	SECOND GENERATION (IF PRESENT)
Oystershell scale	Late May, early June	
Pine needle scale	Mid May	Mid-late July (rare)
Scurfy scale		
Juniper scale	Mid-May through June	
San Jose scale	May, early June (variable)	Late June-July (generation 2); late summer (generation 3)
Black pineleaf scale	June through early July	
Walnut scale	Late June, early July	Early-mid August
Striped pine scale	Early to late June	
Pine tortoise scale	Late May through June	
Cottony maple scale	Late June, July	
Golden oak scale	Late June, July	
Spruce bud scale	Early June through mid-July	
Soft brown scale	Late June, July	
Pit scales	Late June through early August	
Common falsepit scale	Late April through May	
Pinyon needle scale	Early-mid April	
Oak kermes scale	Late September, October	
European elm scale	Late June, July	

OYSTERSHELL SCALE

Lepidosaphes ulmi (L.)
Homoptera: Diaspididae

Hosts: Many hosts are commonly attacked by oystershell scale, making this one of the most destructive insect pests of trees and shrubs. Aspen, ash, cotoneaster, willow and lilac are among the most commonly damaged landscape plants.

Damage and Diagnosis: Oystershell scale feeds the living cells of trunks and branches. As it feeds it often kills the area of the feeding site. Heavy infestations produce stunting, foliage yellowing and bark cracking. Dieback of branches is common result of oystershell scale injury and infested trees are often so weakened they succumb to Cytospora canker. Oystershell scale is the most damaging insect of deciduous trees and shrubs grown in landscape settings in much of the state. It is rarely a concern in natural settings.

Oystershell scales are armored, light to dark brown, elongated and oyster-shaped. On some hosts the scale is covered with a fine powder of wax.

Life History and Habits: There is one generation of oystershell scale per year in the region. Winter is spent in the egg stage under the old cover of the mother scale. The pearly white eggs hatch in late May or early June, and pale yellow crawlers emerge. The crawlers move over the bark in search of sites where they may feed and, if successful, will molt in about a week. Following this molt the scales are legless and are immobile for the rest of their life. Oystershell scales mature in mid-July and mate. Eggs are laid in late summer and early fall. The mother scale dies at the end of the season.

Management: Relatively few natural enemies appear associated with oystershell scale. Parasitic wasps kill a few and a predatory mite has been observed to feed on overwintering eggs. Many scales and overwintering eggs die if areas of bark beneath them are killed.

Vigorous plant growth, provided by proper siting and care, appears to help reduce scale infestations. Often only one or two trunks in an ornamental clump planting are seriously infested.

On smaller trees, old scale coverings and eggs can be destroyed by scrubbing the bark with a soft, plastic pad. Very heavily infested branches may need to be pruned.

The most effective chemical controls are "crawler sprays", applied to coincide with egg hatch. Carbaryl (Sevin), chlorpyrifos (Dursban), diazinon, Orthene, Tempo, Talstar, and Astro are currently available crawler sprays. (Note: Aspen foliage is sensitive to liquid formulations of many insecticides.) Alternatively, insecticidal soaps or oil sprays applied in three to four day intervals during the crawler period can also provide control. Although most treatments are largely ineffective after scale crawlers have molted, summer (foliar) spray oils can control young nymphs for several weeks after they have settled on the bark.

Effectiveness of oils applied as dormant treatments has been more erratic with oystershell scale than with many other scales. This is because the eggs are well protected by the covering produced by the mother scale. Dormant applications of oils are more likely to be effective in spring, after the scale covering has weathered.

Old scales remain in place for several years after the scales have died. In order to determine if controls are effective, old scales should be cleared from at least some of the branch, so that reinfestation can be detected. Also, when crushed, dead scales are dry and flake easily off the bark; scales covering eggs typically will produce some moisture when crushed.

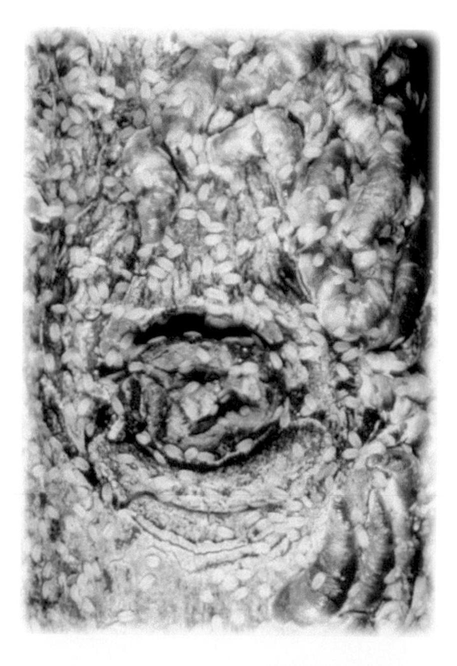

PINE NEEDLE SCALE

Chionaspis pinifoliae (Fitch)
Homoptera: Diaspididae

Hosts: Pine, spruce, and fir. Serious damage is most common on mugho pine.

Damage and Diagnosis: Pine needle scales suck sap from the needles. Usually there is some localized discoloration around the feeding site and high numbers of scales induce premature needle shed. Prolonged outbreaks, which are rare in the region, can kill branches and even young trees. On many trees, outbreaks are often confined to limited areas - usually a single branch or two.

The adult female is almost pure white, elongated and armored. They are slender and slightly yellow at the front end, widening at the rear. Crawler stages are generally oval and light purple. After settling, they molt and are generally translucent yellow, often with a dark center marking.

Some pines, notably bristlecone or foxtail pine, naturally secrete a white wax on the needles. This may superficially resemble pine needle scale, although it does not have the regular form of scale insects. Furthermore, old scale coverings, and more importantly, scales killed by natural causes, do not readily drop off. This makes assessment of infestation severity difficult.

However dead scales can be determined by their easy dislodging with an easy scrape of the fingernail. As they readily dry out after death they flake off more easily.

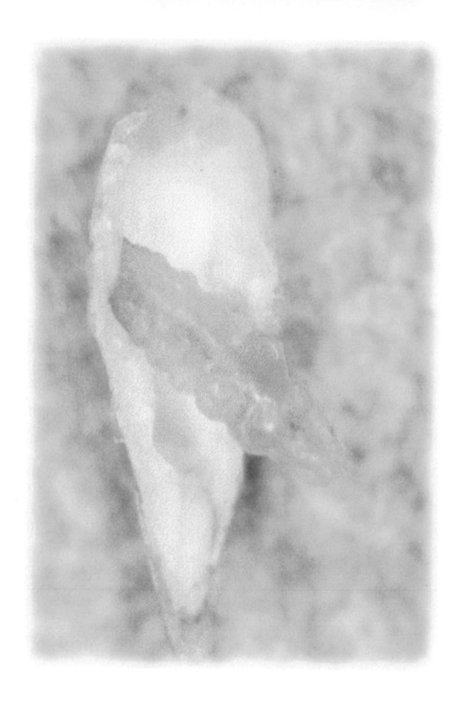

TOP: Oystershell scale crawlers. (F. Peairs)
CENTER: Pine needle scale on spruce.
BOTTOM: Pine needle scale adult exposed from scale covering.

TOP: Young nymphs of pine needle scale on needle.
SECOND: Euonymus scale.
THIRD: Lady beetle larva feeding on pine needle scale. (D. Cooper)
BOTTOM: Scurfy scale with recently settled crawlers.

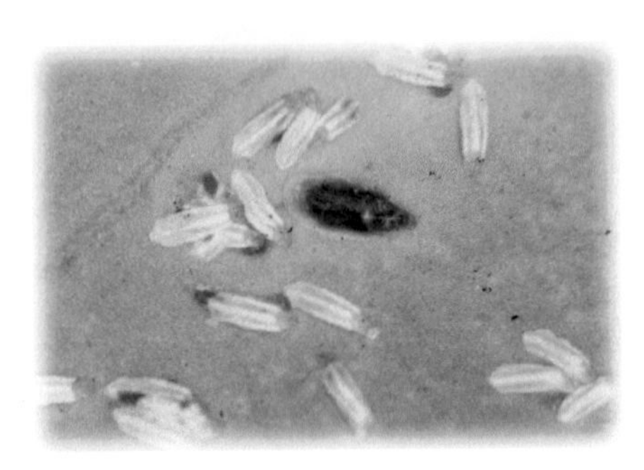

Live scales so disrupted will typically ooze fluids (often reddish purple) upon crushing and/or minute eggs may be discerned.

Life History and Habits: Overwintering stages of pine needle scale occur under the scale cover of the female. A mixture of life stages can be found into midwinter - females maturing eggs, females that have already laid their full complement of eggs and died, and females maturing eggs that will be laid in addition to those already present. Egg production can continue through winter, at low levels. Some females lay eggs only in early spring.

Egg hatch has been noted to be initiated as early as late April in southern areas of Colorado but normally may be expected to begin around mid-May. Despite the mix of pine needle scale overwintering life stages, most egg hatch can be expected to happen over a fairly short period, from late May through midJune.

After settling, the nymphs remain in place throughout their life. Whitish egg tests,produced by mature females, are formed primarily after late July. Males are not known from this area and egg production begins in early fall, extending through winter and early spring.

Related Insects: Scurfy scale (following) is very similar in appearance to pine needle scale, but is associated with deciduous hosts, notably *Populus* and willow. Infrequently euonymus scale, *Unaspis euonymi* (Comstock), becomes established in highly protected sites. This species has very different male and female forms: males are small, white and primarily found on foliage; females more closely resemble oystershell scale and feed on stems and twigs.

Management: Severe cold temperatures appear to very adversely affect survival of overwintering stages and may be the most important factor limiting outbreaks.

Natural enemies of pine needle scale are well represented and often effect a high degree of natural control. The most important biological control species appears to be a small lady beetle, *Coccidophilus atronitens*, with a one-year life cycle well synchronized with pine needle scale. Several small parasitic wasps are common natural enemies of pine needle scale in the region.

SCURFY SCALE

Chionaspis furfura (Fitch)
Homoptera: Diaspididae

Hosts: Aspen, cottonwood and willow are the most common hosts. Apple, hawthorn and mountain-ash, and *Prunus* spp. are among the other reported hosts

Damage and Diagnosis: Scurfy scale feeds on the sap of twigs and branches. During outbreaks, which are quite rare, large dirty-white crusts of the scale will be present and infested areas of bark killed.

The adult female is armored, pear-shaped and flat, and about 1/8 inch long. Color varies from white to dirty gray. They are similar in appearance to the closely related pine needle scale. Males are similar in appearance but much smaller. The crawler stage is dark purple.

Life History and Habits: Scurfy scale spends the winter as reddish-purple eggs beneath the coverings of the dead mother scale on tree branches and trunks. Eggs hatch in May and nymphs settle on branches and trunks to feed. They mature rapidly and adults may be present in late June or early July. A second generation develops in late summer, with adult females of this generation producing overwintering eggs in fall.

BLACK PINELEAF SCALE

Nuclaspis californica (Coleman)
Homoptera: Diaspididae

Hosts: Ponderosa and other pines. Rarely on spruce.

Damage and Diagnosis: The scale develops by feeding on the sap from needles. High populations may contribute to chlorosis or reddening of needles and crown thinning. However, this insect is usually only abundant on trees that are previously stressed due to other factors.

Black pineleaf scale is generally dark and elliptical and are found on the needles. Sometimes in mature scales the dark area is surrounded by white, creating a bullseye appearance.

Life History and Habits: This insect has not been studied under regional conditions. It has one, and probably two generations per season. Overwintering likely occurs as a partially developed scale that matures in late spring. Eggs and crawlers occur in June and July. If there is a second generation, egg hatch would occur in late summer.

Management: Outbreaks of this scale have often been shown to be associated with stressful growing conditions of the host plant such as root injury or compaction, drought stress or smog. Dusty conditions are reported to interfere with natural enemies of this insect and contribute to outbreaks.

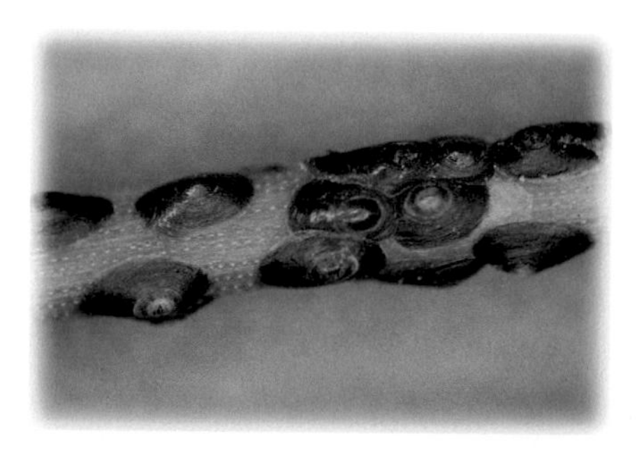

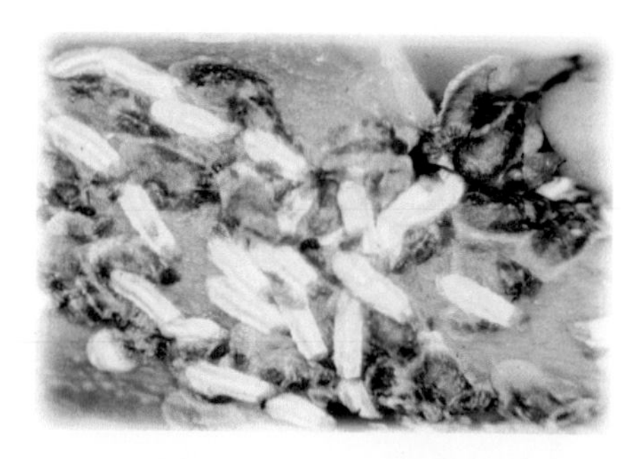

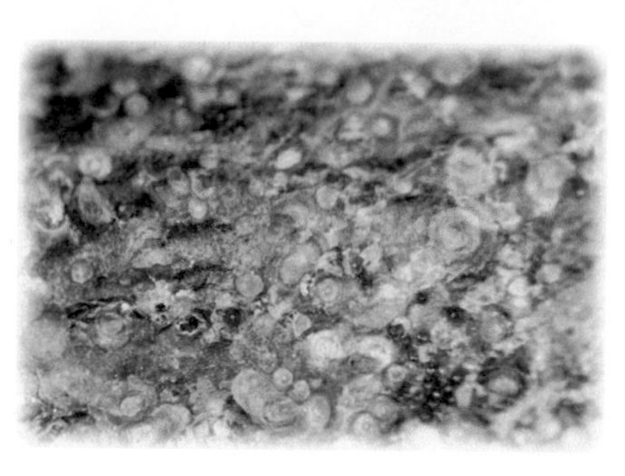

TOP: Black pineleaf scale.
CENTER: Juniper scale.
BOTTOM: Walnut scale on bark.

JUNIPER SCALE

Carulaspis juniperi Bouche
Homoptera: Diaspididae

Hosts: Various junipers, particularly pfitzer junipers

Damage and Diagnosis: The scales suck sap from the needles, reducing vigor and production of new growth. During heavy infestations the entire shrub takes on an off-color and may look as if dusted with snow. Decline and dieback occur during outbreaks. This insect occurs infrequently in Colorado, primarily in the Boulder area.

Adult female scales are circular, about 1/10-inch in diameter, and white. The developing male scales are smaller and more elongated, emerging as butterscotch-colored winged forms in the adult stage.

Life History and Habits: The overwintering stage of the juniper scale are eggs under the old covering of the mother. Eggs hatch in midspring, probably around early May. The newly hatched crawlers move to feeding sites, settle, and soon secrete a waxy covering. They continue to grow and molt until becoming full-grown by mid to late summer. The winged males emerge at this time and mate with production of overwintering eggs following. There is one generation produced per year.

Management: Crawler sprays should be applied before eggs hatch in spring. Horticultural oils are not recommended since many junipers have shown sensitivity to treatment.

WALNUT SCALE

Quadrispidiotus juglansregiae (Comstock)
Homoptera: Diaspididae

Hosts: A wide variety of deciduous trees and shrubs are listed as hosts, with most local infestations associated with linden, silver maple and ash. Despite its name, the only walnut species (*Juglans* spp.) that hosts this insect is Persian walnut.

Damage and Diagnosis: Infested trees lose vigor with thinning and yellowing of foliage. Twig dieback can occur during heavy outbreaks.

Life History and Habits: Life history of the walnut scale in the region is unknown. Based on information in other regions it probably spends the winter either as a partially developed scale in the second instar. Development continues in spring with mature females being present in late June. After mating, the females produce eggs that hatch in early July. The nymphs from these eggs mature rapidly in summer, maturing around early August and produce a second generation.

Related Species: The San Jose scale, *Quadrispidiotus perniciosus* (Comstock), also occurs throughout much of the state, primarily on fruit trees and rosaceous shrubs. The body of the San Jose scale is covered with a darker wax cover and lacks the central indentation of the walnut scale.

Management: Dormant oil applications can control overwintering stages. Because of the extended egg-hatch period, crawler sprays may be difficult to time correctly. Crawler sampling should begin in mid to late June.

ABOVE: San Jose scale on apple fruit. (J. Capinera)

SAN JOSE SCALE

Quadraspidiotus perniciosus (Comstock)

Homoptera: Diaspididae

Hosts: Apple, rose, pyracantha, cotoneaster and crabapple are common hosts but many hardwood trees and shrubs in the rose family (Rosaceae) may be infested.

Damage and Diagnosis: The scales feed on the sap of twigs and small branches weakening and sometimes killing cells around the feeding site. A small reddish area often develops on the twigs around the feeding scale and extends internally to the xylem. Infested trees lose vigor with thinning and yellowing of foliage. Dieback of twigs and branches can occur during heavy outbreaks. San Jose scale also may infest fruit. Fruit becomes spotted around the feeding site of the scale.

San Jose scale is fairly inconspicuous and may build up in huge numbers that appear as a grayish crust on the branches. Individual scales are armored, circular, flattened approximately 1/16-inch diameter when full grown. They are generally a dark gray/brown scale with a lighter center. The body of the insect, observed by removing the wax cover, is bright yellow and circular.

Life History and Habits: San Jose scale overwinters as second instar nymph. It remains dormant until sap flows in the spring, then continues to feed and develop. While feeding, both nymphs and adults secrete hard waxy coating (armor) over their bodies. First stage nymphs ("white caps") are circular and dirty white. Later stages become gray brown and develop a more pronounced nipple in the center.

Coverings of the males are considerably smaller and elongate. Adults are small, gnat-like insects that are winged. They mate with the females who may subsequently produce hundreds of eggs and crawlers over a period of less than a month. Development can be rapid but some may go through a temporary dormant period. There are probably about three generations produced during the growing season in Colorado. As the season progresses there is considerable overlap and all stages may be found during summer.

Related Species: The walnut scale, *Quadrispidiotus juglansregiae* (Comstock), also occurs throughout much of the state. It has a wide host range of non-rosaceous deciduous trees including linden, silver maple, ash and poplars. Appearance is similar to that of the San Jose scale but the body is more orange or orange-red and sunken in the center.

Management: Dormant oil applications can control overwintering stages. Crawler sprays are best applied against the first generation, since the populations are most synchronized at this time. The first generation crawlers often occur shortly after apple petal fall.

Pheromone traps have been developed for this insect. These capture the small, winged male scales and can be used for timing treatments after the first generation. However, since egg hatch becomes increasingly spread out as the season progresses, the value of these later treatments can be marginal.

San Jose scale is attacked by several parasitic wasps. The twice-stabbed lady beetle is an important natural predator.

COMMON FALSEPIT SCALE

Lecanodiaspis prosopidis Maskell
Homoptera: Lecanodiaspididae

Hosts: This insect has been found on ash, mulberry and honeylocust. Other hosts are reported elsewhere in its range. Common falsepit scale is most common in southeastern Colorado but has been identified from the Metro Denver area.

Damage and Diagnosis: Nymphs and adults suck plant sap through the bark of twigs and branches causing reduced growth and/or twig dieback depending on the severity of the infestation. Dead leaves cling to the twig throughout the winter.

Common falsepit scales are about 1/6 inch in diameter, rounded and cream colored. They are typically found in pits formed on the bark of the host in response to their feeding.

Life History and Habits: The common falsepit scale overwinters in the eggs stage within the body covering of the mother. Eggs hatch in May and the crawlers move to the newly emerged shoots. They settle in crevices of the bark and feed at this site for several months, becoming full-grown in midsummer. Males emerge in August to mate with the females and eggs are laid in late summer and early fall. There is one generation per year.

Management: Controls have not been identified. Since the eggs are held within a covering (test) under the mother scale over winter, dormant oils are not likely to be effective. Crawler sprays should be applied in late April or early May.

TOP: Common falsepit scale on mulberry.
BOTTOM: Golden oak scale.

GOLDEN OAK SCALE

Asterodiaspis quericicola (Bouche)
Homoptera: Asterolecaniidae

Hosts: White oaks. Fastigate English oak has been most commonly infested

Damage and Diagnosis: The golden oak scale feeds on the sap of developing twigs, and may also feed on leaves when in high density. Reduced growth, distortion and dieback of twigs may occur from severe infestations. Delays in leafing out in spring and longer retention of old leaves are reported from other pit scales.

Mature females are generally hemispherical and settled in a depression on the twigs of oak. Typical color is golden brown, or darker, sometimes with green. A slight waxy fringe may develop around the scale.

Life History and Habits: Biology of the golden oak scale is not studied under regional conditions so the following description is based on related species. The golden pit scale spends the winter as mature females, on the bark of twigs. They resume feeding in spring and begin to lay eggs, which fill the covering (test) of the female. Crawlers emerge over a period of several months, with peak crawler activity in early summer. They do not move far from the mother scales, settling on first year wood of twigs. As they feed, tissues swell forming the distinctive pit in which they live. They are immobile shortly after settling. Males are unknown. There is one generation per year.

Management: Some parasitic wasps are associated with this insect are appear to provide a high level of natural control. Parasitized scales are most easily identified by the circular exit holes the parasites cut.

Crawler sprays should be based on surveys of crawler activity but probably will be timed for the period around mid-June through mid-July. Dormant oil treatments should

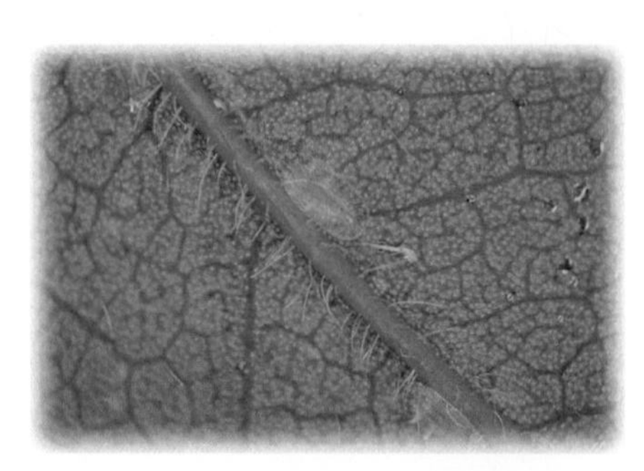

TOP: Full-grown female cottony mapel scale with egg sack.
SECOND: Cottony maple scale nymph on leaves.
THIRD: Cottony maple scale, overwintering females on twigs.
BOTTOM: Soft brown scale.

be effective. For specific recommendations, see the section in the control supplement regarding **Soft Scales**, although this is a species that is not technically a soft scale.

COTTONY MAPLE SCALE

Pulvinaria innumerabilis (Rathvon)
Homoptera: Coccidae

Hosts: Silver maple, honeylocust, hackberry, linden and other hardwoods.

Damage and Diagnosis: Cottony maple scale is one of the largest and the most conspicuous scale insect that occurs in the region. Adult females may swell to over 1/4 inch diameter when producing a large cottony egg sack. Adults and nymphs feed on plant sap of twigs and leaves, respectively. Sustained outbreaks can reduce plant vigor and even cause dieback. However, natural controls are abundant for this species and outbreaks are usually terminated following a single season.

Cottony maple scale also excretes honeydew as it feeds. When high populations are present this can allow sooty mold to grow on trunks and branches and may attract nuisance wasps and bees.

Life History and Habits: The cottony maple scale spends the winter as adult, mated females on twigs and branches. During winter they appear as small, indistinguished dark lumps but begin to grow rapidly when they resume feeding in spring. A very large, wax-covered egg sack is ultimately produced by late spring, somewhat of the appearance of a marshmallow, and may contain over a thousand eggs. Eggs hatch from late June through July and the newly hatched crawlers settle on the undersides of leaves, usually near the midrib. During this leaf-feeding stage, the nymphs are flattened with oval form and nearly translucent.

Nymphs feed on leaves until late summer, migrating to twigs and small branches before leaves drop. They then molt for the last time, after which the females remain in place for the rest of their lives. Males are produced at this time that are mobile and mate with the females, dying afterwards. There is one generation per year.

Management: Abundant natural controls appear to be able to control cottony maple scale in most locations. Several parasitic wasps, predatory plant bugs (*Dereaocoris nebulosus*) and other general predators are particularly important. As the result of these biological controls infestations of cottony maple scale typically are short lived even without applied controls.

Oil sprays applied during the dormant season can help control overwintering cottony maple scale; midsummer foliar oil applications can control stages on leaves. Crawler sprays should be first applied during mid to late June and may need reapplication because of the extended egg-hatch period. Systemic applications of imidacloprid should be effective against summer stages that occur on leaves.

EUROPEAN FRUIT LECANIUM (Brown Soft Scale)

Parthenolecanium (= Lecanium) corni (Bouche)
Homoptera: Coccidae

Hosts: European fruit lecanium has a very wide host range including many deciduous trees and shrubs. Elm and stone fruits (*Prunus*) are the most common hosts.

Damage and Diagnosis: European fruit lecanium is a soft scale that feeds on the phloem of small branches. Under sustained outbreaks they are capable of causing decline and dieback, but generally are under a high level of natural control which largely prevents any significant injury. European fruit lecanium is also capable of producing large amounts of honeydew, which can be a serious nuisance problem and favor the growth of sooty mold.

Coinfestations with other scales can occur, particularly mixed populations with European elm scale. The European elm scale can be identified by having a covering that is dome-shaped, dark brown, and fairly smooth.

Life History and Habits: The European fruit lecanium spends the winter as a second instar nymphs on the twigs, appearing as a small, raised lump about 1/4-inch diameter. They continue growing in spring, molt to the adult stage and develop rapidly. Most feeding, honeydew production, and injury occurs at this time.

The body of the female swells with eggs during late spring. Males may be produced and mating occur at this time but European elm scale often reproduces asexually. The eggs hatch underneath the body of the mother scale, and the crawlers move out continuously for several weeks in midsummer during favorable weather. The mother scale then dies, after having produce several hundred eggs. The newly hatched crawlers move to feed on leaves for the remainder of the growing season. Prior to leaf fall the pale-brown Instar II scales return to the twigs for winter.

Management: Although a common and widespread insect, serious, sustained outbreaks of this species are rare, presumably due to natural controls. Parasitic wasps are one of the more commonly observed natural enemies of this insect. Applied controls, including dormant oils, are similar to those used for other soft scales.

STRIPED PINE SCALE

Toumeyella pini (King)
Homoptera: Coccidae

Hosts: Various pines, but is particularly damaging to Scots pine and lodgepole pine

Damage and Diagnosis: Feeding stunts development of new growth and can induce premature needle drop. Heavy infestation may kill branches and severely detract from tree appearance. Since the mid 1980s problems have greatly increased expanding along the Front Range from the original infestations in the Denver area.

Striped pine scale produces abundant amounts of honeydew, which attracts scavenging yellowjacket wasps in fall and honeybees in spring. Black sooty mold often develops on the honeydew excreted by the insects, covering twigs and branches.

The adult female is hemispherical and attached to pine twigs. General coloring is dark brown or black with reddish brown or cream colored mottling. They have dark red blood. Adult females reach up to 1/4 inch in diameter. Males are much smaller and develop on the needles, producing a winged form. Nymphs are generally orange/brown and globular in form. Males are more elongate and in late stages of development are enclosed in a papery light colored covering, which is later left on the needles as adults emerge. In early season, nymphs are most often observed on needles, but females later move to twigs.

Related/Associated Species: The pine tortoise scale, *Toumeyella parvicornis* (Cockerell) also occurs in Colorado, although it has been less abundant and damaging than the striped pine scale. Biology appears to be similar, except that male stages migrate from needles to twigs to mate. Females are slightly smaller than striped pine scale and are more uniformly colored.

Life History and Habits: Striped pine scale spends the winter in the form of fertilized females on twigs. Feeding resumes in spring, at which time the insects are most conspicuous and produce the largest amount of honeydew. As the females begin to mature as many as 1500 eggs they become greatly enlarged.

Egg hatch occurs beneath the protective covering of the female and begins in late May or early June. Emergence of the crawlers can continue for a month or more, but peak activity typically follows egg laying by about two weeks. Male scales usually settle on

TOP: Mature female striped pine scales.
CENTER: Male striped pine scale on needle.
BOTTOM: Comparison of pine tortoise scale (left) and striped pine scale (right), mature females.

the needles; females cluster on twigs at the base of the needles. Once a suitable location has been found on needles, they insert their mouthparts and remain in place for the remainder of the summer. Winged male stages are produced at the end of the summer and mating takes place at that time. The mated females remain on the twigs throughout the winter. There is one generation per year.

Management: No insect predators or parasites seem to be very important in controlling this insect in Colorado. Warblers and other insect-feeding birds can destroy large numbers of overwintering scales.

Dormant oil sprays applied in early fall or spring can help control striped pine scale. Heavier weight oils have been more effective for control than ultra-fine spray oils. Crawler sprays using contact insecticides or oils should be first applied during early June in most locations. Because of the extended egg-hatch period, a reapplication may be needed to provide control during outbreaks. Horticultural oils, Sevin, or Merit, and Dursban have been moderately effective for control of summer populations. Excellent control has been achieved in Colorado State trials with soil applications of Merit. Treatments should be applied by early summer before scales have significantly matured.

TOP: Fletcher scale.
BOTTOM: Spruce bud scale. (Oregon State University)

FLETCHER SCALE

Parthenolecanium fletcheri (Cockerell)
Homoptera: Coccidae

Hosts: Junipers, arborvitae and yew

Damage and Diagnosis: Fletcher scale is a typical soft scale, of rounded form when full grown and somewhat similar to the closely related European fruit lecanium. Most are caramel-brown and they are attached to needles. During the period when they are most actively growing in late spring they may produce large amounts of honeydew. However, Fletcher scale has been an uncommon insect in the region, apparently well suppressed by natural controls.

Life History and Habits: Fletcher scale overwinters as a small, flattened second instar nymph on the needles. In spring, they resume development and swell rapidly as eggs are produced and mature. Eggs begin to hatch in mid to late June. The yellowish crawlers typically are present over a two or three-week period, after which they settle on needles and produce the yellow-brown summer form. There is little further development through the summer and only one generation is produced per year.

Management: Control is rarely necessary in the region, although Fletcher scale is an important nursery pest in other areas of North America. A single well timed crawler spray should be able to control infestations that do exist. Oil sprays can be useful, but should not be applied to juniper because of phytotoxicity risks. See **Soft Scales** in the recommendation guide.

SPRUCE BUD SCALE

Physokermes piceae (Schrank)
Homoptera: Coccidae

Hosts: Spruce

Damage and Diagnosis: Spruce bud scale sucks sap from the twigs of spruce and can cause needle drop and twig die back during outbreaks. It also produces abundant amounts of honeydew and sooty mold is associated with outbreaks. Infestations tends to be concentrated on the lower branches.

The mature females are reddish-brown and globular. They closely resemble a spruce bud.

Life History and Habits: Spruce bud scale spends the winter as a small, first instar nymph on the needles of spruce. In midspring they resume activity and move to the twigs where they settle and feed. Females become full-grown and swollen with eggs in June and eggs hatch during late June and July. There is one generation per year.

Management: Although this insect has not been studied in the western United States, outbreaks are rare presumably because of the activity of natural enemies such as parasitic wasps.

Crawler sprays, applied in midsummer to coincide with egg hatch, should provide some control.

EUROPEAN ELM SCALE

Gossyparia (= Eriococcus) spuria (Modeer)
Homoptera: Eriococcidae

Hosts: Elm, particularly American and rock elm

Damage and Diagnosis: European elm scale is one of the most widespread and destructive scale insects in Colorado. Prolonged infestations weaken branches, often producing premature leaf yellowing (*flagging*) and leaf drop. Heavy infestations cause dieback of twigs and branches.

European elm scale can cause serious nuisance problems due to honeydew production with peak production during June and early July as the females mature. Sooty mold often grows on the honeydew, blackening tops of branches and root flares. An unusual feature of these insects is that their blood is red.

Associated Insects: The elm leaf aphid, *Tinocallis ulmifolii* (Monell), also is a common honeydew producer on American elm. Its biology is discussed elsewhere.

Life History and Habits: European elm scale spends the winter as second instar nymphs, packed into cracks on twigs and smaller branches. They are oval in general form and pale grey due to the light waxy cover of the body. In spring they resume develop and the females swell greatly, becoming darker with a distinct waxy fringe. During late April and May male scales may also begin emerging from small white cocoons and mate with the females. However, males are not always produced and this species can reproduce asexually.

Eggs hatch within the body of the female and crawlers emerge over a period of several weeks, peaking between mid-June and mid-July. They move to leaves and settle on the leaf underside, the dark yellow nymphs almost always being found tucked next to main leaf veins. In late summer they migrate back to the twigs where they overwinter.

Management: Several species of parasitic wasps commonly kill some scales, causing the affected scales to be stiffened and puffed in appearance (i.e. scale "mummy"). Several general predators such as, predatory plant bugs, predatory mites, and spiders also kill many of the nymphs on leaves. On Siberian elm, competition with defoliators, such as elm leaf beetle, probably kill many of the leaf inhabiting stages of the scale.

Horticultural oils, particularly more viscous formulations, can control overwintering stages. Treatments should be applied before bud break since older scales are difficult to control with oils.

Crawler stages can be controlled by applications of various insecticides made in late June, at onset of egg hatch. Repeat applications may be needed to maintain coverage through the extended period of egg hatch. Leaf feeding stages of the scale, during late July and August, as well as nymphs that are moving to twigs in late summer, can be controlled with horticultural oils and several insecticides (e.g., Sevin, Talstar). Insecticidal soaps have been minimally effective against nymphal stages. Soil injected treatments of imidacloprid (Merit) have been very successful for control of this insect.

TOP: European elm scale full-grown females and males.
CENTER: European elm scale nymphs on elm leaf.
BOTTOM: European elm scale crawlers on adult females.

KERMES SCALE

Allokermes gillettei (Cockerell)
Homoptera: Kermesidae

Hosts: Pin oak is the most commonly infested host in Colorado. Different species of Kermes scale infest other species, including Gambel oak.

Damage and Diagnosis: Reduced growth and vigor of the branches results from feeding by this scale. Heavy infestations cause leaf flagging, dieback of twigs, and twig abscission. Oozing sap and a sticky wax is a common and unusual symptom produced around infested areas of twigs.

Females are immobile, globular, tannish scales specked with brown. They are about 1/4 inch in diameter, the largest scales present in the region, and are usually found at the base of leaf stems.

Yellowjacket wasps, which feed on the mature scales and eggs, can become a nuisance problem on scale infested trees.

Life History and Habits: The overwintering stage of oak kermes scale is a very small, first instar nymph on branches of oak concentrated arond the buds. Following emergence of the new growth most remain concentrated at point of new shoot growth; others move from these areas in spring to the new growth and feed on sap, often settling at the base of the petioles. They continue to feed for several months becoming full-grown in late June and early July.

At maturity the females mate with the minute, winged males. The females then begin to mature eggs, becoming very large in late summer, almost the size of a small marble. The eggs hatch in September and October, but the crawlers may remain under the protection of the mother scale of several weeks. Newly hatched scale "crawlers" move to the overwintering sites on the branches and trunk for shelter. There is one generation per season.

Management: Controls have not been developed for this species. Dormant oils should be effective on overwintered stages. Crawler sprays applied in early fall are also likely to be effective.

Fox squirrels and yellowjacket wasps have been observed to feed on mature female scales.

PINYON NEEDLE SCALE

Matsucoccus acalyptus (Herbert)
Homoptera: Margarodidae

Host: Pinyon

Damage and Diagnosis: Feeding by adult females and nymphs causes needles to turn yellow and prematurely fall. Most defoliation occurs on older needles, producing a tufted appearance with younger needles primarily persisting on infested trees. Repeated yearly attacks can kill young trees and weaken large trees, which then become susceptible to bark beetle attacks. This insect is most common and damaging in southern Colorado.

TOP: Oak kermes scales on oak twig.
SECOND: Kermes scale and gumming reaction to infestation.
THIRD: Pinyon needle scale, "bean stage".
BOTTOM: Pinyon needle scale injury.

The stage most commonly observed are the second instar nymphs, often described as the 'bean stage' found attached to needles in early spring. Adult females that subsequently are produced are black, armored, mobile scales, about 1/16 inch long that migrate to the trunk where they produce a cottony mass of eggs in spring. Adult males, present in early spring, are winged, fly-like, and rarely seen.

Life History and Habits: Pinyon needle scale overwinters as a second instar nymph, which is legless and resembles a small, black bean. They resume development in early spring, molting to the mobile adult form. Mating occurs in early April. Eggs are laid in masses around collar, branch crotches, and underside of larger branches. Egg masses are covered by a white cottony wax, which can be quite conspicuous on heavily infested trees. Newly hatched nymphs settle on the previous year needles. The second stage is formed in late summer and overwinters attached to the needle. There is one generation per year.

Management: Drenching trunk sprays of dimethoate (Cygon) are specifically registered for use of this insect. These should be applied against the adult stages in early spring before egg laying. Sprays used against other scale insects should also be effective.

INSECTS THAT OVIPOSIT IN TWIGS

BUFFALO TREEHOPPER

Stictocephala bisonia Kopp and Yonke
Homoptera: Membracidae

Hosts: Apple, peach, ash, elm and many other deciduous trees

Damage and Diagnosis: The buffalo treehopper injures plants not during feeding but when it inserts eggs into twigs. Small rows of scars appear on injured twigs that become scabby and may be sites for plant pathogens to become established. Serious injury is uncommon but has occurred in young orchards grown amongst legume cover crops that favor this insect.

The buffalo treehopper is a rather unusual insect in appearance. It is generally triangular-shaped with the sides of the front developed into small points, somewhat resembling a miniature bison. They are grass green colored and about 3/8 inch long.

Related and Similar Injuries: Certain leafhoppers also oviposit in twigs, producing similar injuries.

Life History and Habits: The buffalo treehopper overwinters in the egg stage, under the bark of twigs. The eggs hatch in late spring and the nymphs usually feed on grasses and broadleaf weeds around the base of trees on which eggs were laid. They become full-grown in late July or August and females lay eggs during from August until a killing frost occurs. In the process of egg laying, the female cuts two slightly curved slits in the bark and lays a small group of up to one dozen eggs under the loosened bark.

PUTNAM'S CICADA

Platypedia putnami (Uhler)
Homoptera: Cicadidae (Platypediidae)

Hosts: Putnam's cicada is native to shrub lands of mountain-mahogany or Gambel oak and pinyon/juniper habitats. They have adapted to landscapes and most commonly are associated with crabapple, honeylocust, maple, mountain ash, and oak.

Damage and Diagnosis: Damage is produced not by feeding, but instead occurs when the females insert eggs into twigs (oviposition wounds). This causes a gouging and splintering of twigs that frequently causes them to break, producing conspicuous "flagging" of wilted foliage. Nymphal damage from root feeding is not considered to be significant for this or any regional cicada species.

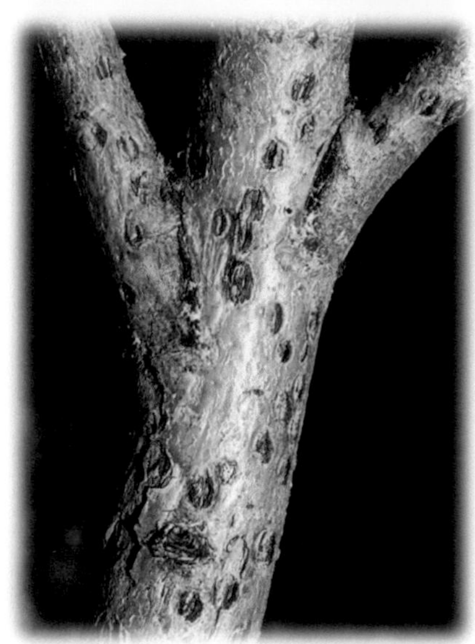

TOP: Buffalo treehopper.
SECOND: Buffalo treehopper egg laying scars in peach.
THIRD: Twig breakage resulting from oviposition wounds by Putnam's cicada.
BOTTOM: Old egg laying scars of buffalo treehopper.
(David Leatherman)

TOP: Putnam's cicada recently emerged from nymphal skin. (L. Bjostad)
BOTTOM: Emergence holes made by Putnam's cicada. (David Leatherman)

Putnam's cicada produces a clicking noise, likened to the striking of two dimes. Their large size and ability to produce noise results in many concerns related to this insect. They are large, dark-colored insects with bulging eyes and membranous wings, folded tent-like over the body.

Life History and Habits: Life history of Putnam's cicada is poorly understood but is thought to take three to five years to complete. The entire immature stage occurs underground, with the nymphs feeding on the plant roots. When full-grown, the nymphs emerge from the soil and transform into adults, leaving behind their cast nymphal skins on the lower trunk of their host.

Adults are present for about four to six weeks from June through early July. After mating, the adult females lay eggs in slits in the twigs of various hosts. Upon hatching, nymphs drop to the ground, burrow beneath the soil surface, and spend the next two to five years feeding on the roots of plants.

Related Species: Over 30 different cicadas occur within the Central Rockies, although the periodical cicadas (*Magicicada* species), such as the "17-year locust", are not present. Putnam's cicada is the only species which has had associated plant damage in Colorado. The high-pitched buzzing of the dog-day cicadas (*Tibicen* species) in shade trees often attracts attention.

Management: No controls have been developed to prevent egg-laying wounds by this insect. Control of other cicadas with insecticides has been very disappointing since the adult cicadas are highly mobile and are present over an extended period (weeks). Very high value plants, particularly younger trees that are still getting established, may be protected by covering with netting to exclude the adults.

TIP MOTHS, TWIG AND TERMINAL FEEDING INSECTS

Several species of insects attack terminal growth of woody plants, usually causing dieback of terminal growth. Most common are the larvae of various moths, although the white pine weevil can be very damaging to spruce at higher elevation areas.

Tip moth control requires that the insecticide cover susceptible terminals at the time of egg laying and egg hatch. Often this coincides with shoot elongation, although it varies with different insect-host plant combinations.

For some species pheromone traps are available to identify flight and egg laying periods. Because the insecticide is being applied to expanding tissue, repeat applications are often required. Insecticides with systemic activity (e.g., Orthene, Cygon) or persistence on bark (e.g., Dursban, Astro) are standard for the treatment.

Some measure of control could be gained by mechanically destroying tip moth pupae of species that overwinter just beneath the ground line. Typically these species, such as the common southwestern pine tip moth, have pupae that attach themselves to the root collar of the extreme lower trunk by means of plaster like cocoons. Use care not to injure the bark of the tree, as wounds in these areas are ideal for invasion by harmful fungi.

Tip moth injuries are typically much more conspicuous than they are actually damaging. As the plants recover, there is often little long-term injury, other than slight increased bushiness at the injury site or curve in the stem as it continues to grow. However, tip moths do occur as serious pests of landscape plants in some areas (e.g., Nantucket pine tip moth in the Albuquerque area) and their effects on tree growth can be significant in forest production and in Christmas tree plantations.

Tip moth populations tend to be highly cyclic. Several natural enemies, primarily parasites, act to reduce the survival of the tip moth larvae and can be highly effective.

Specific control suggestions appear in the supplement with section related to tip moths and other insects affecting terminal growth detailed under **Southwestern pine tip moth**, **Pinyon tip moths/Pinyon nodule maker**, and **White pine weevil**.

SOUTHWESTERN PINE TIP MOTH

Rhyacionia neomexicana (Dyar)
Lepidoptera: Tortricide

Hosts: Several species of pines, particularly Austrian and mugho. Scots, ponderosa, and bristlecone (foxtail) are occasionally attacked.

Damage and Diagnosis: Larvae develop by tunneling under the bark of new shoots, producing girdling wounds that cause them to brown and crook. Heavy infestation for consecutive years may retard growth, leaving trees short and bushy.

Adults are small moths with mottled reddish patterns on forewings and light tan-colored hind wings, wingspan of about 3/4 inch. The larvae, found under the bark of infested twigs are reddish-orange to yellow caterpillars, about 1/2 inch long when mature. The head capsule and anal plate are light brown.

Life History and Habits: The southwestern pine tip moth spends the winter as a pupa, in plaster-like cocoons attached to the base of the trunk. Adults emerge from early to mid spring, typically laying eggs as new needles emerge (candling stage). Eggs hatch in 14 to 21 days, and small larvae first feed inside a needle. Later feeding occurs inside needle sheaths or buds and eventually hollows out growing shoots. Larvae mature by midsummer and then usually drop to the ground to pupate. There is one generation per year.

Related Species: At least eight other species of *Rhyacionia* are found in Colorado including the western pine tip moth, *R. bushnelli* (Busck), and the ponderosa pine tip moth, *R. zozana* (Kearfott). The western tip moth has somewhat different habits than the southwestern pine tip moth in that pupation occurs within the damaged shoot, rather than around the root collar, and they may have two generations per year.

Management: Several parasites attack the tip moths, and outbreaks tend to be highly cyclical due to these natural controls.

Insecticide treatments should be applied during the period when eggs are being laid and newly hatched larvae and eggs are exposed. This typically occurs during the early candling stage, when needles are emerging from new shoots. Chlorpyrifos (Dursban), Orthene, and dimethoate (Cygon) are effective for control of this species.

TOP: Southwestern pine tip moth larva in Austrian pine.
CENTER: Southwestern pine tip moth adults.
BOTTOM: Cottonwood twig borer larva.

COTTONWOOD TWIG BORER

Gypsonoma haimbachiana (Kearfott)
Lepidoptera: Tortricidae

Hosts: Cottonwood and other poplars

Damage and Diagnosis: The larvae tunnel into new shoots causing branch tip dieback. Long-term infestations lead to a distorted branching pattern, particularly characterized by bushiness. Excessive twig shedding can also become a significant nuisance problem. Internal tunneling of the stem can be used to identify cottonwood twig borer as the source of twig shedding rather than squirrels or other causes. The larvae are dirty gray caterpillars with brown heads and may be found associated with tunneled twigs.

Life History and Habits: The overwintering stage is apparently a very young larva, which makes small pits in the bark, often near old leaf scars or tunneling wounds. Some older larvae may overwinter within hollowed out terminal buds.

Larvae resume activity in spring, boring into actively growing shoots, tunneling down the pith. When full-grown (late May-early June), they emerge and crawl down the trunk, pupating in protected sites on trunks and branches. Adults emerge in about

TOP: Pitch nodule of the ponderosa pitch nodule moth.
CENTER: Ponderosa pitch-nodule moth larva exposed from wound site.
BOTTOM: Pinyon pitch nodule maker wound.

eight to ten days and females typically lay eggs on the upper leaf surface. The very young larvae first feed on the midrib or vein of leaves. Later, they migrate to twig tips and begin tunneling. The twig-tunneling phase takes about a month to complete. There are probably two or three generations produced per year.

Management: Several species of birds feed on the larvae within tunnels. Presumably, this species is attacked by various insect parasites.

No chemical controls have been identified. Dormant applications of oils should provide some control of overwintered larvae on bark. Foliar sprays of insecticides effective against **Leaf feeding Caterpillars** (see supplement) applied shortly before bud break should also provide sufficient residue activity to kill larvae as they first become active. Because of the multiple, overlapping generations of this species, it will be difficult to precisely time treatments during the growing season.

PINYON PITCH NODULE MOTH

Retinia (= Petrova) arizonensis (Heinrich)
Lepidoptera: Tortricidae

Host: Pinyon

Damage and Diagnosis: Larvae feed on terminal growth, causing twig dieback. A characteristic round nodule of purple-red pitch is produced around the feeding site. The larvae are reddish-yellow caterpillars with a black head and dark area behind the head, found within the pitch nodule.

The adult is a rusty brown moth with a wingspan of approximately 3/4 inch. Forewings are mottled with brown, white, and silver scales.

Related and Associated Species: Another related tip moth, *Retinia metallica* (Busck), attacks terminal growth of ponderosa pine and is found statewide. In pinyon, the pinyon tip moth, *Dioryctria albovittella* (Hulst), often kills terminals but does not make the same pitch nodule.

Life History and Habits: Winter is spent as a partially grown larvae within the characteristic nodules of pitch. Feeding resumes in spring followed by pupation in late spring. Adult flights typically peak in late July and early August. Eggs are laid on needle sheaths and the newly hatched larvae feed first on the young needle tissue before tunneling into shoots. There is one generation per year.

Management: Individual larvae can be fairly easily located in pitch nodules during late spring, picked and killed from small plantings. Plants that are being used for new landscaping should be checked and receive some such treatment, since this species can cause significant damage in some settings.

Preventive insecticide applications should be applied during midsummer, the only time when the larvae are exposed on the exterior of the twigs.

PINYON TIP MOTH

Dioryctria albovittella (Hulst)
Lepidoptera: Pyralidae

Host: Pinyon

Damage and Diagnosis: The larvae feed underneath the bark of twig terminals, producing girdling wounds that cause twig dieback. Often there is some oozing of pitch at the wound site, but it does not form the smooth, purplish nodule of the pinyon pitch nodule moth (above). Larvae may also tunnel cones. Older larvae, found within the tunneled terminal in late spring and early summer are light golden brown in color with a dark brown head capsule. Full-grown larvae reach approximately 3/4 inch in length. Adults are small grayish moths with forewings marked in a zigzag pattern and a wingspan of about 1 inch. Their hind wings are generally grayish white.

Related and Associated Species: Other damaging species of *Dioryctria* in Colorado include the pinyon pitch mass borer (*D. ponderosae* Dyar) and Zimmerman pine moth [*D. zimmermani* (Grote)]. These trunk and branch borers are discussed in a separate section. Non-damaging species also feed on cones.

Life History and Habits: Overwintering stage is a first instar larva found within a small silk cocoon (hibernacula) on the bark. In mid to late May the insects begin to feed by tunneling into base of unopened buds. As insects grow they mine the pith of the terminal growth. Large amounts of pitch and silk may collect at the wound site. Large larvae often leave the original wound site and move to new tissue, cones, or new shoots. Pupation takes place in the terminals and cones. Adult flights and egg laying occur primarily from late June through August. There is one generation per year.

Management: Control should be timed for periods when the larvae are exposed on the plant. This occurs both in midsummer, after egg laying, and again in May as they become active and begin to enter stems. For specific controls see the sections on the supplement on **Tip moths**.

TOP: Pinyon tip moth larva.
CENTER: Pinyon tip moth injury.
BOTTOM: Peach twig borer damage. (Clemson University)

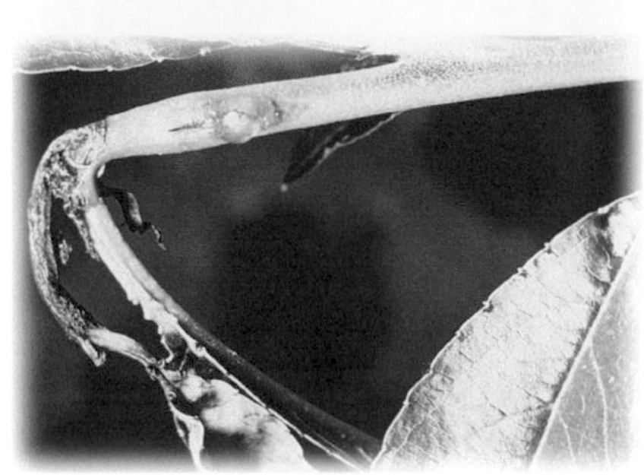

PEACH TWIG BORER

Anarsia lineatella Zeller
Lepidoptera: Gelechiidae

Hosts: Peach, primarily. Other stone fruits and apple are rarely infested.

Damage and Diagnosis: Larvae burrow into the developing twig tips, killing them back three to four inches. Midsummer borers tunnel into peach fruit. This insect is only known from the West Slope.

Life History and Habits: The peach twig borer overwinters as a partially grown caterpillar, protected in a small silk-covered cell on the bark of fruit trees. In early spring, the caterpillars become active, migrate to the twigs and tunnel the buds and emerging terminal growth. The damaged new growth typically wilts and dies (flagging). The caterpillars then pupate in the tree and emerge as moths in May.

The emerged moths lay eggs on the twigs, small leaves, and developing fruits. The caterpillars usually first feed on terminal growth of twigs, later moving to fruits shortly after the pits start to harden. After becoming full-grown, these caterpillars pupate on the trunk and larger branches. Moths that emerge from this second generation lay eggs on the fruit. Caterpillars restrict their feeding to the fruit at this time, causing most crop injury. (Apricots, which fruit earlier than peaches, are damaged more by the first generation.)

After the caterpillars become full-grown, there is a third generation. This produces that overwintering larvae, and no significant injury is caused by it.

Management: Several different species of parasitic wasps commonly attack peach twig borer and may kill 50 percent or more of the caterpillars. In addition, caterpillars exposed on leaves or bark are susceptible to general predators such as green lacewing larvae and damsel bugs.

Dormant oil or lime-sulfur sprays used for control of other insects and diseases also can kill many of the overwintering larvae on the tree. Sprays of several other insecticides, applied just prior to bloom can effectively kill the caterpillars as they begin to emerge from their overwinter shelters and move about on the tree.

The moths are small (about 1/4 inch), and ash-grey. They are rarely observed but can be easily captured in pheromone traps baited with the peach twig borer lure. This can help to better time sprays used for control of the insect later in the season.

TOP: White pine weevil injury to spruce terminal.
CENTER: White pine weevil larvae in spruce terminal.
BOTTOM: Magdalis weevil. (Whitney Cranshaw)

There have been promising research developments indicating that the 'male confusion' method, involving permeating the air with the female insect's sex pheromone to inhibit mating, might be useful to control peach twig borer in isolated orchards or plantings. Some products have recently been marketed for this purpose.

WHITE PINE WEEVIL

Pissodes strobi (Peck)
Coleoptera: Curculionidae

Hosts: Colorado blue spruce, Engelmann spruce. White pine is the common host in eastern states.

Damage and Diagnosis: White pine weevil is sometimes the most serious insect pest of Colorado blue spruce in landscape plantings, particularly at the higher elevations. In other parts of the U.S. this insect is also an important pest of white pine. Feeding by the developing insects causes the top terminal (leader) to suddenly wilt and die in early summer. Upper branches are affected less frequently. This damage can subsequently result in a bushy, deformed tree that may be considered aesthetically undesirable.

Immature stages (larvae) of the white pine weevil feed underneath the bark of the spruce leader, girdling the plant. When sufficient damage has been done, the top growth will wilt and curl, becoming completely dead in a few weeks. Only the top leader and upper branches are affected by the insect and damaged leaders have an overall characteristic appearance involving a "shepherd's crook" appearance with the needles turning a gray-blue color. Often, at the base of the damaged growth, there are small (1/8-in) round exit holes in the bark made by the emerging insects.

Once the top leader is killed, some side branches will change their growth habit and begin to grow upwards to take the place of the killed leader. If successful, these new leaders will form main trunks and multiple main trunks will occur above the damaged area. This changes the form of the tree from its normal tapering growth to one that is more densely bushy. This can be considered to detract from the appearance of the tree.

The adult stage of the white pine weevil is a small (1/4-in) snout beetle flecked with brown and white patches. A feature it shares with other weevils is the long, curved snout with elbowed antennae arising from it. Larvae are almost identical to those of bark beetles, looking like a grain of cooked white rice. The larval head is brown. The larvae are found within the terminal growth of spruce.

Life History and Habits: The insect overwinters in the adult stage, under leaf litter and in other protected areas. After snow melts and temperatures begin to warm, (mid-March to early May) the weevils become active and females seek out spruce trees. They feed on the cambium of main branches near the leader and insert eggs into the feeding cavities that are formed. Small points of oozing pitch on the main leader are indicators of this feeding and egg laying activity.

Eggs hatch in one to two weeks and the young grubs (larvae) tunnel downward underneath the bark. Damage increases as the insects grow and wilting starts to become noticeable in June and July.

When full-grown, the white pine weevil larvae tunnel deeper into the stem and form a cocoon made of wood chips in which they pupate. In about two weeks the adult beetles emerge through small holes they chew through the bark. The chip cocoons remain behind and are a useful means of diagnosing old white pine weevil injury.

Adult weevils feed on the needles, buds and twigs of spruce for several weeks before going into a dormant condition for overwintering. Some minor chewing injury to buds may result if infestations are severe.

Related and Associated Species: A closely related weevil species, *P. terminalis* (Hopping), occurs in the tops of lodgepole pine, producing similar injury.

Magdalis lecontei Horn is a weevil associated with terminal growth and small branches of ponderosa and, occasionally, lodgepole pine. Larvae make tunnels in a similar manner to the white pine weevil, but leave fine-packed borings in the tunnels rather than coarse shredded material. Adults of *M. lecontei* are usually blue-black, somewhat metallic and have a prominent snout. Attacks are almost always confined to recently dead or dying branches.

Management: Insecticides applied in spring when adult weevils feed on trees and lay eggs can provide control. Standard bark beetle or wood borer insecticides should provide control if used at rates labelled for the above insects. These include the insecticides carbaryl (Sevin), chlorpyrifos (Dursban), and permethrin (Astro). Timing of these treatments will vary by location and year but usually should be made in late March or early April. Only the upper areas of the tree need to be treated.

Mechanical removal of infested terminals while the insects are still present can provide some future control if other sources of weevils are not in the area. For there to be any control value, this should only be done in June or July before adult emergence. Terminals should only be cut as far down as necessary to remove the weevil larvae. Pruned terminals should be destroyed, as the insects will be able to complete development in the prunings.

If top growth has been killed, proper training of a single side branch as a replacement leader can help to salvage the future appearance of the tree. Often several side branches will begin to grow upward and the healthiest of these should be favored. Temporarily binding competing shoots or pruning the tips of competing shoots will allow for a single leader to again be established. This new leader should be annually protected with insecticides until the tree is no longer highly susceptible to attack. Such a decision can be based on the incidence of attacks to hosts in the area.

TWIG BEETLES

Pityophthorus species, *Pityogenes species, and certain other minor genera of bark beetles*
Coleoptera: Scolytidae

Hosts: Pines, spruce, true firs, Douglas-fir and certain other ornamental conifers (but not junipers - see **cedar bark beetles**).

Damage and Diagnosis: The larvae tunnel the inner bark and girdling kills branches, tops and small trunks. Damaged plants show distinctive 'flagging' of terminal growth.

In natural situations these beetles function as pruners of shaded-out or broken twigs and branches. They also often serve as indicators of conditions that favor outbreaks of other, more damaging bark beetles such as mountain pine beetle. In urban settings they are major contributors to the death of recently transplanted or stressed pines. Certain species of *Pityophthorus* also occur under trunk bark of large trees under attack from bark beetles in the genera *Ips* and/or *Dendroctonus*. In such situations they are of minor importance compared to the other bark beetles present.

Adults are tiny bark beetles, 1/32 to 1/6 inches long, generally dark brown. Most species have rounded rear ends but some male *Pityogenes* have a pair of curved, fingerlike projections. Larvae are small, legless, "C"-shaped grubs with a caramel-colored head and are found within the twigs.

Life History and Habits: Adult emergence, flight and attack activities occur throughout the warm months and may begin by mid-March following mild winters. They find suitable hosts and produce broods just under the bark. Attacks produce much tan or reddish sawdust but little pitch. The pattern of galleries made beneath the bark by adults and feeding larvae is generally star-shaped and lightly etches the wood. Most species of twig beetles have two to four generations per year.

TOP: White pine weevil adults.
CENTER: Old "chip cocoons" of white pine weevil.
BOTTOM: Twig beetle tunneling of ponderosa pine.

TOP: Twig beetles.
BOTTOM: Nest cells of *Pemphredon* sp. hunting wasps in pith of ash.

Management: These beetles are usually secondary pests, breeding in dying wood and can be indicative of substantial stress on the plant. Improvement of cultural conditions to allow more vigorous growth should be fundamental to control. Where freshly cut wood or damaged trees occur, these beetles can build to high populations. Such breeding sites should be eliminated near landscape plantings.

If transplanted trees are being threatened, standard bark beetle preventive sprays during the first season should provide control. Treatments should be persistent or frequent enough to maintain coverage through the warm months, since multiple, overlapping generations occur with this group of bark beetles. Note twig beetle protection may require more applications (two to three evenly spaced through the period of April-October) than is needed for protection from larger bark beetles, which only normally attack the trunks of mature trees). Susceptible tree species with a trunk diameter greater than one-half inch transplanted during this period should receive protective sprays either slightly before, or the day of, being placed in their new site. Waiting a few days or week to preventively spray may be too late.

PEMPHREDON WASPS

Pemphredon spp.
Hymenoptera: Sphecidae

Hosts: Pemphredon wasps nest in cavities they excavate from pithy plants. Ash, elder, sumac, rose, and caneberries are among the plants most often used.

Damage and Diagnosis: Pemphredon wasps, and some other hunting wasps, excavate the pith from plants to establish nesting chambers. No plant injury results, as the insects restrict their nesting to the pith. As these hunting wasps prey on aphids, they may provide modest beneficial effects. Nesting is restricted to pith that has been exposed by pruning cuts or breakage.

When colonized branches are split a series of cells is usually evident. Each of these may contain aphids, aphid fragments and/or the grubs of developing wasps. When full-grown they are small (less than 1/4 inch) black wasps.

Related and Associated Insects: Scores of species of wasps develop by hunting other insects and returning the prey to their nest. Unlike the *Pemphredon* wasps most do not excavate pith and instead use existing cavities or tunnel into soil when constructing nests. Others make nests of mud and certain social species, such as hornets and yellowjackets, construct paper nests using chewed wood. Most species of hunting wasps tend to specialize in use of certain prey, which may include leafhoppers, crickets, spiders, weevils or many other insects. The largest hunting wasp in the state is the cicada killer, which paralyzes dog-day cicadas and stores them in underground chambers that they dig.

Life History and Habits: These are hunting wasps that use aphids for prey. Female wasps establish nests by excavating the pith from plants; these species only use plants with pith intact. In most plants, particularly those with a small diameter pith, the individual nest cells are built sequentially on top of each other. However, a system of branching chambers may be made in large diameter twigs.

Prey are located by the female who grasps a single aphid, paralyzes it by stinging and returns to the nest. She then repeats this process until the cell is adequately provisioned, usually with about two dozen aphids. She then lays an egg on the aphids she has stored and seals the cell with a plug of chewed pith. A series of cells are subsequently provisioned and sealed off.

Eggs hatch in about three to five days and the developing wasp larva feeds on the aphids for about two weeks, before becoming full-grown. They then remain in this prepupal stage for a variable amount of time, depending primarily on what the season is. Late in the year most remain as a prepupa, overwintering before resuming development in spring. However, generations that develop in spring or early summer remain as a prepupa for only a short time and then transform to the pupa. Regardless of the season, pupal stages last about three weeks. Adults then emerge, in reverse order of when eggs were laid, i.e., those reared in the last cells produced are first to exit. There are probably two generations annually produced.

Other insects may be found in the cells and cavities these wasps produce. Several parasites attack the developing wasps. Also, other species of hunting wasps that utilize existing cavities may colonize nests of pemphredon wasps, sometimes destroying in the process the developing stages of pemphredon wasps.

Management: No controls are recommended. Sealing pruning cuts to prevent access to the plant pith can prevent nesting.

conditions generally associated with *larger branches and stems*

Vascular Wilts

True vascular wilts of trees are relatively uncommon in the region except for a few notable examples. Tree wilts are caused by fungi that affect the vascular system causing the tree to wilt. Some fungi cause vascular damage in the root system such as black stain root disease of pinyon, while others mainly affect the smaller twig leaf areas such as Dutch Elm disease while others cause dysfunction in the stem and branches such as Verticillium wilt. Trees can show wilt symptoms from extreme heat, drought, root damage, and cankers girdling a stem or branch. The diseases listed in this section are those caused by organisms that invade the vascular system -xylem or the "true wilt diseases"

Diagnosis Hints: Wilting of leaves is the main symptom of wilt diseases but terminal dieback and decline is also common. Discoloration of the xylem is common and cross sections of the branch or stem will show the discoloration in the current year's vessels. The discoloration needs to be running vertically in an annual ring and not moving radially. Radial staining pattern is from "blue" stain fungi that live in the ray parenchyma cells and are usually associated with various bark beetles.

TOP: Dutch elm disease discoloration in elm wood. (W. Jacobi)
BOTTOM: Dutch elm disease flagging in elm tree. (W. Jacobi)

DUTCH ELM DISEASE

Sexual: *Ophiostoma* or *Ceratocystis ulmi*
Asexual: *Pesotum ulmi*

Hosts: All species of elms native to North America are attacked by the fungus that causes Dutch elm disease, as well as other genera in the elm family. In Colorado it is most damaging to the American elm. Siberian elm is tolerant to Dutch elm disease and is an important potential breeding site for insect vectors of the pathogen.

Diagnosis and Damage: The first evidence of fungal infection is wilting or "flagging" in one or more of the upper branches of a tree. Leaves yellow, wilt and eventually turn brown, but remain on branches. The symptoms then spread to adjacent branches. Eventually the entire tree wilts and dies, taking from several weeks to several years. When bark on infected branches is peeled back, light to dark brown streaks or blue to gray discoloration in the wood indicate the presence of a vascular infection. This symptom is not entirely diagnostic and positive diagnosis requires a laboratory test.

Trees affected with Dutch elm disease do not survive and this disease has all but wiped out native stands of American elm as well as urban and farmstead plantings.

Biology and Disease Cycle: This fungus is spread from declining or dead elms to healthy elms via insect vectors or root grafts. In this region one insect transports the fungus from tree to tree, the lesser European elm bark beetle, *Scolytus multistriatus*. The beetles breed in declining or dead trees or logs infected with the fungus. Sticky fungal spores adhere to the insect's body and are carried to healthy trees where adult beetles go to feed on twig crotches. The fungus is consequently introduced into the host tree and invades and grows in the water-conducting vessels, causing wilting and death of the tree. Once in a host tree, the fungus can travel to nearby (35 to 50 feet) elm trees via root grafts.

Conditions Favoring and Management of Disease: Locations containing dead elm wood or declining elm trees are prime breeding grounds for the beetle vectors, increasing the likelihood of disease spread. Susceptible elms of the same species planted within close proximity to each other favor the spread of the pathogen through root grafts.

There is no cure for Dutch elm disease and the primary emphasis in management is prevention. Systematic inspection of every elm within a community to detect the early symptoms of infection, isolation by disruption of root grafts between infected and healthy trees, and prompt removal and disposal of all dead and dying elm material with intact bark are key to a good management program. Early infections can be removed from elm trees by pruning. A minimum of 8 to10 feet of streak-free wood (no vascular discoloration) must be removed. Individual high value trees may be protected by injection of chemical fungicides at labeled intervals.

VERTICILLIUM WILT

Verticillium spp.

Hosts: Predominantly affects ash, catalpa, elm, sumac and maple, and occasionally found on linden, and fruit trees.

Diagnosis and damage: Foliar symptoms typically include wilting, curling, yellowing, marginal or interveinal browning, or death. Often these symptoms may look like water stress. Other symptoms may include dieback of branches or a portion of the tree. Furthermore, wood under the bark may exhibit discolored streaks or bands. This streaking may be scattered within previous year's growth rings or be confined to current year's sapwood. The color of the streaks can range from light tan in ash trees, grayish to olive-green in maples, or brown-black in elms. Other diseases may produce symptoms similar to verticillium wilt. Positive identification is best determined by laboratory tests.

The disease can be lethal, causing sudden and total collapse of the entire plant or often show progressive or intermittent symptoms. Yellowing and defoliation often progress upward. Sometimes, only a branch or portion of the plant is affected. Trees may suddenly recover if conditions become favorable for plant growth and defense mechanisms. The disease is found throughout the Rocky Mountain region at a low incidence and is usually seen in trees under stress from poor growing conditions.

Biology and Disease Cycle: Verticillium wilt is caused by a soil inhabiting fungus, which affects the plant's vascular system. Infection occurs through roots or where damage to the stem has occurred near the soil line. Once the fungus invades the plant, it spreads into water conducting tissues (xylem), disrupting water movement and normal plant functions. The pathogen survives in roots and trunk of killed trees remaining viable for several years and as resting structures (microsclerotia) persisting in soil for years.

TOP: Verticillium on sumac. (W. Jacobi)
BOTTOM: Verticillium stain in maple wood. (W. Jacobi)

Condition Favoring and Management of Disease: Trees under moderate to severe water stress on other root related stresses are more susceptible to an attack from this fungus. Damage to trees also increases their susceptibility. Optimum temperature for growth of the fungus in plants is 65 to 72 F.

There is no chemical cure for verticillium wilt. When the disease is suspected, increasing a tree's vigor may help defend against the disease. Infected trees should be fertilized and watered throughout the growing season. High nitrogen fertilizers increase wilt severity. Balanced fertilizers such as 10-10-10 are recommended.

When infections have been verified within a portion of the tree, remove all dead branches or prune back to wood showing no vascular streaking. All dead branches or dead wood should be disposed of properly. A note of caution on pruning: branches showing slight wilting of leaves should not be removed immediately; they may recover in response to water and fertilizer treatments. Avoid construction or lawnmower damage around tree bases since the fungus can also enter through wounds at or near the soil line.

Resistant or tolerant species should be used in replacement plantings to reduce the chance of infection from contaminated soils. Conifers are immune to the pathogen. Overall, the best long-term solution to avoid the disease in landscape plantings is to utilize resistant varieties and maintain healthy plants. Plants listed as resistant usually can withstand infection or possibly avert the disease all together.

Canker Diseases

Cankers are dead areas on the stem bark and cambial area caused by fungi, bacteria, or other living agents. Cankers can kill plants especially if the plant is stressed and can not defend itself well. Drought, freeze damage, dormancy, soil conditions, other diseases and insects are common stresses that allow canker organisms to damage trees. Canker causing organisms spread from tree to tree by wind and occasionally by insect vectors. Canker causing organism usually enter the tree through wounds or branch crotches. Canker causing organisms grow in the bark and kill the tree by killing the cambium.

Diagnosis Hints: **Cankers** maybe sunken if the canker has been active for several years killing the cambium while surrounding tissue continued to expand. Rapid killing of bark and cambium of a branch or stem by an aggressive canker organism does not produce the typical sunken canker seen in many picture guides. Some cankers may be oval, elongate or round. The resulting damage of a canker is usually dieback of the portion of the tree above the canker. Determining what canker is on your tree is important because it could be caused by a slow growing organism that will not kill your tree rapidly and you can do something about it. Physical wounds/injuries differ from cankers in that canker causing organisms are expanding the affected area over time killing more bark each year. You can tell physical wounds or injuries from cankers by looking for evidence of expansion of the canker. Old callus ridges, several layers of sunken wood, target appearance of damage are evidence of canker expansion. Physical wounds have a smooth wood wound surface and usually have callus around the edge indicating the tree is trying to grow over the wound. Insect borers are often found associated with cankers since both organisms are found on stressed trees.

Management: Disease prevention is usually obtained by improving the growing environment, reducing stresses and planting disease resistant species. Some cankers can be removed if the plant is healthy and can defend itself. See Thyromectria canker for detailed instructions on canker removal. There are limited chemical treatments for canker problems.

Canker Diagnosis Guide

HOST	SYMPTOMS AND SIGNS	DISEASE
Aspen only	Large black callus ridges around sunken canker	Ceratocystis canker
Aspen only	White flecks in black-sooty bark - long elongate canker	Cryptosphaeria canker
Aspen only	Sooty/black bark with small cup shaped fruiting bodies on bark	Black canker
Aspen and many hardwoods	Discolored bark with small pimples spaced about 1/4 inch apart	Cytospora canker
Aspen only	Checker board pattern of bark cracking in canker	Hypoxylon canker
Many hardwoods - elms, honeylocusts,	Orange/pink to black pin head sized fruiting bodies sticking out from bark or just under thin bark layer	Tubercularia canker
Honeylocusts only	Irregular brown to black fruiting mass in lenticels on main stem or small pin head sized pycnidia on branches	Thyronectria canker

ABOVE: Siberian elm canker - circled area of fruiting bodies. (W. Jacobi)

SIBERIAN ELM CANKER (BOTRYODIPLODIA CANKER)

Botryodiplodia hypodermia

Hosts: Susceptible hosts are American, Siberian, English and smooth-leafed elms. Other elm species have not been tested for susceptibility.

Diagnosis and Damage: Infected bark on host becomes reddish-brown to black and often splits longitudinally. Outer bark may become loose and coil back. Inner bark tissues turn reddish to blackish brown, becoming water-soaked and soft. Cambium and sapwood beneath infected bark discolors reddish brown. The fungus can girdle a stem and cause foliage to wilt and the stem above that point to die. This girdling often causes sprouts to develop below this point, giving the infected tree a bushy appearance. Yellow foliar symptoms on American elm can superficially resemble those of Dutch elm disease. Fruiting bodies (pycnidia) produce spores (conidia) that are initially single celled and colorless, but eventually become brown, and some two celled. The pycnidia are small and tightly spaced on the bark with only a small pimple protruding through the bark. Pycnidia are usually over looked since they are so numerous and so small that they look like "normal" bark.

Botryodiplodia is one of three damaging canker pathogens (*Cytospora, Tubercularia, Botryodiplodia*) of Siberian elm in the Great Plains. It causes dieback and death of infected trees, and has severely limited the usefulness of this species in windbreaks.

Biology and Disease Cycle: Pycnidia develop on infected, dead bark in the autumn, causing a roughening of the bark surface. These fruiting structures contain spores (conidia) that turn from colorless to brown when mature. Mature spores are present year round, and masses of them exude from pycnidia with moisture present. Splashing or running water disperses them.

Conditions Favoring and Management of Disease: Botryodiplodia canker causes significant damage only to drought-stressed elms and infects through wounds. Winter-injured twigs on drought-stressed trees are common infection points for this fungus. Observations have also indicated that herbicides, especially 2,4-D, may predispose Siberian elm to infection, but experimental evidence is lacking.

BLACK CANKER ON ASPEN

Ceratocystis fimbriata (Ceratostomella fimbriata)

Hosts: Predominantly affects aspen in the mountains of the region but may affect other hardwoods.

Diagnosis and Damage: Large, black, swollen callus ridges with target-like appearance of concentric rings. No other canker on aspen has large concentric callus ridges. Cankers grow at a greater rate vertically (2.8 cm annually) than horizontally (1.3 cm annually) causing them to become elongate.

This canker can kill twigs, branches and trunk by girdling, and it affects all sizes and ages of aspen. Trees lose value for timber and become hazardous by breaking at the canker site. Decay fungi enter the tree at cankers and weaken the tree in addition to the damage from the canker. Cankers seldom girdle or kill large trees, but can enlarge for 30 to 40 years. The canker mainly stresses trees, allowing insect borers and other cankers (Cytospora, Sooty bark) to kill the trees. This is the most common canker on aspen after Cytospora canker throughout western states

Biology and Disease Cycle: *Ceratocystis fimbriata* overwinters in cankers. It is spread from tree to tree primarily by insects visiting wounds. Though it can infect through leaves, petioles and young stems, the primary site of infection is trunk wounds. The first symptoms are a circular necrotic area around the wound or branch junction. The tree lays down fresh cambium around the infection in the spring. The fungus grows through this callus in the dormant season. This process is repeated annually, giving the ringed or target appearance to the canker. Eventually dead bark will fall off exposing rings of dead wood. Perithecia, the fruiting bodies of the fungus, form in dead wood or bark along the canker perimeter in the spring. Ascospores ooze out in a sticky mass. Perithecia are hard to detect so are not a good diagnostic factor.

Conditions Favoring and Management of Disease: No specific conditions are known that promote infection or disease development. Prevention - avoid wounding, insect damage and stress. Prune out affected branches during winter. Plant aspen in compatible sites.

CRYPTOSPHAERIA CANKER ON ASPEN

Sexual Stage: *Cryptosphaeria lignyota (Diatrype lignyota, C. populina, Eytypa populina)*

Asexual Stage: *Libertella* sp. (*Cytosporina* sp.)

Hosts: Cryptosphaeria affects primarily aspen in the mountains of the region, and occasionally cottonwoods (black, eastern and yellow) and Lombardy poplar.

Diagnosis and Damage: Long, narrow, inconspicuous cankers (5-10 cm x 3 m) that follow the wood grain. Small fruiting bodies (perithecia) imbedded in bark and closely spaced and can be felt almost easier than seeing. Dead bark is stringy, and sooty-like sooty bark canker but has white flecks in the sooty bark. The fungus grows within sapwood before moving out to and killing cambium. Small trees may die before the trunk is girdled. Branch cankers girdle the branch and extend into the trunk. Cytospora canker is commonly found on trees with Cryptosphaeria canker. Mortality from this canker is high (26 percent) and is found in 80 percent of aspen stands in Colorado

Biology and Disease Cycle: Infection takes place in inner bark and wood exposed by fresh wounds. Perithecia form under the outer layer of bark that has been dead at least one year. Orange fruiting bodies (acervuli) occasionally form around the edge of the canker. It is believed that ascospores are wind dispersed and conidia are water splashed. The fungus invades the wood first like a decay fungus and then moves out to the bark and kills it in long narrow cankers.

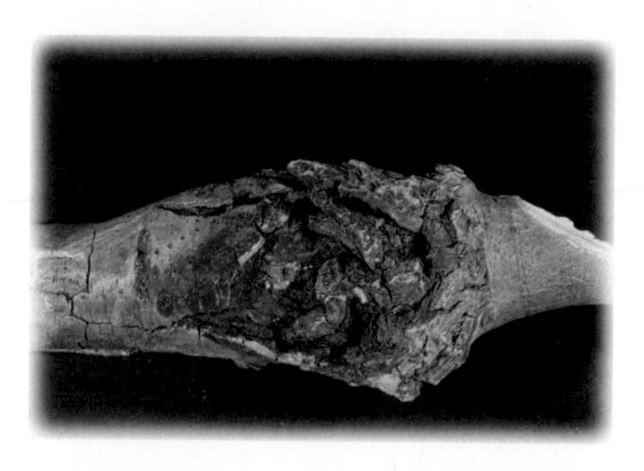

TOP: Black canker on aspen showing target-like concentric rings. (USDAFS)
SECOND: Black canker on aspen. (W. Jacobi)
THIRD: *Cryptosphareria* canker on aspen. (W. Jacobi)
BOTTOM: *Cryptosphareria* canker on aspen, closeup of sooty-like bark with white flecks. (W. Jacobi)

TOP: Cytospora canker gummosis symptoms on peach. (Luepschen)
SECOND: Cytospora canker on spruce. (W. Jacobi)
THIRD: Cytospora canker on small branches.
FOURTH: Cytospora canker on willow. (W. Jacobi)

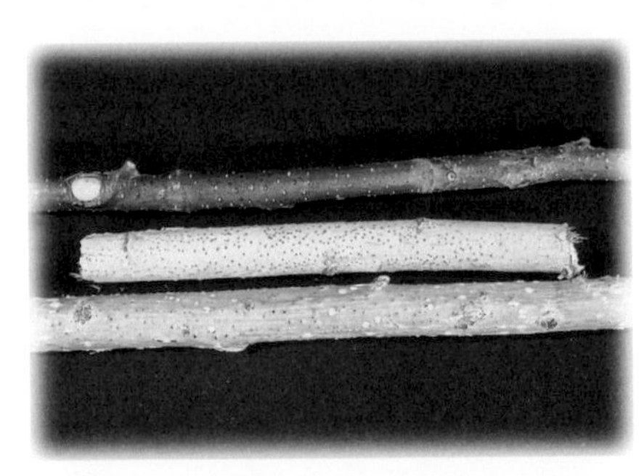

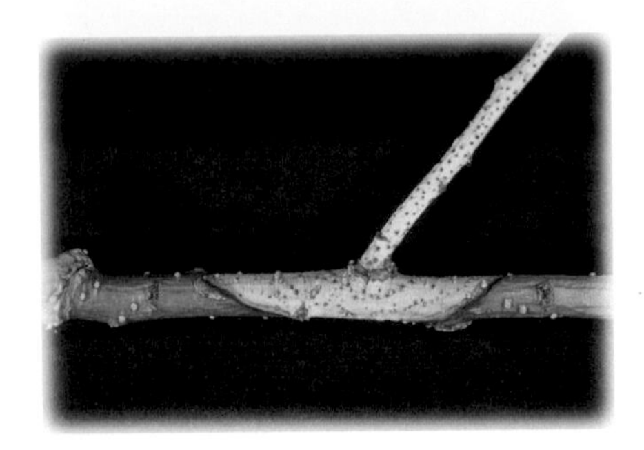

Conditions Favoring and Management of Disease: No specific conditions are known to favor this disease. Prevention - avoid wounding, insect damage and stress. Prune out affected branches during winter. Plant aspen in compatible sites.

CYTOSPORA CANKER

Sexual stage: *Valsa* spp., *Leucostoma* spp.
Asexual stage: *Cytospora* spp., *Leucocytospora* spp.

Hosts: Affects many species of trees in Colorado including aspen, cottonwood, poplar, elm, willow, birch, maple, mountain ash, linden, honeylocust, ash, sycamore, mulberry, oak, apple, stone fruits, and spruce.

Diagnosis and Damage: Symptoms vary depending on host and species of Cytospora. For most hardwoods in the region definite sunken cankers are not common. The fungus is able to damage mainly stressed trees so the cankers are often extensive, not sunken, and not easily noticed unless the canker girdles the stem causing leaf death. Cankers are usually irregular in shape and more or less elongate, appear on trunks or limbs. Discoloration of the outer bark may be yellow, brown, red-brown to gray or black depending on host species affected. Pimple-like fruiting structures (pycnidia) often develop in canker areas, are space some distance apart. Under moist conditions, pycnidia ooze long orange, coiled thread-like spore tendrils. Oozing of liquid on aspen and oozing of gums on peach, cherry and plum is common. Sexual and asexual spores are diagnostic, as they look like curved rods or macaroni noodles.

On spruce dying or dead branches may call attention to canker development. Old branches are more susceptible than young ones. Lesions appear as sunken areas surrounded by swollen callus tissue resulting in a gall-like deformation. Small, black fruiting structures may be evident on the canker. Large amounts of clear amber resin flows from infected areas and may obscure the canker location.

Damage is associated with the decline or death of planted or native trees. Cytospora is an opportunistic pathogen, and attacks plants that have been predisposed by various stresses such as fire, drought, flooding, winter damage, or infection by other pathogen. It can be a limiting factor in the establishment and growth of some poplars and willows.

Biology and Disease Cycle: The fungus overwinters in cankered bark and spreads by means of spores dispersed by rain, wind, insects or birds. Infection occurs only through bark wounds, dead tips of twigs or branch stubs. The fungus grows in the bark until limited by defense processes of the tree. Fruiting bodies form in infected bark to complete the life cycle.

Conditions Favoring and Management of Disease: Predisposing environmental or cultural factors are paramount in determining vulnerability to attack. Trees weakened

BELOW: Spore tendrils of Cytospora canker.

BELOW: Cytospora canker on alder. (W. Jacobi)

by other stresses, biotic or abiotic are most subject to infection. Temperatures of 70 to 85 F (20 to 30 C) favor rapid canker formation.

Management includes preventing stress. The most common stresses to avoid are drought, flooding of soil, and infection by other pathogens. Proper preparation of planting area when planting trees, proper watering, pruning and fertilizing, and planting resistant varieties will go a long way in helping a tree resist infection. Some species of fruit trees may benefit from the use of fungicidal wound dressings on pruning cuts. Once infection has occurred, the best treatment is to increase plant vigor and sanitation by removing infected limbs and other areas, sterilizing tools between cuts.

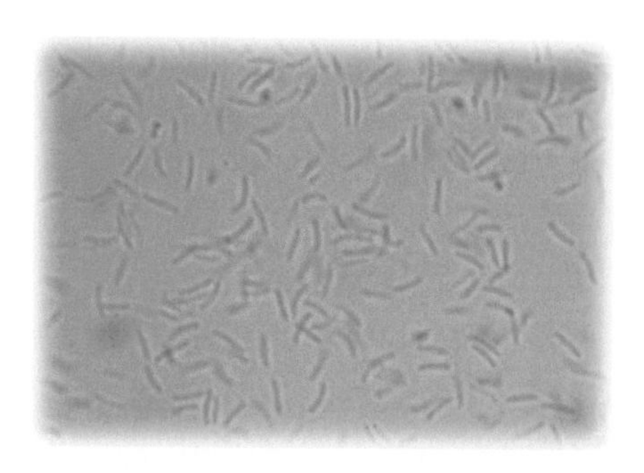

TOP: Conidia of Cytospora.
SECOND: Cystospora canker on left, Dothiora canker on right. (W. Jacobi)
THIRD: Dothiora canker on aspen, note closely spaced pycnidia at base of branch on the left. (W. Jacobi)
BOTTOM: Endothia canker on Gambel oak. (M. Schomaker)

DOTHIORA CANKER

Dothiora polyspora

Hosts: Observed on aspen and willow.

Diagnosis and Damage: This disease is an important killer of young aspen trees (1 to 10 yr.) in the mountains of this region. The fungus invades and girdles injured twigs killing them in the process. Stem wounds created by animals also are points of infection where canker development can occur. Fruiting bodies (pycnidia), which appear as small, closely spaced (Cytospora has more widely spaced pycnidia), pimple-like structures on infected twigs, produce conidia that are oval and non-distinct. Ascospores from perithecia are multicelled and teardrop shaped. Dothiora canker kills twigs by girdling, if entering wounds in main stems, girdling will occur in two or three years.

Biology and Disease Cycle: The biology of this fungus has not been studied. Pycnidia probably release spores during wet conditions in spring and early summer.

Conditions Favoring and Management of Disease: Conditions that cause stress, or create an adverse environment for host plants are favorable to infection and canker development. Drought, soil flooding, and frost are implicated in predisposing aspen sprouts to the fungus. Stem wounds caused by browsing and trampling by domestic livestock, deer, and elk appear to be particularly susceptible to infection. Preventing stresses and wounds are the best control measures for dothiora canker. Cankers on high value trees sometimes can be excised by carefully cutting away the infected bark and adjacent healthy tissue.

ENDOTHIA CANKER OF GAMBEL OAK

Sexual: *Endothia singularis*
Asexual: *Endothiella singularis*

Hosts: Affects Gambel oak in the mountains and foothills of Colorado and New Mexico.

Diagnosis and Damage: Cankers, which form on twigs, branches or trunks, exhibit specialized vegetative hyphae (stromata) that erupt through the bark. These irregular-shape stromata, which may be scattered or clustered, are mahogany red on the exterior and scarlet to orange within. The stromata become brittle with age and when bumped, break open and set free a powdery scarlet mass of conidia, which are one-celled, hyaline, cylindric. They are formed in nearly spherical cavities in the stroma . The sexual fruiting structure, a brown to black slender necked perithecia, embedded in the stroma, releases ascospores that are hyaline, cylindric and one-celled.

This canker can kill twigs, branches and trunk by girdling, and it affects all sizes and ages of Gambel oak. Decay fungi enter the tree at cankers and weaken the tree in addition to the damage from the canker.

Biology and Disease Cycle: Infection is through wounds and this initiates fungal spread through inner bark and cambium. Two spore stages are formed on the bark in cushion-like stroma.

TOP: Hypoxylon canker on aspen, older canker with checkerboard pattern. (W. Jacobi)
SECOND: Hypoxylon canker on aspen. (W. Jacobi)
THIRD: Hypoxylon canker on aspen. (W. Jacobi)
BOTTOM: Common canker on rose cane. (W. Feucht)

Conditions Favoring and Management of Disease: Disease development is favored on stressed or wounded trees. To prevent, avoid wounding, insect damage and stress. Prune out affected branches during winter.

HYPOXYLON CANKER

Hypoxylon mammatum

Hosts: Aspen

Diagnosis and Damage: Young infections appear as slightly sunken, irregular, yellowish-orange areas around wounds. The fungus invades new tissue each year, causing the canker to elongate. Diseased bark appears laminated or mottled black and yellowish white, and white mycelial fans are formed near the canker margin under the bark. Gray fungal pillar-like fruiting bodies form beneath loose, blistered bark. Older cankers appear as roughened gray-black areas where the **dead bark** then cracks and falls off in small patches **in a checkerboard fashion**. Hypoxylon canker results in broken stems, forked tops and dead tops and may kill trees within three to eight years.

Biology and Disease Cycle: Fungal spores are wind disseminated and enter host trees through sapwood wounds. Aspen bark contains fungitoxic compounds that inhibit mycelia growth, so the fungus invades the bark from within. The fungus produces a toxin that causes bark death and collapse in advance of the fungus. Consequently, the fungus is well established in the sapwood before canker symptoms appear on the bark. A year after infection, the fungus produces pillar-like fruiting structures that cause blistered areas in the center portion of the canker. The outer layer ruptures and allows for dispersal of spores.

Conditions Favoring and Management of Disease: Conditions that favor wounding of host species favor infection. Some clones have exhibited resistance. Low density stands , mixed stands, and thinned stands appear to have more infection, as do trees on the edges of, rather than within, stands. No control measures are known for hypoxylon canker and as yet cannot be prevented except by preventing wounds.

ROSE CANKERS

BROWN CANKER

Sexual Stage: *Cryptosporella umbrina*
Asexual Stage: *Diaporthe umbrina*

COMMON CANKER

Sexual *Stage: Leptosphaeria coniothyrium*
Asexual Stage: *Diapleela coniothyrium*

Host: Rose

Diagnosis and Damage: Brown canker symptoms begin as small red to purple spots on current year's canes. Affected areas enlarge and turn tan to whitish surrounded by purple margins. Fruiting structures, or pycnidia, protrude through the tan to whitish tissue. During wet weather yellow spore tendrils exude from pycnidia. Symptoms of **common canker** begin as yellow to red spots initially at the union of stock and scion. As the canker enlarges, tissue within the center becomes light brown and shrinks or dries out, giving the appearance of a sunken area with color change. Small pycnidia may be produced in the cankered area. Both cankers may cause stem girdling, resulting in wilting and death of plant parts above affected tissue.

The diseases occur on a limited basis and are usually of little importance in the Rocky Mountain Region. However, in wet years die back of scattered canes may occur. The disease is most serious on roses in storage and on newly transplanted roses, especially if planted and maintained incorrectly.

Biology and Disease Cycle: The fungi overwinter on diseased canes. In moist weather spores ooze from pycnidia and are dispersed in water droplets or moist wind. The pathogens rapidly colonize wounded or weakened areas on rose stems. Stem cankers typically develop at the cut end of canes when stubs are left after pruning, around tissue damaged by insect feeding, thorn scars, or abrasions.

Conditions Favoring and Management of Disease: Improper pruning, stem or cane injury, and moist weather will promote the diseases. Avoid injury to rose canes. Pruning cuts should be made immediately above the node (area with bud formation) at an angle to leave a minimum amount of dead wood (no stubs). Rose stubs usually die back to the first node, leaving weakened, dying tissue that favors colonization by the fungi. Rose canes with cankers and dying stems should be removed as soon as possible. Sharp pruning tools should be used to avoid injury or damage. Fungicides are usually not necessary. However, if canker is damaging, fungicide sprays may be used to cover pruning wounds.

ROUGH-BARK OF ASPEN AND ASPEN GALL

Curcurbitaria staphula or various fungi and *Diplodia tumefaciens*

Hosts: Susceptible hosts are aspen, cottonwood and poplar, most often in high altitude, forest sites.

Diagnosis and Damage: Symptoms of rough bark of aspen are rough oval spots and fissured bands of grayish-black, corky bark that often extend all or part way around the trunk or branches. The damage caused to the trees by rough bark is unknown but presumed to be unimportant, although some of the fungi affect the bark periderm, cortex, and phloem. Lichens and one or more fungi often are found fruiting on the corky ridges; consequently, a microscopic examination is necessary to identify them. Aspen gall (*D. tumefaciens*) produces galls when infecting aspen branches and rough bark when infecting large stems. Other cork-bark diseases of aspen have been identified in Canada, some are wide spread in the West. *Rhytidiella baranyayi* causes angular, rough bark without forming a band around the trunk and is frequently initiated around branch stubs. *Seimatosporium etheridgei* forms cushion-like swellings, more circular in outline and smaller that the other rough barks, the central portion later assumes a cork-bark appearance. These diseases do not kill trees, but heavy infection does make ornamental trees unsightly and thus reduces their value.

Biology and Disease Cycle: Spores that gain entrance to the tree by wounds or natural openings such as lenticels cause most infections. In the case of *Diplodia tumefaciens*, penetration is no deeper than the outer layer of the cortex, because it apparently stimulates the formation of a protective periderm. The fungus invasion and formation of a protective barrier against the pathogen continues for years, resulting in the formation of rough bark and branch galls.

Conditions Favoring and Management of Disease: No control guidelines have been established for these diseases, although management could include wound avoidance of host trees, maintaining high tree vigor with appropriate cultural practices, and removal of heavily infected trees.

SOOTY-BARK CANKER ON ASPEN

Encoelia pruinosa (*Cenangium singulare* , *C. pruinosa*, *Dermea pruinosa*, *Phibalis pruinosa*)

Hosts: Predominantly on aspen, but found on black cottonwood and balsam poplar.

Diagnosis and Damage: Black stripes or areas that give a "barber pole" appearance on the main stem. Dead bark crumbles to a sooty powder. There is a black leopard pattern

TOP: Aspen gall on aspen. (W. Jacobi)
CENTER: Sooty bark apothecia fruiting bodies on aspen. (W. Jacobi)
BOTTOM: Sooty bark on aspen. (W. Jacobi)

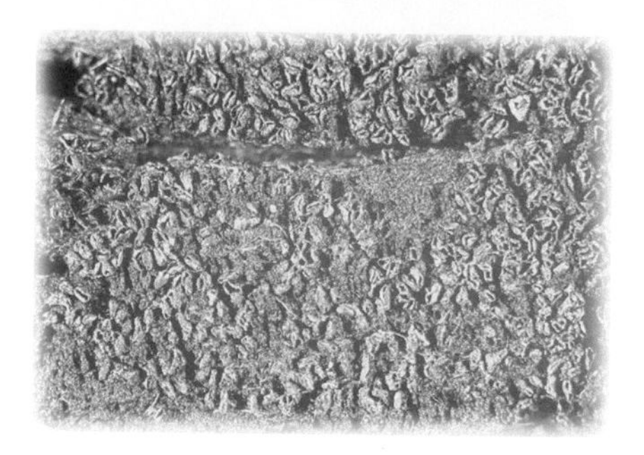

on the wood surface when the bark has fallen off. Small gray cup like fruiting bodies (apothecia) are diagnostic of this canker and easy to see on older cankered bark.

Cankers grow rapidly - 16 cm x 45 cm in one year in Colorado. The disease is one of the most common causes of mortality in Colorado and the Rocky Mountain region at middle elevations in large, old dominant trees (>60 yr.).

Biology and Disease Cycle: Infection occurs in superficial or deep wounds (those reaching the xylem). The fungus then grows through the inner bark and cambium. Little or no callus forms. Annual zones of color variation in the bark can be seen. Apothecia form in the blackened bark throughout the year. Ascospores are shot out and dispersed by the wind.

Condition Favoring and Management of Disease: Conditions favoring infection and damage are unknown. Avoid wounding, insect damage and stress. Prune out affected branches during winter. Plant aspen in compatible sites.

TOP: Fruiting bodies on Thyronectria canker on honeylocust.
BOTTOM: Thyronectria canker on honeylocust.
(W. Jacobi)

THYRONECTRIA CANKER

Thyronectria austro-americana

Hosts: Honeylocust species

Diagnosis and Damage: Disease symptoms include dieback of affected branches, reduced foliage, yellow foliage, premature fall coloration and early leaf drop. Cankers are found at the base of trees, at branch crotches, around wounds or on branch stubs. Cankers can range from slightly flattened surfaces to distinctly sunken areas with large callus ridges at the canker margin. Areas of stems and branches with thin bark may have a red-yellow discoloration. The condition of the bark and cambium (the tree's growth tissue, between bark and wood) can indicate the presence of a canker. Infected bark and cambium will be loose and wood beneath them may have a dark discoloration (wine-red to yellow) instead of a normal white or light color. The reddish color associated with the center of honeylocust stems is not related to this disease. In areas of the bark that have been dead for a year or less, the fungus produces specialized small cushion-like vegetative hyphae (stromata) where spores are produced, they are first brownish, gradually blackening with age. These cushions may be seen in bark lenticels or in the bark surfaces of branches and stems. Cankers at the tree base usually are fatal. Death of the tree or affected parts occurs because of cambial death. Main stem or branch crotch cankers may completely girdle depending on the tree's health. Stressed trees cannot stop the fungus whereas healthy trees may be able to stop canker expansion and recover.

Biology and Disease Cycle: The fungus overwinters as vegetative material (mycelium) and fruiting structures on infected trees. Since the fungus also can live in dead tissue, it can become established or produce spores on dead wood such as branch stubs, wound edges or firewood. High humidity and wind-driven rain favor spore release and spread infection. Infections may take place through branch crotches, pruning wounds, or other physical wounds in the bark. The fungus grows in the bark, cambium, and outer wood where it eventually kills the cambium and surrounding cells. Fruiting bodies can form within one month after the tree bark is killed and are abundant on dying or dead trees. The fungus is orange when grown in a culture.

Conditions Favoring and Management of Disease: A variety of stresses predispose honeylocusts to infection by canker fungi. Avoiding stress due to improper planting practices, drought, overwatering, and insufficient area and oxygen for root growth can help prevent the infection. In general, initially planting small trees (1 to 2 inches) will assure a better chance of success than planting large trees (4 to 10 inches) that are stressed and may become cankered. Trees should be watered adequately (about one inch/week) but not overwatered. Long watering, less often, rather than short periods, more often, and watering once a month during dry winters will increase tree vigor.

Management options include prevention of wounds and promotion of high tree vigor. Any injury to the base of a honelocust is potentially an entry point for the fungus. Lawnmowers, weed trimmers and construction work commonly cause basal injuries. Injuries to the stem and trunk, such as those caused by squirrel gnawing, pruning of branches and sunscald should be minimized. Should physical damage occur, remove loose bark and allow the wound to dry. Prune out dead or infected branches to reduce the chances of subsequent infections. Cankers on limbs may be pruned out during cool, dry weather. Cut at a branch junction and at least one foot below the visible margin of the canker. Small cankers on main stems may be cut out and dead or dying bark and discolored wood should be removed, extending one inch into healthy tissue. If the tree appears to be callousing, **do not** cut into healthy tissue. Wound dressings are not recommended. All tools used in pruning and cutting should be disinfected and allowed to dry after each cut by spraying with Lysol or dipping in 70 percent rubbing alcohol or 1 to 9 ratio bleach and water solution. Prompt removal of all infected tree parts or dead trees reduced the chances for infecting nearby trees. Since the fungus can grow on dead wood and produce spores that can infect nearby trees, the wood must be kept dry, buried in a land fill or burned within three weeks of cutting. Research shows Sunburst honeylocust is the most susceptible to cankers while Moraine and Skyline are moderately resistant and Imperial, Holka and Shade Master are more resistant.

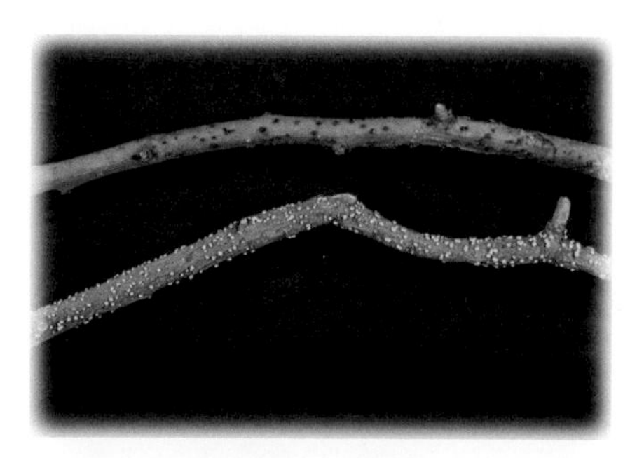

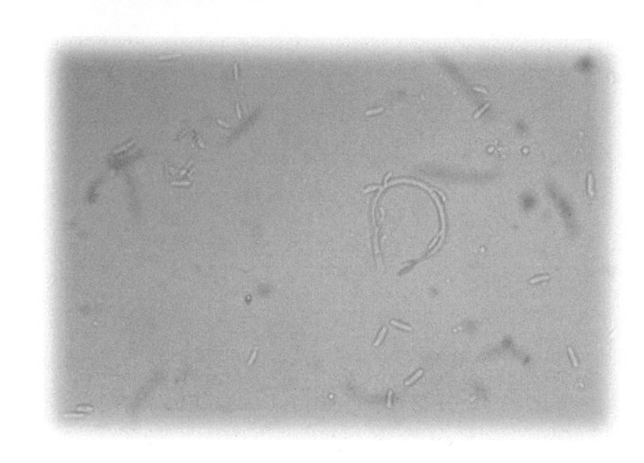

TOP: Tubercularia sp. young (light) and old sporodochia, (dark) (asexual) fruiting structure on small branch. (W. Jacobi)
SECOND: Tubercularia canker in branch crotch of honeylocust. (W. Jacobi)
THIRD: Nectria canker on hardwood. (W. Jacobi)
BOTTOM: Tubercularia canker, spores and curved conidiophores. (W. Jacobi)

TUBERCULARIA AND NECTRIA CANKERS

Sexual stage: *Nectria cinnabarina*
Asexual stage: *Tubercularia ulmea* and *T. vulgaris* (asexual stage of *N. cinnabarina).*

Hosts: Tubercularia and Nectria fungi cause cankers and dieback on a number of hardwoods. Tubercularia canker caused by (*Tubercularia ulmea*) is the most common in the region and it does not appear to have a sexual stage. Occasionally *Nectria cinnabarina* is found causing cankers with the sexual stage (perithecia) visible on the tree. The most common hosts of Tubercularia canker in Colorado include Siberian elm, shrub dogwood, honeylocust, linden, maple, mulberry, and Russian olive.

Diagnosis and Damage: Cankers appear in spring and are of two forms, girdling ones that cause twig or branch dieback, often leaving branches with dead leaves still attached, and oval or elongate, slightly sunken non-girdling cankers. Hosts species may discolor in canker area, exude gum, or not show until cracks form at the margins of the areas. Vigorous sprouts may develop below cankered areas. Small cushion-like fruiting structures (sporodochia) that are first pink-orange and blacken with age may be growing out of lenticels on the bark surface or present under a papery thin layer of bark. Tubercularia conidia are diagnostic, as they are produced on a curved conidiophores.

Cankers at the base of the host are usually fatal, main stem or branch crotch cankers may completely girdle depending on the tree's health. Stressed trees cannot stop the fungus whereas healthy trees may be able to stop canker expansion and recover.

Biology and Disease Cycle: Tubercularia and Nectria fungi overwinter as fruiting bodies and mycelium in cankered bark. Spores released from fruiting bodies (sporodochia or perithecia) travel via wind and rain to dead trees, or live trees that have been wounded. Tubercularia can only infect live trees through wounds such as those caused by hail, wind, snow, animals, machinery, etc. It is a weak parasite and often infects stressed trees. Although infection often will result in small annual cankers that callus over during the next season, it occasionally causes perennial cankers.

Conditions Favoring and Management of Disease: Conditions that cause weakening or create an adverse environment for host plants and that contain sources of fungal spores favor disease development. Rapid drops in temperature below -20 C,

TOP: *Daldinia concentrica* conk on aspen. (W. Jacobi)
SECOND: *Ganoderma applamatum*, conk on right shows top, still attached to tree, left conk shows underside. (M. Schomaker)
THIRD: *Peniophora rufa* fruiting structures on aspen. (M. Schomaker)
BOTTOM: *Perenniporia fraxinophila* conk on green ash. (USDAFS)

as may happen after relatively mild weather in late autumn or winter has been shown to result in girdling cankers on some host species.

Management options include removing and destroying affected twigs and limbs, pruning back to the nearest living branch beyond the canker. Disinfect pruning tools with alcohol or Clorox and water mix (1to10 ratio) after each cut. Prune during dry weather near the end of dormant period. Manage trees for optimum vigor through proper soil conditions, watering and avoid wounding.

Stem Decay Fungi

STEM DECAY FUNGI OF HARDWOODS AND CONIFERS

A wide array of fungi attack and decay living and dead trees. Fungi that decay dead trees and wood are, in general, beneficial organisms that recycle carbon by breaking down wood cell wall components. This process returns billions of tons of carbon to the atmosphere and allows new plants to utilize the carbon as carbon dioxide. Decay fungi also play an important role in soil fertility by adding organic matter and contributing to aeration of the soil by decaying roots.

Recognition of decay in trees is not always easy in the Western U.S. because fruiting bodies (conks or mushrooms) do not always form. Decay recognition is best done by looking for other indicators of possible decay such as, open wounds, animal nests in the trees, broken branches, and swollen areas on the stem or branches. On living trees it is not always necessary to identify the fungus decaying the tree since it is the recognition of the decayed wood that is the main objective. Trees weakened by decay are prone to windthrow or breakage and are thus called hazard trees. Infected trees which are situated such that they may cause property damage or potentially injure people should be evaluated by a professional arborist to determine the extent of damage, and should be removed if there is any question regarding the stability of the affected parts.

Infection by decay fungi is usually accomplished through wind-borne spores, which enter the tree through wounds on the trunk, branches, root crown, or roots. Older trees are generally considered to be more susceptible to infection by decay fungi than younger, more vigorous trees. In the majority of decays, the disease is not treatable once it is established in the host tree. The most effective preventive measures are promoting tree health, keeping younger trees when removing trees, and avoidance of wounding.

Decay fungi are often classified by the type of decay they cause. Brown rots are those in which the fungus breaks down the cellulose and hemicellulose components of the wood's cell walls leaving the brownish lignin component. Brown rotted wood tends to be dry and fragile, readily breaking into cubes or crumbling. Brown rots cause rapid structural weakness with initially little visible change in the wood so are considerable more dangerous than white rots. White rot fungi are able to breakdown both the lignin and cellulose components leaving behind white to yellow, spongy, stringy, laminated, or crumbly material.

Listed below are a few decay fungi common to the region with a brief description of the fungus

DECAYS OF HARDWOODS

Bjerkandera adjusta. Common decay on many hardwoods, particularly on aspen. Causes a white rot and colonizes the sapwood of wounded or diseased living trees. Thin, shelf-like fruiting bodies are produced annually on the stem of the infected tree. The conks are often found in overlapping clusters, and appear slightly velvety with a white to tan or gray upper surface and a gray lower surface. Small angular pores.

***Collybia velutipes:* (*Flammulina velutipes*)** Common decay on elm, poplar, willow and aspen. Causes a brown mottled white rot. Fleshy mushrooms, usually seen in clusters, caps and stems are orangish-brown to reddish-yellow, smooth and slimy, and stems are 1 to 3 inches tall.

Daldinnia concentrica Found on several species of hardwoods including aspen and ash. Causes decay on dead wood. Fruiting bodies are conspicuous dark jet black, round, usually not stalked, and emergent from the bark of infected trees. Cutting open fruiting body shows concentric rings like at tree.

***Ganoderma applanatum* (Artists' Conk)** Found in living and dead aspen and cottonwood. Major cause of tree failure, "blow down" with older aspen. Fruiting bodies is shelf-like and woody, light brown to gray on the upper side, has a cream-colored undersurface, and the conk is found at or near the bottom of the tree. The fungus causes decay of sapwood and heartwood in the trunk and roots of infected trees, but does not cause death of living tissues. Decaying wood appears light in color and exhibits mottling in the early stages, becoming increasingly spongy over time. Typically a dark colored band separates decaying wood from adjacent healthy sapwood.

Peniophora polygonia One of the most common decay fungi on aspen in Colorado. This fungus causes a white brittle rot and is characterized by a pink to brownish pink stain found throughout large portions of the stem. Fruiting bodies are pinkish red, crust-like patches with white margins.

Perenniporia fraxinophila Commonly found on living ash and occasionally on elm, hawthorn, and sycamore. Causes a white mottled heart rot that in advanced stages appears yellow to yellowish white, soft, and crumbly. Fruiting body is a perennial, bracket shaped, and grayish on the upper surface, which darkens and cracks with age.

Phellinus everhartii Found on living Gambel oak. Causes a white rot of the heartwood that appears golden brown or lighter and is associated with trunk cracks. Conks are perennial and hoof shaped with the upper surface yellowish to reddish brown when young, blackening and cracking with age.

***Phellinus igniarius* (Hardwood Trunk Rot)** Hosts include live birch, ash, elm, poplar, willow, black walnut, and buckthorn. Decay is a yellowish white spongy rot that contains black zone lines running throughout and surrounding the decay column. Fruiting bodies are perennial, woody, and hoof shaped with the lower surface close to horizontal. The upper surface of the conk is dark brown to black and rough; the lower surface is brown.

Phellinus pomaceus Produces an uniform white rot of living fruit trees, particularly in *Prunus* species of plums, peaches, and cherries. Fruiting bodies are perennial, woody, and hoof shaped. The upper surface of the conk is light grayish brown and slightly velvety when young, becoming black and cracked with age; the lower surface is shiny and light yellowish to reddish brown.

Phellinus punctatus Attacks both live and dead hardwoods including green ash, lilac, willow, black locust, American plum, and common buckthorn. Decay is a uniform white rot of the heartwood of living trees. Fruiting bodies are perennial, woody, and hoof shaped with the lower surface close to horizontal. The upper surface of the conk is dark brown to black and rough; the lower surface is dull, smooth, and yellowish to grayish brown.

***Phellinus tremulae* (Phellinus Decay of Aspen)** Found only in living aspens. One of the most common decay of aspen in the mountains. Causes a yellowish white spongy, sweet smelling decay that is generally confined to the central core of the stem and surrounded by black zone lines. Fruiting bodies develop at branch stubs and scars and on wound scars. The conks are perennial, woody, and triangular shaped in longitudinal section – the upper and lower surfaces are at approximately 45° angles from

TOP: *Perenniporia fraxinophila* conk on green ash. (USDAFS)
SECOND: *Phellinus everhartii* conk on gambel oak. (M. Schomaker)
THIRD: *Phellinus pomaceus* conk. (W. Jacobi)
BOTTOM: *Phellinus punctatus* conk on cherry. (M. Schomaker)

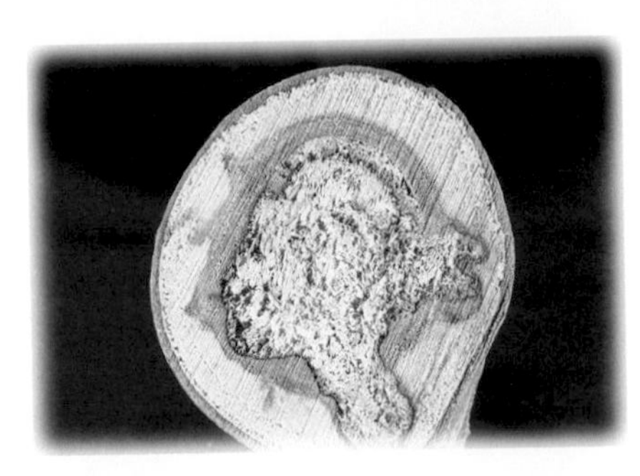

horizontal. The upper surface of the conk is dark brown to black and rough, becoming light brown near the margin.

***Pholiota populnea* (Yellow Laminated Butt Rot of Cottonwood)** Found on cottonwoods and poplars. Decay appears early as brown streaks through the heartwood, progresses to a white mottled appearance, and in advanced stages is yellowish brown and laminated. Fruiting bodies are large, light brown gilled mushrooms with white scales, often found in clusters on living trees.

***Pleurotus ostreatus* (Oyster Mushroom)** Causes a flaky white rot of the sapwood and heartwood of various hardwoods. In this region it is found on cottonwoods and aspens. The fruiting bodies have a short off center stalk, are fleshy gray to white, with gills extending onto the stalk. Mushroom clusters can be found at the base of the infected tree or growing from branch stubs.

Trametes hirsuta – Causes a white rot of dead trees on several species of hardwoods including apple and elm. Fruiting bodies are soft, annual shelf-like conks. The upper surface is brown to gray, banded, and coarsely hairy; the margin is yellowish brown; the lower surface is white to tan.

DECAYS OF CONIFERS

***Cryptoporus volvatus* (Pouch Fungus)** Found on pine and Douglas fir. Causes a soft grayish-white rot of sapwood in recently killed trees within one to two years of death. Attacks outer sapwood while tree is standing and has been found on dead portions of living trees. Fruiting bodies are cream to tan colored, annual, pouch-like structures with a layer of tissue covering the pore surface with a small hole that allows dispersal of spores by wind and insect vectors.

***Dichomitus squalens* (Red Rot)** Found on pine, spruce, true fir, and Douglas-fir. Important decay in Colorado on ponderosa pine. Causes a white pocket rot in the heartwood of living conifers and in dead conifer wood. Decayed wood is crumbly and disintegrates easily. Fruiting bodies appear as irregularly shaped white to light brown crusts on downed wood and rarely on live trees.

***Echinodontium tinctorium* (Indian Paint Fungus)** Hosts are subalpine fir and occasionally Douglas-fir and Engelmann spruce. The fruiting body, which develops only after the fungus has caused significant decay, is thick and woody, with a gray to black, rough, cracked upper surface, a dark undersurface from which tooth-like spines protrude, and a rust colored interior. Early decay is indicated by a water-soaked, brown appearance in the wood, which eventually becomes stringy and mottled. Margins between decayed wood and adjacent sapwood are typically stained red to brown.

***Fomitopsis officinalis* (Quinine Fungus)** Found on Douglas-fir, spruce, and pine. Causes a brown cubical rot found in the trunk and upper bole of the tree. Fruiting bodies are perennial, woody, and pendant often attaining large dimensions. When young, conks are leathery and white, becoming chalky white and crumbly with age. The fungus acquires its name from the bitter taste of fruiting bodies and thick, white mycelial felts present in the shrinkage cracks of decayed wood.

***Fomitopsis pinicola* (Red Belt Fungus)** Commonly found on pine, Douglas-fir, spruce, true fir, and also found on several species of hardwoods including cottonwood and aspen. Causes a brown cubical sap and heart rot that is soft and crumbly in texture. Early decay appears as a yellowish brown stain. Occurs on dead, downed, and living trees.

TOP: *Phellinus tremulae* conk on aspen. (W. Jacobi)

SECOND: *Phellinus tremulae* heart rot on aspen. (W. Jacobi)

THIRD: *Pholiota squarrosa* mushrooms. (M. Schomaker)

FOURTH: *Pleurotus ostreatus* mushrooms on willow. (M. Schomaker)

BOTTOM: *Coriolopsis gallica (Trametes hispida)* on apple. (Luepschen)

Fruiting bodies are perennial, woody brackets. The upper surface is usually zoned and ranges in color from dark brown to grayish-black and has a distinctive reddish margin that gives this fungus its name. The lower surface is white to cream colored

***Hericium abietis* (Coral Fungus)** Commonly found on true fir and Engelmann spruce. Early decay appears as a yellow to brown heartwood stain. Advanced decay is evident by elongated pits (~ 1 cm long), which may be filled with yellow to white mycelium and are oriented longitudinally within the wood. Fruiting bodies are annual, soft, fleshy, white, and corral-like with many downward pointing spines on a very branched stalk.

***Inonotus circinatus* and *Inonotus tomentosus* (False Velvet Top Fungus)** Common hosts are Douglas-fir, true fir, pine, and spruce. These decay fungi cause a white pocket rot of the heartwood of the roots and butt of living conifers that is characterized by white elongated spindle shaped pits surrounded by a pink to reddish stain indicative of incipient decay. Fruiting bodies are found on the base of infected trees or on the ground at the base of the tree. They are annual with a velvety, yellow brown to reddish brown upper surface that darkens with age and a buff to yellowish brown lower surface. Differentiation between these species is difficult; however, fruiting bodies *of I. tomentosus* are smaller, thinner, more commonly found on the ground, and usually found in groups while those of *I. circinatus* are larger, thicker, more commonly grow on the stem, and found singly.

***Lentinus lepideus* (Scaly Cap Fungus)** Hosts include pine, Douglas-fir, and true fir. Causes a brown cubical rot of the heartwood from roots to upper bole. Incipient decay appears as a yellowish stain. Fruiting bodies are large, leathery, stalked mushrooms with whitish caps covered with darker tan to brown scales.

***Phaeolus schweinitzii* (Velvet Top Fungus)** Hosts include Douglas-fir, pine, true fir, and spruce. Causes a cubical brown rot of the heartwood that is confined to the roots and lower 3 m of the stem. Annual fruiting bodies are usually compound and either bracket like when growing on the trunk of infected trees or stalked when growing on the ground. The upper surface is red-brown and velvety, the margin rounded and yellowish, and lower surface yellow-green to brown.

Phellinus nigrolimitatus Commonly found on spruce, also on other conifers. Causes a white pocket rot with distinctive large pits (up to 2.5 cm long) surrounded by firm wood in living conifers and downed woody material. Fruiting bodies are perennial woody conks with blackish-brown to yellow or reddish-brown upper surface and smooth cinnamon color lower surface.

***Phellinus pini* (Decay of Pine)** Nearly all species of conifer are affected including pine, spruce, true fir, and Douglas-fir. Affected wood will appear as a typical white pocket rot, exhibiting a pinkish to purplish red discoloration interspersed with small, well- defined, longitudinal, spindle shaped pockets of whitish fibers. The decay produced can be distinguished from other local white pockets rots: *Phellinus nigrolimitatus and Hericium abietis* form larger pockets with distinct margins; *Dichomitus squalens* and *Inonotus tomentosus* produce pockets with indistinct margins. The bark surface above decayed areas may exude resin, often from swollen knots, bulges, or cracks. The fruiting bodies, which usually arise from knots or branch stubs, generally appear as perennial, hoof-shaped, woody conks up to 30cm in diameter, gray to brown in color with wavy

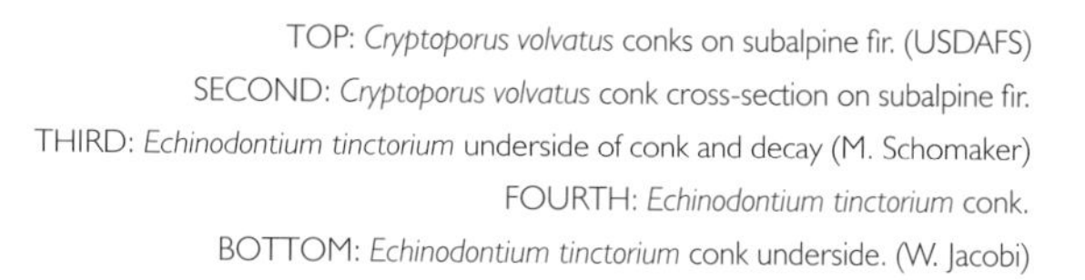

TOP: *Cryptoporus volvatus* conks on subalpine fir. (USDAFS)
SECOND: *Cryptoporus volvatus* conk cross-section on subalpine fir.
THIRD: *Echinodontium tinctorium* underside of conk and decay (M. Schomaker)
FOURTH: *Echinodontium tinctorium* conk.
BOTTOM: *Echinodontium tinctorium* conk underside. (W. Jacobi)

concentric rings on the upper surface and rusty brown on the lower surface. In some cases, fruiting bodies may instead appear as small, flat incrustations on the bark surface.

***Pholiota adiposa* (Yellow Cap Fungus)** Found on conifers including true firs, pines, and spruce and on hardwoods. The fungus produces a white mottled rot of the heartwood. Advanced rot appears as widely scattered pockets filled with yellowish-brown fungal mycelium. Fruiting bodies consist of a yellow stalked mushroom with a yellow to yellowish brown cap and may be found growing in clusters from the base of the infected tree or from the bole.

***Pyrofomes demidoffii* (Juniper Pocket Rot)** The fungus produces a white trunk rot of living junipers and is common in pinyon-juniper ecosystems. Advanced decay appears as numerous large pockets filled with yellow-white fibers and buff mycelial felts. Fruiting bodies are perennial, woody, hoof shaped, and often become columnar. The upper surface of conk is dark brown to black with deep cracks; the lower surface is light yellowish brown.

Bark Beetles

Bark beetles are insects that develop feeding on the cambium under the bark of trees. Typically the adult beetles first make distinctive galleries in which they mate and lay eggs. The larvae tunnel outwards from these galleries and can girdle the tree.

Another very important aspect of bark beetles is their involvement with various fungi, particularly those in the genera *Ophiostoma* and *Ceratocystis.* These include the fungi involved in Dutch elm disease and the 'blue stain' fungi of conifers. Bark beetles are very important in the distribution of these fungi to new trees. In many instances the two organisms - bark beetle and fungus - are co-dependent. Bark beetles spread the fungus and the fungus helps to provide declining hosts that are optimal for beetle development.

Most bark beetle species successfully breed in trees that are severely stressed or dying. Newly transplanted trees are particularly susceptible to attack by these insects. Proper cultural practices to promote tree vigor are the most important means of preventing most bark beetle problems. Often, these techniques are sufficient for control.

A few bark beetles are capable of killing healthy trees, either through coordinated "mass attacks" (mountain pine beetle for example) and/or through introduction of disease causing fungi into healthy trees (elm bark beetles). Sanitation and insecticide applications are of great importance in managing these pest situations.

An active area of research develop involves the behavioral chemicals (pheromones) used by bark beetles to regulate their attacks. Both pheromones that are involved in attraction (aggregation pheromones) and pheromones that are repellent to bark beetles (anti-aggregation pheromones) have been identified for some species. These have promise in managing beetles when populations are not high, but have been generally ineffective in preventing attacks during outbreaks.

Bark beetle preventive insecticide applications involve a thorough wetting of the bark (point of run-off) prior to the egg laying period. Carbaryl (Sevin, Sevimol), chlorpyrifos (Dursban) and permethrin (Astro) are currently used for control of bark beetles.

TOP: *Fomitopsis officinalis* conk. (W. Jacobi)
SECOND: *Fomitopsis pinicola* conk on Douglas fir. (M. Schomaker)
THIRD: *Phaeolus schweinitzii*, bottom of basidiocarp. (W. Jacobi)
FOURTH: *Phaeolus schweinitzii*, top of basidiocarp. (W. Jacobi)
BOTTOM: *Phellinus pini* conk. (W. Jacobi)

Specific recommendations for bark beetle control are included in the supplement in the sections: **Mountain pine beetle**, **European elm bark beetle**, and **Engraver (Ips) beetles**.

ASH BARK BEETLES

Hylesinus species (= *Leperisinus species)*
Coleoptera: Scolytidae

Hosts: Green and white ash. Autumn purple ash is a particularly favored host.

Damage and Diagnosis: Adult bark beetles cut egg galleries under the bark and larvae tunnel perpendicular to the gallery. These injuries can girdle and sometimes kill branches. On rare occasions entire trees are killed by these insects. Injured limbs and heavily shaded branches in the interior of the tree are most commonly attacked. Transplanted trees can be at special risk. Ash bark beetles may infest almost the entire tree, from finger-diameter branches to the main trunk.

Egg galleries run across the grain and often have two "arms" with a central chamber in the middle. Also characteristic of these insects are that small "ventilation holes" perforate the bark above the egg galleries. The tunnels are almost invariably colonized by fungi that stain the wood a rich brown color around the feeding sites. Sap may ooze from wounds in twigs, staining the bark.

Ash bark beetle have become increasingly important within the region during recent years. This is likely due to the general increase in ash plantings and, in particular, the increase in ash that has been damaged or is in decline.

Life History and Habits: At least two species occur in Colorado, *Hylesinus californicus* (Swaine) appears to predominate in the western areas and *H. aculeatus* (Say) in the east. Life history of both species is poorly understood under local conditions but the following is based on information from more northern areas and Colorado observations.

Overwintering can occur as either late-instar larvae under the bark or as adults that winter within niches cut into green bark of the outer trunk. Adults begin to become active in early to mid-spring and females construct girdling tunnels under the bark. During this tunneling small ventilation holes are also constructed that are exteriorly visible and sap may ooze from the wounds. These tunnels are the main egg galleries characterized by running at right angles to branch length.

The larvae are pale, legless grubs that develop by feeding under the bark, often extensively scoring into the sapwood. Those developing from spring eggs become full grown in late spring or early summer and pupate within the tunnels. Adults emerge from the branch and feed on green wood, causing little damage.

There is evidence that a partial second generation is produced in some situations. These may not complete development and overwinter as larvae. Bark beetles that have reached the adult stage move to the trunks at the end of the season to cut hibernation chambers within which they winter.

Management: Ash bark beetle outbreaks often are, at least in part, related to conditions of growing stress. Well-sited, vigorously growing ash should be at much less risk of attack by this insect. Good cultural practices are fundamental to ash bark beetle management.

Pruning and disposal of infested branches can limit population development of the beetle. This needs to be done in spring, before adults emerge. Infested branches can be identified by showing wilting.

Preventive insecticide applications should be applied to coincide with periods when adults begin constructing egg galleries and laying eggs. In general this should occur by mid-spring. A repeat application may be needed later in the season for high risk trees

TOP: Phellinus pini conk. (W. Jacobi)
CENTER: Ash bark beetle adult and galleries under bark of green ash.
BOTTOM: Ash bark beetle galleries.

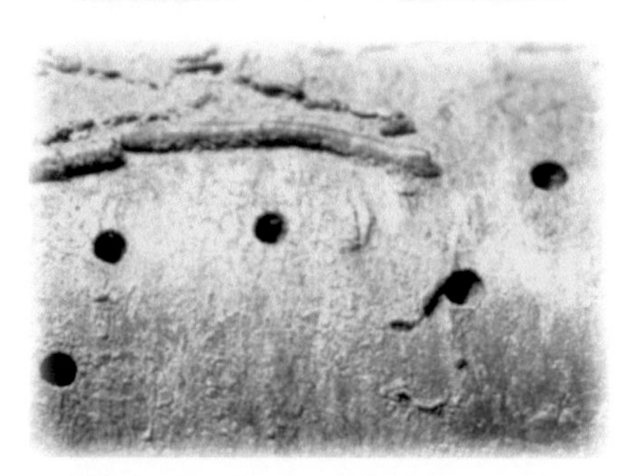

TOP: Oozing from shothole borer wounds in plum.
SECOND: Tunneling by shothole borer in Nanking cherry twig. (David Leatherman)
THIRD: Exit holes produced by shothole borer. Photograph courtesy Clemson University Cooperative Extension.
BOTTOM: Shothole borer. (Ken Gray)

where a second generation is present. Large, 2-inch caliper transplants, are at particular risk of attack and often will benefit from preventive insecticide treatment during the establishment years. Infestations of lilac/ash borer also seem to predispose trees to ash bark beetle attack.

Insecticides applied to the lower trunk in late summer can kill overwintering beetles which cut into the trunk for winter shelter. It is unknown if preventive insecticide sprays of branches, applied in early spring as for elm bark beetle control, may also be an effective control of adults as they move to trees for egg laying.

Several natural enemies of ash bark beetle have been observed. Clerid beetles and chalcid wasps attack the larvae and jumping spiders prey on the adults.

SHOTHOLE BORER

Scolytus rugulosus (Muller)
Coleoptera: Scolytidae

Hosts: Fruit trees (particularly *Prunus* species) and a few other hardwoods like mountain-ash, hawthorn and, in rare cases, elm. Most infestations involve overmature or damaged plum and cherry.

Damage and Diagnosis: Shothole borer is the most common bark beetle affecting fruit trees in the region. The larvae stage develop under the bark, in tunnels they excavate parallel to the egg gallery of the female. This produces girdling wounds that can weaken, and sometimes kill, the plant beyond the damaged area. Trees in poor health are much more susceptible to shothole borer damage.

Oozing gum often occurs on *Prunus* species where beetles enter the wood to lay eggs. When the adult beetles emerge through the bark they chew small exit holes, the most commonly observed evidence of shothole borer activity.

Life History and Habits: The shothole borer spends the winter as a grub-like larva under the bark of trees. They continue to develop the following season, cutting a chamber into the sapwood in spring to pupate. The adults are small (1/10-inch), gray-black beetles that may begin to emerge in late April or May but can subsequently be found throughout the growing season. After mating the females seek out tree branches in poor health and chew out a one to two inch long egg gallery under the bark. Eggs are laid along the gallery and the newly hatched larvae later feed under the bark, making new galleries away from the central egg gallery. (The pattern made by the egg and larval galleries is useful for diagnosing shothole borer infestations.)

There are likely to be two generations per year in the region. The larvae of the first generation become full grown by midsummer and a give rise to a second generation period of adult activity and egg laying. However, emergence of adult beetles extends over a considerable period and there are no distinct generations.

Related Insects: The fir engraver, *Scolytus ventralis* LeConte, develops under the bark of true firs. Perhaps the most notorious of the *Scolytus* species is *S. multistriatus*, the smaller European elm bark beetle, vector of Dutch elm disease. *Scolytus fagi* Walsh has been recovered from hackberry in eastern Colorado.

Management: Shothole borers rarely attack, and survive poorly in, trees that are actively growing. Serious damage by shothole borer can be almost entirely avoided by growing trees under favorable growing conditions. Trees stressed by drought, winter injury, poor site conditions, wounding injuries, or other problems are at greatest risk of shothole borer damage.

TOP: European elm bark beetle tunneling base of bud. (Ken Gray)
SECOND: European elm bark beetle. (Ken Gray)
THIRD: Lesser European elm bark beetle galleries under bark.
BOTTOM: Lesser European elm bark beetle galleries.

Regularly prune dead or dying branches limbs in which shothole borers breed. Since the insects can continue to develop in pruned wood, remove or destroy it before adult beetles emerge.

On non-bearing trees, some insecticides are registered to control shothole borers (and other bark beetles). These need to be applied to the branches before the adult beetles have tunneled into the tree and laid eggs, typically before late May to early June.

SMALLER EUROPEAN ELM BARK BEETLE

Scolytus multistriatus (Marsham)
Coleoptera: Scolytidae

Hosts: Elm

Damage and Diagnosis: This introduced beetle is the principal U.S. vector of the Dutch elm disease fungus, *Ophiostoma (=Ceratocystis) ulmi*. (See **Dutch Elm Disease**) By breeding in dead or dying (diseased) trees, and then flying to healthy trees to feed, the beetles effectively spread the disease.

Where extremely high populations occur, the beetles may sometimes directly kill trees. This occurs from cambium girdling following mass attacks by beetles. Most breeding, however, occurs in recently killed or dying limbs and trees.

Trees under stress are at much greater risk from European elm bark beetle attacks. Emergence holes indicating prior infestations appear as small pin holes in the bark, and the bark ultimately loosens from effects of the larval tunneling. The galleries produced by this insect under the bark consist of a vertical egg gallery with radiating larval galleries.

Life History and Habits: The smaller European elm bark beetles overwinters in the larval stage, under bark. Larvae mature in the spring and pupate. Adult beetles typically emerge around mid-May, although earlier emergence may occur. European elm bark beetles are generally dark, about 1/8 inch long with reddish-brown to reddish-black wing covers. The posterior end of the beetle is concave and has a small projection or spine, typically of members of the genus *Scolytus*.

Beetles emerging from Dutch elm disease-infected trees may become contaminated with spores of *O. ulmi*. The adult beetles then fly to healthy elm trees where they feed on the crotches of two to three year old twigs. Most transmission of Dutch elm disease occurs during this twig feeding period, as spores can be transferred to the scored xylem of the feeding wound.

After feeding, adults seek out diseased and weakened trees, construct brood galleries under the bark, and lay eggs. (Egg galleries run parallel to the wood grain.) The larvae develop over the course of about two months and adults emerge in mid to late summer. Adults again produce a second round of twig feeding and egg laying, with the subsequent larval generation overwintering. New infections of Dutch elm disease are considerably rarer following second generation feeding, since the xylem vessels of the summer wood are smaller and restrict movement of the fungus.

Management: Management of Dutch elm disease requires the use of several different control approaches, applied in an integrated manner. Fundamental to any Dutch elm disease management plan is a program of strict sanitation. Prompt identification and removal of diseased elm trees prevents the development of large populations of elm bark beetles carrying the fungus. American elms should be regularly checked for symptoms of infection during the growing season. A tree infected late in the growing season may live through the winter. Early season (June) surveys can detect such trees and allow for prompt removal before the disease spreads to other elms. Historically, the Colorado State Forest Service has been in charge of processing twig samples from trees suspected of Dutch elm disease to verify infection.

Trees determined to be infected with Dutch elm disease should be promptly removed, at least within 20 days. If trees cannot be removed immediately, they should be girdled by a two inch deep cut into the wood to prevent the fungus from moving to the roots where infections can move to adjacent trees. However, girdling does not prevent beetle development so beetle transmission is not inhibited by this method.

Because Dutch elm disease can move from tree to tree through root grafts, without the insect involvement, trenching should be done immediately around infected trees to prevent root graft transmission. Trenches should be 18 inches deep and two inches wide.

Beetle transmission of Dutch elm disease can sometimes be further reduced by spraying tree crowns with insecticides to kill the beetles. Sprays should be made before late April, to catch the first emerging beetles. Since insecticides can persist for fairly long periods on bark, these applications can be made as early as November.

FIR ENGRAVER

Scolytus ventralis LeConte
Coleoptera: Scolytidae

ABOVE: Fir engraver tunnels.

Hosts: True firs

Damage and Diagnosis: This insect attacks many pole-sized and mature trees each year in the region. Stressed trees with trunk-diameters over 6 inches are preferred. Attacks may or may not kill the host. A brown-staining fungus is introduced into infested trees. The gallery system under the bark is easily identified by having the main egg gallery deeply score the wood and run perpendicular to the trunk (similar to the ash bark beetles). The larval galleries run with the long dimension of the trunk. Thus, the pattern and scoring of the wood by fir engraver separates it from balsam bark beetle and others likely to be encountered in fir. The adults are the largest of the genus *Scolytus*, averaging about 3/16 of an inch long. They are shiny black and males have a long tubercle on the underside of abdomen tip.

Life History and Habits: Trees in poor vigor are selected for attack, such as those suffering from defoliation by Douglas-fir tussock moth or western spruce budworm, root disease (such as from *Fomes annosus*), soil compaction, and wounding. Throughout most of the region there is one generation a year, with two years being required in some northern parts. Flight can occur any time during the late spring to early fall period. Some or all of the trunk may be colonized and result in tree death, top-kill or just patches of dead cambium. Fir engravers are often accompanied by other bark beetles or wood borers.

Management: Direct control of infestations is usually not practical because the local population usually contains a high percentage of individuals harbored in live trees. Actions which improve and maintain the vigor of potential hosts probably have the greatest value, such as the prevention of wounding and soil compaction and attention to defoliators. Standard conifer bark beetle preventive sprays could be used to protect individual trees.

MOUNTAIN PINE BEETLE

Dendroctonus ponderosae Hopkins
Coleoptera: Scolytidae

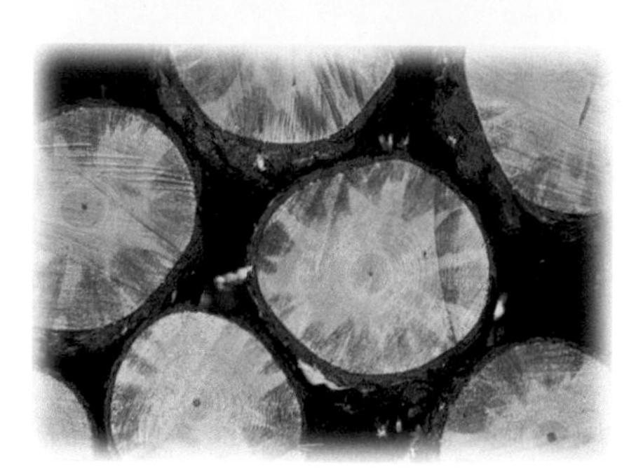

TOP: Pitch tubes from mountain pine beetle attack.
SECOND: Mountain pine beetle adult pitched out of wound.
THIRD: Mountain pine beetle.
BOTTOM: Blue stain associated with Mountain pine beetle.
(David Leatherman)

Damage and Diagnosis: Mountain pine beetle (MPB) is an insect native to the forests of western North America. Previously called the Black Hills beetle or Rocky Mountain pine beetle, periodic outbreaks of the insect can result in losses of millions of trees. Outbreaks develop irrespective of property lines, being equally evident in wilderness areas, native forests, mountain subdivisions, and back yards. Even windbreak or landscape pines many miles from the mountains can succumb to beetles imported in infested firewood.

Mountain pine beetles develop only in pines. Epidemic cycles occur about every 10 to 30 years, depending on forest condition, weather and other factors poorly understood. During low population periods or early stages of an outbreak, attacks are primarily seen in trees under stress from injury (such as by lightning, fire or mechanical insults), poor site conditions, overcrowding, root disease, or old age. However, as beetle populations increase, attacks may involve most trees of suitable size in the outbreak area, irrespective of their apparent health.

Mature trees over 8 inches in diameter are most susceptible to attack and, once infested, they usually die. A complex of bluestain fungi that turns the sapwood grayish-blue and disrupts the vascular system is introduced into the tree by attacking beetles.

Signs and Symptoms of Mountain Pine Beetle Attack:

- Popcorn-shaped masses of resin, called 'pitch-tubes' on the trunk where beetles attacked and tunneling begins. Pitch tubes may be brown, pink, yellow or white.
- Reddish boring dust in bark crevices and on the ground immediately adjacent to the tree base.
- Evidence of woodpecker feeding on the trunk (patches of bark are removed and bark flakes lie on the ground or snow)
- Reddish needles (crowns of successfully attacked trees usually turn off-color be ginning in May or June following attacks the previous summer).

Mountain pine beetle adults are cylindrical, stout bodied beetles, brown to black in color and about 1/4 inch long. The larvae are yellowish-white, legless grubs with dark heads found within tunnels under the bark.

Related Species: Several other *Dendroctonus* species are associated with pines. Largest is the red turpentine beetle, *Dendroctonus valens* LeConte, which attacks the base of trees, particularly following fire injury. Also found in the region are the roundheaded pine beetle, *D. adjunctus* Blandford, the larger Mexican pine beetle, *D. approximatus* Dietz, and the western pine beetle, *D. brevicomis* LeConte.

Hosts: Ponderosa, lodgepole and limber pines are the primary hosts. Scots pine is occasionally damaged. Bristlecone, pinyon and other pines are rarely attacked.

Life History and Habits: Mountain pine beetle has a one-year life cycle in Colorado. In summer, adults leave the dead, yellow to red-needled trees in which they developed. Females seek out living, green trees that they attack by tunneling under the bark. Coordinated mass attacks by many beetles are the norm. If successful, each beetle pair mates, forms a vertical tunnel (egg gallery) under the bark, and produces about 75 eggs. Following egg hatch, larvae (grubs) tunnel away from the egg gallery, producing a characteristic pattern.

MPB larvae spend the winter under the bark. They continue to feed in the spring, transforming to pupae in early summer. Emergence of new adults can begin in early July and continue through September. However, the great majority of beetles exit trees during late July (lodgepole pine) and mid-August (ponderosa pine).

TOP: Douglas-fir beetle tunnels.
BOTTOM: Douglas- fir pole beetle galleries.

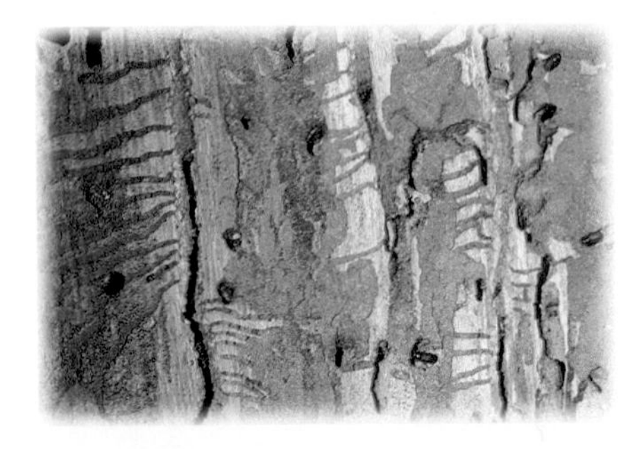

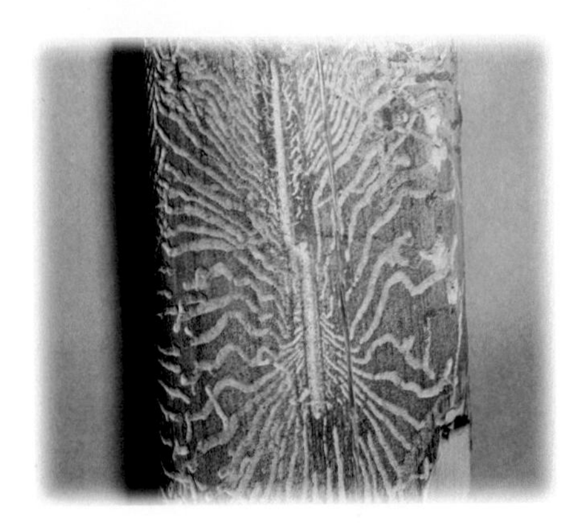

A key part of this cycle is the ability of mountain pine beetle (and other bark beetles) to transmit bluestain fungi. The body exteriors of adult beetles carry spores of these fungi and introduce them into trees during attack. They grow within the tree and, together with beetle feeding, weaken and rapidly kill the tree. While physiologically dead within a period of weeks, successfully infested trees do not show red, faded foliage until 8-10 months after attacked.

Management: Natural controls of mountain pine beetle include woodpeckers and insects such as clerid beetles that feed on both mountain pine beetle adults and larvae. Extremely cold temperatures also can reduce MPB populations. However, during outbreaks these natural controls often fail to prevent additional attacks.

Perhaps the most important natural influence is tree vigor. Healthy trees are less attractive to beetles than trees under stress. Vigorously growing trees also have better defenses that allow them to 'pitch out' pine beetles. Cultural controls which promote tree health, open spacing, and diversity of both tree species and size are the primary means of preventing MPB outbreaks. The forest management practice of thinning is the best long-term way to minimize MPB losses. In general, mature forests from which fire has been excluded over many decades are too dense. They would be much more resistant to both heavy beetle losses and catastrophic wildfire if every other tree were removed. Consult a professional forester to select the best cultural practices for your land.

Mountain pine beetle-infested logs can be treated in various ways to kill developing beetles before they emerge as adults in summer. Logs may be burned to kill the larvae under bark. Intense solar radiation that dries out the cambium and raises temperatures to lethal levels (110F+) can kill mountain pine beetle larvae. Beetles also die if the bark is removed by peeling or milling. Burying is another option to kill the developing insects in infested logs. In some cases, hauling infested logs to "safe sites" a mile or more from susceptible tree hosts is also practiced. Following beetle emergence, wood can be used without threat to other trees.

Chemical control options for mountain pine beetle have been greatly limited in recent years. Former treatments involving ethylene dibromide fumigation have been banned. Cacodylic acid (Silvisar products) and lindane are, or soon will be, unavailable or Restricted-Use. These treatments were primarily used to kill larvae in trees or adults as they emerge.

Certain formulations of carbaryl (Sevin) are registered for use in preventing attacks on individual trees. More recently permethrin (Astro) has also become available. These sprays are applied to living green trees in early summer (before mid-July) to repel or kill attacking beetles. This preventive spray is quite effective through one flight (one year) in ponderosa pine areas. In lodgepole pine areas, evidence indicates one spray of carbaryl may provide satisfactory protection through two flights (two years).

DOUGLAS-FIR BEETLE

Dendroctonus pseudotsugae Hopkins
Coleoptera: Scolytidae

Host: Douglas-fir

Damage and Diagnosis: Gallery construction and feeding on the inner bark of trunks can kill trees, which show foliage fading to a deep reddish brown color. Outbreaks usually occur in areas of wind-thrown trees or at sites damaged by fire or defoliation by western spruce budworm.

The adult beetles resemble mountain pine beetle, about 1/4 inch long. Some individuals are all black, others have black head and thorax with reddish brown wing covers. During attacks on trees, reddish or yellowish boring dust often accumulates in folds of bark or around the base of the tree. While pitch tubes are not usually produced, clear "streamers" of pitch may run down the bark of certain moist trees under attack.

Larvae are whitish, legless grubs with brown heads found within the galleries. Galleries primarily occur in the bark, but also etch the sapwood. The central egg gallery runs with the grain. Because the eggs are often laid in groups on alternate sides of the egg gallery, the resultant larval tunnels which run perpendicular to the egg gallery make a characteristic pattern.

Associated Insects: The Douglas-fir pole beetle, *Pseudohylesinus nebulosus* (LeConte) fills a similar niche in smaller diameter Douglas-fir. The insect is widespread and often found in areas infested with Douglas-fir beetle, sometimes in the same tree.

Life History and Habits: The Douglas-fir beetle appears to occur in two situations: apparently healthy, large-diameter, host stands where isolated, infested dead-tree groups increase to several to a few dozen trees over a few years and then decrease; and in seriously stressed stands such as those suffering from insect defoliation or exposure to fire that did not ruin the phloem tissue. In these stressed-tree situations, trees killed by the bark beetle may eventually outnumber those damaged by the initial problem. In Colorado, areas of trees blown down by strong winds are not as important in the development of Douglas-fir beetle outbreaks as in other parts of the West. In all situations the largest diameter trees present seem to be preferred, but Douglas-fir beetle may attack stems as small as six inches in diameter. Adults and large larvae overwinter.

Douglas-fir beetles spend the winter as either large larvae or adults under the bark. Adults typically begin emerging in late April and May (earlier than mountain pine beetle), but over 75 percent of the population emerges the last three weeks of June. Thus, the emergence period can extend from April into August.

Attacking beetles bore under the bark, mate there and the females lay eggs in small groups along a vertically-oriented gallery. These attacks result in accumulations of reddish boring dust at the tree base and streams of clear pitch running down the bark. The larvae feed outward in the inner bark area killing the tree. The needles of infested trees turn red a few to several months after initial attack. There is one generation per year, but more than one developmental stage often occurs in any one tree, because the attack period can extend over three to four months.

Management: Controls for Douglas-fir beetle are generally similar to that for mountain pine beetle. However, since adults can emerge earlier, developing beetles must be destroyed and preventive sprays applied by late April.

RED TURPENTINE BEETLE

Dendroctonus valens LeConte
Coleoptera: Scolytidae

ABOVE: Red turpentine beetle tunnelling into lower trunk.

Hosts: Large-diameter pines, rarely spruce.

Damage and Diagnosis: Pines and, rarely, other conifer species are sometimes attacked by the red turpentine beetle. Trees scorched near the base by fire or injured during construction are particularly susceptible. Turpentine beetle attacks are characterized by large, pinkish-purple pitch tubes confined to the lower few feet of the trunk. Beetles may be active throughout the warmer months, peaking in midsummer.

Inner-bark feeding by red turpentine beetles does not produce a branched gallery pattern typical of most bark beetles. Rather, the adults make short, irregular tunnels and developing larvae feed as a group, excavating an irregular, round-edged patch under the bark (resembles the outline of a puffy cloud). Trees can survive attack, but resultant weakening can make hosts more susceptible to other bark beetles.

The red turpentine beetle is the largest bark beetle found in the region. They generally resemble other *Dendroctonus* species (e.g., mountain pine beetle) but are wine red and about 1/4 to 3/8 inch in length. Larvae, found under the bark, are legless cream-colored grubs with brown heads, somewhat C-shaped and up to 3/8 inch long.

Life History and Habits: Turpentine beetles favor large declining or injured trees. Pines scorched by fire, wounded by construction equipment or suffering from soil compaction and grade changes are common targets. They bore into the lower trunks, where large pink masses of resin, called pitch tubes, form at attack points. The adult beetles lay clusters of eggs within their tunnels, and when these hatch the larvae feed in groups, consuming irregular areas of cambium/phloem just under the bark. These areas may join each other and girdle the tree depending on the density of attacks. Adult attacks occur through the warm months. There are two to three generations per year.

Management: Cultural practices promoting tree vigor and preventing basal trunk wounds (hay bales surrounding trees on construction sites, for example) can help prevent attacks by red turpentine beetle. Preventive insecticide sprays (carbaryl) applied before adult attacks, are also effective.

SPRUCE BEETLE

Dendroctonus rufipennis (Kirby)

Coleoptera: Scolytidae

ABOVE: Spruce beetle tunnelling.

Hosts: Spruce, primarily Engelmann spruce

Damage and Diagnosis: Larvae develop under the bark of spruce and girdling wounds produce cut the flow of water and nutrients within the tree. Foliage fades/discolors about a year after the tree is infested, turning pale green before shedding. However no pitch tubes are produced in response to attack by the adults. Boring dust collecting are entry points may be observed.

Spruce beetle is a forest species that usually occurs in low numbers. However historically it has produced some extensive tree kills during outbreaks. Over mature stands or situations where there is a large amount of wind thrown timber greatly increase risk of outbreaks.

The adult beetles are 1/6 to 1/4 inch long, dark brown to black with reddish wing covers. It is similar in appearance to the Douglas-fir beetle.

Life History and Habits: Completion of the life cycle of the spruce beetle takes at least two years and possibly three years at higher elevations. Adults spend the winter around the base of the trunk, in small chambers cut into the bark in late summer. They emerge shortly after snow melt and are usually most active during June and July. Most egg laying is done at this time.

The egg gallery is vertical and the base is often filled with pitch but pitch tubes, characteristic of mountain pine beetle, do not form. Larvae feed perpendicularly to the egg gallery. During the first year the grub-like larvae become partially grown, resuming feeding the following season. Pupation occurs in small chambers cut just beneath the bark. Adults emerge in summer and move to lower trunks to form the protective overwintering chambers.

Management: "Trap trees" felled shortly before flights, can concentrate large numbers of beetles. These logs then need to be either sprayed before flights or removed and destroyed before emergence of the insects.

ENGRAVER BEETLES

Ips species
Coleoptera: Scolytidae

Hosts: Pines and spruce, very rarely Douglas-fir

Damage and Diagnosis: A yellowish- or reddish-brown boring dust in bark crevices or around the base of tree is indicative of *Ips* attack. Bluestain fungi are introduced into trees by adult beetles. *Ips* beetles usually attack weakened trees of all sizes such as recent transplants, tops and limbs of mature trees already attacked by other bark beetles, trees injured by construction or topped during powerline clearing, drought-stressed or over watered trees.

Adults are small beetles, 1/8 to 1/4 inch long, reddish brown to black in color. They have a pronounced cavity at the rear end, which is lined with three to six pairs of tooth-like spines. Larvae are found within the galleries. They are small grubs, about 1/4 inch long when mature, white to dirty gray in color, legless, with dark heads.

Life History and Habits: Adults overwinter and begin to attach weakened trees in the spring. They bore into the bark and construct "Y" or "H"-shaped egg galleries, pushing the boring dust out of the entrance hole as they work. These 'cleared out' galleries have a different appearance than the frass-filled galleries of *Dendroctonus*. Eggs are laid along the gallery and young larvae soon hatch and begin tunneling small lateral galleries that lightly etch the sapwood. From two to four generations of these beetles may develop per year.

Management: Actions which prevent mechanical injury and improve tree vigor reduce the likelihood of *Ips* attack. Freshly-cut material, such as that resulting from pruning or thinning operations (called 'slash'), should be removed from the vicinity of valuable trees, chipped or treated in some manner that destroys the inner bark area. Treatments could include scattering slash to promote rapid drying or piling and burning the material. New transplants should be protected for at least the first year in their new site by application of preventive sprays (see discussion under twig beetle management).

"SPRUCE IPS"

Ips hunteri Swaine
Ips pilifrons Swaine
Coleoptera: Scolytidae

Hosts: Spruce

Damage and Diagnosis: A yellowish- or reddish-brown boring dust in bark crevices or around the base of tree is indicative of *Ips* attack. Blue-stain fungi are often introduced into trees by adult beetles. *Ips* beetles usually attack weakened trees of all sizes such as recent transplants, tops and limbs of mature trees already attacked by other bark beetles, trees injured by construction, drought-stressed or over watered trees.

Adults are small beetles, 1/8 to 1/4 inch long, reddish brown to black in color, with a pronounced cavity at the rear end lined with three to six pairs of tooth-like spines.

Life History and Habits: Ips beetles associated with spruce have at least two generations per year. Along the Front Range flights typically occur in April, followed by a second generation in late July. However, adult flights have been observed into October during periods of warm fall weather.

Attacks on trees are concentrated at, and often limited to the tops. Spruce ips are also commonly associated with wind-thrown or felled spruce and may out compete spruce bark beetles for breeding sites in these trees.

TOP: Engraver beetle tunnelling under bark.
SECOND: Galleries produced by spruce ips.
THIRD: Larvae tunneling by spruce ips. (Whitney Cranshaw)
BOTTOM: Symptoms of top dieback from spruce ips.

TOP: Cedar bark beetle tunnels.
BOTTOM: Balsam bark beetle tunneling of subalpine fir.

WESTERN CEDAR BARK BEETLES

Phloeosinus spp.
Coleoptera: Scolytidae

Hosts: Found commonly in all native junipers. Rocky Mountain juniper is the most common host. It is rarely, if ever, associated with prostrate juniper varieties.

Damage and Diagnosis: Adults feed on and may girdle small twigs, typically several inches back from the tip. The adults are 1/8-inch long, reddish brown to black shiny beetles of typical oblong bark beetle form. The wing covers are marked by lengthwise rows of minute puncture marks.

Most damage is done as developing bark beetle larvae feed under the bark. If only certain branches are killed, this results in conspicuous flagging. More commonly entire trees are killed. The gallery system has a central tunnel running parallel to the branch or trunk, with numerous side tunnels coming off it at right angles. These galleries are usually free of extensive sawdust. Larvae are minute, legless white grubs, found within their tunnels during active infestations.

Trees or branches subject to attack are almost always under severe stress. Girdling roots and drought have frequently been associated with attacks. Isolated bark beetle flagging on individual branches does not necessarily indicate the entire tree is at risk of dying.

Trees dying of cedar bark beetle trunk attacks turn bright red. In this respect, they look no different from a distance than those which die of other causes such as root problems, acute herbicide drift, and basal bark rodent gnawing. Inspecting the lower trunk of suspect trees on the plains may require pulling away packed accumulations of tumbleweed and other debris.

Life History and Habits: The life cycle of western cedar bark beetles are little studied in the Rocky Mountain region. Over wintering stage is as larvae found under the bark. Adults usually emerge in late spring or early summer and feed on twigs. However heavy late-summer flights have been observed in mid-July through early September, and probably indicate two generations are the norm for certain species.

The egg gallery is somewhat widened into an arrowhead shape at its bottom and then extends upward for a few inches parallel to the long-dimension of the trunk. Larvae feed perpendicular to the central egg gallery and girdle the branch or trunk. There can be either one or two generations produced per year, depending on weather.

Management: Implementing cultural practices which improve tree vigor can usually eliminate most threat of future attacks. Girdling roots have been in some infestations and dry, open winters may increase infestation risk unless plantings are provided supplemental water. Dead branches and trunks, which still harbor developing beetles, should be removed prior to their emergence to reduce local populations.

Chemical controls have not been reported. Standard bark beetle preventive treatments (e.g., chlorpyrifos, permethrin, carbaryl) applied to maintain coverage during the period of adult activity (June through September) should be effective.

WESTERN BALSAM BARK BEETLE

Dryocoetes confusus Swaine
Coleoptera: Scolytidae

Hosts: Primarily true firs, such as subalpine, corkbark and white. Engelmann spruce and lodgepole pine are occasional hosts.

Damage and Diagnosis: This insect, in concert with other factors such as adverse weather and root disease, is capable of causing widespread tree mortality in high-elevation forests. During epidemics large areas of subalpine fir turn deep red. Despite hundreds of thousands of trees dying, the remote location of damage results in little public attention. Only when infestations involve ski areas, nearby residential areas or recreation sites, does concern arise.

Adult beetles are about 1/4-in, dark brown and inconspicuously covered with sparse hairs. The profile of the rear wing covers is without distinct spines but is somewhat flattened rather than being evenly rounded. Female beetles have a dense circle of reddish-blond hairs on the front of the crown. The egg galleries are roughly star-shaped, with four or five short, winding 'arms' coming off the central mating chamber. All the galleries are shallow, largely restricted to the cambium and rarely etch the wood (unlike the fir engraver).

Life History and Habits: The biology of this insect is poorly understood. The life cycle may require one to two years, with studies in Colorado showing peak adult flight and attacks in midsummer. Once under the bark, adults construct egg galleries. Unlike most bark beetles, egg-laying is split between fall and spring, with the over wintering adults hibernating in galleries which are off-shoots or continuations of the egg galleries. This, combined with the larval galleries, leads to a distinctive but somewhat confusing (see scientific name of this beetle) total pattern. Other bark beetles, including a few very small species, are common associates of the western balsam bark beetle. Their roles and interactions are not well understood.

Management: Very little is known about the cause of outbreaks and even less is known about how to prevent them. Apparently, long-term periods of warmer-than-average weather, mild winters, etc. stress high-elevation forests. Balsam bark beetle problems tend to develop along the lower (drier?) edges of the host range, but problems elsewhere may occur over a wide range. Because low-value fir naturally intermingles with higher-value spruce, some management schemes advocate conversion of spruce-fir forests prone to balsam bark beetle to stands which are predominately spruce. Carbaryl and permethrin preventive sprays should also be effective for individual tree protection. Sprays should be applied annually in late spring as soon as snow melt allows and before adult beetle flight commences. Trees that have had roots disturbed due to road or pipe work and other insults are good candidates for preventive spraying.

BELOW: Galleries produced by cedar bark beetles. (David Leatherman)

Borers of Trunks and Larger Branches

Several insects develop by tunneling into the trunks and/or branches of trees and shrubs. Among the most important are the 'clearwing borer' moths (Family: Sesiidae) the metallic wood borers/flathead borers (Family: Buprestidae) and the longhorned beetles/roundheaded borers (Family: Cerambycidae). Most damage typically results from girdling of the cambium causing decline and die back. Some borers, notably the roundheaded borers, can tunnel extensively through the wood and cause structural weakening.

Most borers are only capable of successfully attacking dying or stressed trees. Proper watering and care is the first-line approach for prevention of borer injuries. Good cultural practices can also allow an infested tree to better tolerate borer injuries.

Insecticidal control approaches require applications just prior to egg laying and egg hatch. The occurrence of these life stages vary with different borer species and often last a month or longer. Consequently, repeat applications are often necessary for control. After eggs hatch and young borers have moved underneath the bark, insecticide treatments are ineffective. Pheromone traps are available to aid in the timing of treatments for control of clearwing moth borers (peach crown borer, lilac/ash borer).

Currently, chlorpyrifos (Dursban), permethrin (Astro) or carbaryl are recommended for control of wood borers. These materials are used as coarse sprays applied to the trunk until point of run-off. Lower branches and the crown area may need protection, depending on the habits of the borer. Spray deposits typically last for several several weeks to months; insecticide degradation is much slower on coarse bark (e.g., elm) than on smooth bark (e.g., birch). Soil and trunk injected insecticides provide irregular control of wood borers, at best, particularly where injuries have already occurred which affect translocation of the insecticides to the borer feeding sites.

Specific recommendations for control are included in the supplement with sections related to wood borers including: **Zimmerman pine moth**, **Pinyon pitch mass borer**, **Lilac/Ash borer**, **Peach tree borer**, **Bronze birch borer**, and **Wood borers (General)**.

Flight Periods and Hosts of Common Woody Plant Borers in Colorado

SCIENTIFIC NAME	COMMON NAME	COMMON HOSTS	TYPICAL FLIGHT PERIODS
METALLIC WOOD BORERS (Coleoptera: Buprestidae)			
Chrysobothris femorata	Flatheaded appletree borer	Apple, maple, *Populus*, other hardwoods	June-August
Agrilus anxius	Bronze birch borer	Birch	June-July
Agrilus difficilis	Honeylocust borer	Honeylocust	June-July
Agrilus politus		Willow, Populus, many hardwoods	June
Agrilus aurichalceus	Bronze cane borer/ Rose stem girdler	Caneberries, rose, currant	late May-June
Chalcophora species		Pines	June-August
Dicerca species		Aspen	June-August
LONGHORNED BEETLES (Coleoptera: Cerambycidae)			
Plectodera scalator	Cottonwood borer	*Populus*, willow	July-August
Megacyllene robiniae	Locust borer	Black locust	August-September
Saperda calcarata	Poplar borer	*Populus*, willow	June-August
Saperda inornata	Poplar gall borer	Poplar, cottonwood	July-September
Parandra brunnea	Pole borer	Maple, other hardwoods	July-September
Neoclytus acuminatus	Redheaded ash borer	Ash, other hardwoods	April-June
Monochamus species	Pine sawyers	Pines, spruce, fir	May-Sept.
Callidium antennatum	Blackhorned pine borer	Pines, other evergreens	May-June
Callidium texanum		Juniper	May-June
Atimia huachuchae	Juniper borer	Juniper	June-July
Dorcaschema alternatum	Small mulberry borer	Mulberry	June-July
Moneilema spp.	Cactus longhorn	Cholla cacti	June-July
WEEVILS (Coleoptera: Curculionidae)			
FALSE POWDERPOST BEETLES (Coleoptera: Bostrichidae)			
Cryptorhynchus lapathi	Poplar and willow borer	Willow, poplar	July-August
Pissodes strobi	White pine weevil	Spruce, some pines	March-May
Amphicerus bicaudatus	Apple twig borer	Honeylocust, apple, grape, ash and many other deciduous trees	June-July
CLEARWING BORERS (Lepidoptera: Sesiidae)			
Podosesia syringae	Lilac/ash borer	Ash, lilac, privet	April-June
Sesia tibialis	Cottonwood crown borer/ American hornet moth	Cottonwood	Mid June-July

Synanthedon exitiosa	Peach tree borer	*Prunus*	Mid June-September
Synanthedon tipuliformes	Currant borer	Currant, gooseberry	Late May-June
Synanthedon viburni	Viburnum borer	Viburnum	June-July
Synanthedon novaroensis	Douglas-fir pitch moth	Spruce, some pines, Douglas-fir	June-July
CARPENTERWORMS (Lepidoptera: Cossidae)			
Prionoxystus robiniae	Carpenterworm	Elm, maple, ash	June-July
***DIORYCTRIA* BORERS** (Lepidoptera: Pyralidae)			
Dioryctria ponderosae	Pinyon pitch mass borer	Pinyon, ponderosa pine	June-August
Dioryctria zimmermani	Zimmerman pine moth	Austrian, Scots pine	August-September
NOCTUID BORERS (Lepidoptera: Noctuidae)			
Achatodes zeae	Elder shoot borer	Elderberry (*Sambucus*)	July, August
HORNTAILS (Hymenoptera: Siricidae)			
Tremex columba	Pigeon tremex	Maple, elm other hard woods	Late July-August
STEM-BORING SAWFLIES (Hymenoptera: Tenthredinidae)			
Hartigia trimaculatus	Stem-boring sawfly	*Rosa, Rubus*	Late April-May

LILAC/ASH BORER (Ash Borer)

Podosesia syringae (Harris)
Lepidoptera: Sesiidae

Hosts: Ash, particularly white ash. Lilac and privet are other reported hosts.

Damage and Diagnosis: On ash, the larvae tunnel under larger branches and trunks, making rough gouging wounds under the bark and riddling the wood to a depth of about 2 inches. Repeated injuries can seriously weaken trees making them susceptible to breakage. Also wounding induces canker-like swellings to form and interferes with normal sap flow inducing sucker growth and a gnarled growth habit.

On smaller diameter branches, such as is common with lilac and privet, larval injuries may girdle and kill main stems. Unlike many borers, lilac/ash borer may push sawdust frass from the entrance hole as the larvae mature and begin to prepare the exit hole. This coarse material accumulates in bark crevices or at the base of infested trees. As adults emerge the pupal skin may be retained partially extruded at the exit hole.

Adult lilac/ash borers are mimics of paper wasps, quite similar in both size and coloration. The larvae are creamy white grubs with a small dark head. Prolegs on the abdomen are highly reduced but small hook-like crochets are present at the tip of the prolegs, which allows separation from the roundhead borers that are also associated with ash.

Related and Associated Insects: Several longhorned beetles also attack ash (e.g., redheaded ash borer, banded ash borer), although they limit attacks primarily to dying trees and are much less important than the lilac/ash borer in landscape plantings.

The banded ash clearwing, *Podosesia aureocincta* Purrington and Nielsen, has not been confirmed from Colorado. However, isolated infestations appear to be present in some areas of eastern Colorado. This species emerges from ash and lays eggs in late summer and late summer emerging borers from ash should be suspected as this species. Banded ash clearwing moths are not attracted to the standard "clearwing borer" pheromone lure.

BELOW: Lilac/ash borer larva exposed beneath bark.

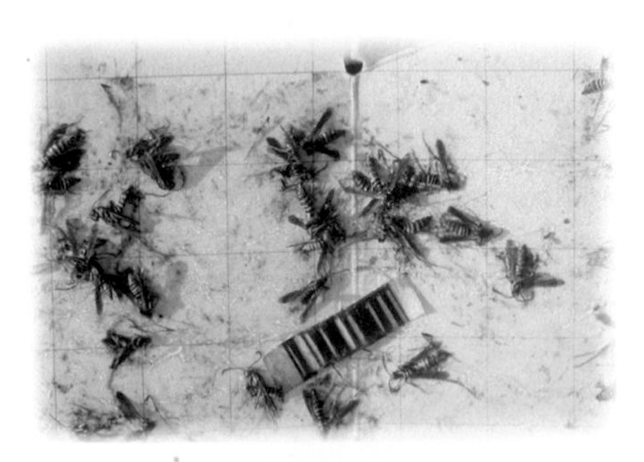

TOP: Pupal skins of lilac/ash borer extruding from trunk of ash tree.
SECOND: Mating pair of lilac/ash borer.
THIRD: Lilac/ash borer adults trapped in pheromone trap.
BOTTOM: Damage to base of trunk of plum by peach tree borer.

Life History and Habits: The ash borer spends the winter as a partially grown larva within tunnels under the bark. It resumes feeding and larval development in early spring, pupating just under a thin cover of the bark. The adults are brown and black "clearwing borers" that superficially resemble paper wasps in both size and color. Adults emergence may begin by early April during warm springs but usually is later and may extend for several weeks with cool overcast weather. Warm temperatures (above 60F) and sunny conditions appear critical for adult emergence which takes place during morning. Frequently the old pupal skin out is only partially extruded from the emergence hole and remains attached to the tree.

Adult activity typically extends for about four to six weeks after initial emergence, but can be substantially more prolonged with unfavorable weather. After mating, females lay eggs on bark, typically near wounds or bark cracks. Eggs may be laid singly or in small groups, and each female lays eggs for about a five day period. Eggs hatch in about a week and a half after which time the larvae tunnel into the cambium and phloem. As they continue to feed and develop they move into the sapwood, tunnel upward and later return to feed under then bark.

Management: Lilac/ash borer is most severe in plantings on marginal sites where stressful conditions occur. Proper siting to relieve future stresses can greatly limit attacks.

Fresh pruning wounds are highly attractive to the egg laying moths. It is important to avoid pruning prior to periods when moths fly.

Pheromone traps can be very useful for monitoring flight periods of this insect. The standard 'clearwing borer' lure is attractive to this species and lilac/ash borers are usually the only clearwing borer moth trapped during the early season. Trunk sprays of insecticides (e.g., permethrin, chlorpyrifos, carbaryl) should be applied about two or three weeks after first moths are captured. If heavy flights continue a month after application, reapplication may be needed if the plants have previously sustained injury.

Several parasitic wasps are recorded as natural enemies of the larvae of this insect.

PEACH TREE BORER (Crown Borer)

Synanthedon exitiosa (Say)
Lepidoptera: Sesiidae

Hosts: Peach, cherry, plum and other stone fruits.

Damage and Diagnosis: Larvae cause extensive damage by burrowing into the sapwood of the tree trunk, usually at or below the soil line. Girdling injuries weaken and frequently kill trees. This is the most important insect pest of stone fruits (*Prunus* spp.) in Colorado.

Infestations of peach tree borer result in production of clear gum at wound areas, mixed with light brown wood fragments. Almost all injuries occur at, or slightly below, the soil line. Often the pale larvae can be found within the wounded area and/or cast skins from the pupae be observed on the soil surface.

Gumming produce by peach tree borer is mixed with wood particles and is dark colored. It is different from other gumming produced on stone fruits caused by factors such as drought stress, mechanical injury, or infection with *Cytospora* fungi. These produce a clear, amber gum and can occur throughout the upper areas of the tree. Shothole borers, a small bark beetle, is common in above ground parts of the tree.

Life History and Habits: Females lay eggs on the bark of the lower trunk or even on the soil near the trunk during July and August. Eggs hatch in about a week and the larvae immediately burrow through the bark into the sapwood of the tree. Tunneling under the bark continues until late fall, with the insects mining down the trunk as cold weather approaches. In the spring, the larvae again feed extensively and complete development in June and July. Pupation occurs within the trunk and lasts about two weeks.

Adults typically emerge in late June and July. They live for several weeks, the females laying eggs over a period of one month or more. Eggs are most commonly laid under bark flaps or near wounds on the lower trunk.

Related Insects: Other *Synanthedon* species can attack shrubs in Colorado including the viburnum borer, *Synanthedon viburni* Engelhardt, dogwood borer, *Synanthedon scitula* (Harris), and the currant borer, *Synanthedon tipuliformis* (Clerck). *Synanthedon novaroensis* (Hy. Edwards) is occasionally associated with wounds of spruce.

Management: Egg laying is concentrated around wounds. Practices that avoid wounding, and control of existing peach tree borer infestations, can decrease later attack by the insect. White paint applied to the trunk can seal bark cracks used by female moths for laying eggs.

Treatment of the lower trunk and crown area with a drenching insecticide application can protect trees from new attacks. These applications should be timed to coincide with the onset of egg laying, which typically begins in late June or early July. Pheromone traps are very useful for determining the time of adult activity and egg laying. The standard "clearwing borer" lure is attractive to adult males of this species.

Individual larvae can be dug out or killed by a sharp wire. However, this needs to be done with care to avoid excessive tree wounding.

Larvae can not be controlled by insecticides after tunneling has progressed. Preventive treatments of insecticides should be applied to the trunk during periods when moths are active and eggs are being laid. This typically occurs in early July and August. Since mulch around the base of trees can provide some protection of the larvae, this should be cleared from the base of the trunk before treatments are applied.

Moth crystals (paradichlorobenzene) can be used to fumigate larvae within the trunk. The crystals are applied around the base of the trunk and are temporarily mounded with soil to retain the fumigant gas in the area of the borer infestation. The crystals should not directly touch the trunk and applications rates vary by trunk diameter. Fumigation "rescue" treatments are best applied during warm periods in fall, after harvest. However, future registration of paradichlorobenzene fumigants for fruit trees may become more restricted, so always check labels to insure that it remains a legal treatment.

Insect parasitic nematodes (e.g., *Steinernema carpocapsae*) have been reportedly used successfully to control peach tree borer larvae on ornamental *Prunus*. This has not been confirmed yet in Colorado trials. If attempted, suspensions of nematodes should be applied as a drench in fall or spring, when soil temperatures are above 55 F.

TOP: Peach tree (crown) borer larva.
BOTTOM: Peachtree borer male.

CURRANT BORER

Synanthedon tipuliformis (Clerck)
Lepidoptera: Sesiidae

Hosts: Currant, gooseberry

Damage and Diagnosis: Larvae tunnel canes, particularly near the base of the plant. Leaves on infested canes show signs of dieback including yellowing foliage. The larvae are usually found within the pith of infested canes and possess rudimentary prolegs that allow them to be separated from the bronze cane girdler, another common borer of Ribes.

Associated Species: The bronze cane girdler, *Agrilus aurichalceus* Redtenbacker, is a metallic wood borer that frequently attacks *Ribes*. However, attacks are limited to the upper canes and generally are limited to stressed or over mature wood. *A. polistus* (Say) also attacks currants as well as other woody plants.

Life History and Habits: The currant borer spends the winter as a nearly full-grown in the base of canes of currant, gooseberry, sumac, or black elder. They feed for a brief period in spring, but cause little damage. They then pupate and later emerge as adults in late May or early June. The bluish-black adult 'clearwing borer' moths appear similar to wasps, and can be seen resting or mating on leaves of the plants.

Eggs are laid on the bark during June and early July and the caterpillar stage larvae bore into the plant. They move downwards, tunneling the pith and wood. These feeding injuries may girdle or weaken the plant causing cane dieback in late summer.

Management: Cut out and discard all canes that show evidence of wilting and dieback in spring, before adult moths emerge. Increasing plant vigor through proper culture can reduce the severity of infestations.

Insect parasitic nematodes, applied as a drench to the crown area of the plants, can control currant borer larvae that are tunneling canes. Control of currant borer with insecticides has had only marginal success. If attempted, treatments should be applied to coincide with periods when the moths are flying and laying eggs, typically early June. However, very few insecticides are registered to permit use on edible currants and gooseberries.

The standard "clearwing borer" pheromone lure is not attractive to this species.

COTTONWOOD CROWN BORER/AMERICAN HORNET MOTH

Sesia tibialis (Harris)
Lepidoptera: Sesiidae

Hosts: Cottonwood and poplars are the most common host; willows and aspen are occasionally attacked

Damage and Diagnosis: Larvae tunnel the cambium and sapwood of trees, usually at or just below the soil line. Small trees can be girdled and killed. American hornet moth is widespread in the state but not a serious pest. Larvae can be separated from the other wood borers found in Populus by the presence of very short prolegs, tipped with a series of hook-like crochets, running along the underside of the abdomen.

Life History and Habits: The adult stage of the American hornet moth is a close mimic of yellowjacket wasps, a day flying moth that can be found mid June into August. Peak flights tend to occur in the latter half of July during which time the females lay eggs around the base of trees. Eggs hatch in about three weeks and the larvae seek injured sites or other easily penetrated areas of bark, usually around the root collar. Early feeding by larvae is restricted to the inner bark and cambium, but they later extend galleries into the heartwood. Larval development is quite extended and requires two years to complete. Pupation occurs within the burrows, just under the bark. Adults typically live for about one week following emergence.

Management: This insect has not proved to be a seriously damaging species. Controls, if necessary, would involve basal trunk treatments of insecticides timed to coincide with egg laying and egg hatch peaks. The adults, which appear similar to large yellowjacket wasps, are attracted to traps baited with the standard "clearwing borer" lure.

BELOW: Carpenterworm larva.

CARPENTERWORM

Prionoxystus robiniae (Peck)
Lepidoptera: Cossidae

Hosts: Elm, ash, cottonwood and other hardwoods

Damage and Diagnosis: Larvae tunnel into sapwood and heartwood of trunks and large branches. Heavily infested trees may be break in high winds and chronically infested trees appear gnarled and misshapen. Carpenterworm is most common in eastern Colorado shelterbelts, usually with some associated growing stress.

Unlike most other wood borers, the larvae maintain an exterior opening through which they continuously expel sawdust. Large exit holes and flaking of bark occurs along damaged areas of trees. Carpenterworm larvae are pinkish-white caterpillars with a dark head and dark brown tubercles on the body, up to three inches long at maturity.

Life History and Habits: Carpenterworms winter as larvae in the tunnels produced in infested trees. Pupation occurs in spring and adult moths appear around May, leaving behind their large purplish pupal cases sticking out of the exit hole. Adults are large stout-bodied moths with grayish-mottled forewings.

The females then lay eggs in clusters in bark crevices or near wounds. Eggs hatch and young larvae bore directly to the inner bark and feed. When about half-grown the larvae bore into sapwood, tunneling upward into sapwood and heartwood. It takes an average of three years to complete development.

Management: Because of the extended period during which the larvae develop within the tree, control requires a considerable period. Trunk sprays of insecticides, similar to that for other borers, can prevent new attacks. Pheromone traps are available for this species that can be used to best time sprays so that they will coincide with adult activity and egg laying.

Unlike most borers, the larvae create an opening to the outside as they develop and they continually push out sawdust leaving smooth tunnels. Moth crystals or injections of insect parasitic nematodes into these active tunnels can control larvae within the wood.

ELDER SHOOT BORER

Achatodes zeae (Harris)
Lepidoptera: Noctuidae

Hosts: Various species of elder (*Sambucus*), particularly golden elder. Corn and dahlia are also reported to be uncommon hosts.

Damage and Diagnosis: Larvae tunnel the new canes, causing them to die back in late spring. They are generally yellowish caterpillars with dark bands on the head, first, and last body segments. When full-grown they reach a length of between 1 and 1 1/2 inches. Adults are a fairly typical cutworm-type moth, with rusty red wings mottled with gray and a tawny spot along the wing tip. Wing span about one and 1/4-inches.

Life History and Habits: The elder shoot borer spends the winter in the egg stage, within an egg mass attached to bark of the old canes. Eggs hatch about the time that the new shoots start to emerge, typically late April. The young larvae first feed on the unfolding leaves for several days before boring into the new shoots. They feed within the shoot for several weeks. They often remain in they original shoot, but if food is exhausted, they will migrate to a new lateral shoot, usually entering near the base. Caterpillars feed and develop for about six to eight weeks, pupating in late June or early July. Pupation may occur within the damaged shoot, but often occurs in the pithy center of dead, dry branches, or on the ground.

Adults emerge in July. Females lay eggs under loose bark of old canes. Eggs are laid in masses averaging about 18 eggs and numerous egg masses may be produced. Eggs do not hatch until the following spring. There is one generation per year.

Management: Several parasitic wasps are reported to attack the elder shoot borer, as well as a tachinid fly. Downy woodpeckers sometimes dig the caterpillars out of elder shoots.

Since the eggs over winter on older canes, cutting and removing old wood before spring is a highly effective control.

BLUE CACTUS BORER

Melitara dentata (Grote)
Lepidoptera: Pyralidae

Hosts: Prickly pear cactus (*Opuntia* spp.)

Damage and Diagnosis: Larvae are caterpillars of vivid dark blue color that tunnel into the cactus pad at base of needles. They live and develop within the pads, producing large amounts of frass at the openings.

Life History and Habits: Adult female moths lay eggs in an unusual "eggstick" mass along the spines of prickly pear cactus during midsummer. The eggs hatch and the larvae tunnel into the cactus pad at the needle base. They spend the winter as a partially grown larva and continue to feed with the return of warm weather in spring. As the larvae increase in size the group scatters throughout the plant, attacking adjacent pads. Occasionally they are abundant enough to kill the pads of the cactus or even the entire plant. The larvae are an unusual blue color, giving the insect its common name, the blue cactus borer.

When full grown, in late June or July, the larvae leave the pad and wander a considerable distance away from the plant. They pupate within a cocoon under debris. Adults typically begin to emerge in mid July and can be found through August. After mating the eggstick masses are laid on the spines. There is one generation per season.

Increased populations have been associated with periods of moist weather that stimulates growth of grasses around the cactus. Tachinid flies and parasitic wasps are natural enemies, emerging during the pupal stage of the blue cactus borer.

TOP: Blue cactus borer larva. (J. Capinera)
BOTTOM: Pinyon pitch mass borer larva in trunk of pinyon.

PINYON "PITCH MASS" BORER

Dioryctria ponderosae Dyar
Lepidoptera: Pyralidae

Hosts: Pinyon, primarily. Ponderosa pine and other pines may be occasional hosts.

Damage and Diagnosis: The pinyon pitch mass borer has been a serious pest of pinyon in landscape plantings throughout much of the state. Ponderosa pine is another recorded host and there are some indications that they may occur in mixed infestations with Zimmerman pine moth in Scots and Austrian pines. Larvae tunnel into the cambial area causing large gouges which ooze a light pinkish pitch. Damage occurs on trunks and large branches. Wounding disfigures and weakens the tree. Heavily infested branches may break.

The larvae creamy or pale pink caterpillars with a dark head, found underneath the pitch mass. Adults are gray-brown snout moths with white zigzag markings, approximately 1/2 to 3/4 inches in length, but difficult to distinguish between other members of the genus.

Life History and Habits: Observations of the life history of the pinyon pitch mass borer in Colorado and Nebraska suggests that most require over one year to complete development. Adult, moth stages are active and laying eggs from late June through August. Larvae, the damaging stage, are pale yellow or pink in color with a light brown head. They tunnel under the bark and feed actively during the warmer months, undergoing four molts before changing to the pupal stage. Pupation occurs in the chamber of pitch and silk produced earlier by the larvae.

In Nebraska, where it is a pest of ponderosa pine, some of the insects can develop within 12 to 14 months. Likely these include individuals produced early in the season (June, early July) that emerge in August of the following year.

Related Species: Several other species of *Dioryctria* affect pines including *D. tumicollela* in branches of pinyon, the pinyon tip moth in pinyon twigs, and Zimmerman pine moth in trunks and branches of various pines.

Management: Several cultural factors appear to be associated with injury to pinyon pine by the pinyon pitch mass borer. Problems tend to be greatest in irrigated settings where pinyon receives far more than their natural requirement for water (6 to 10 inches/year). Such over irrigation promotes succulent growth and branch cracking that should favor the insect. Also, pinyons that are closely planted and overly crowded have had a history of increased problems with pinyon borer in the Denver area. Properly siting pinyon is considered important in managing problems with the pinyon pitch mass borer.

Pinyon borers lay most eggs near existing wounds. When possible, pruning of pinyons should be made in a manner to allow wound closure prior to periods when adult moths are active (late June-August).

Individual borers in trees can be punctured and killed by "worming" with a flexible wire. However, larvae can be difficult to find since they often tunnel for several inches under the bark. Alternatively, moth crystals containing paradichlorobenzene may be inserted into borer pitch masses and used to fumigate the larvae. Some formulations of these crystals are sold with labeling that allows use in shade trees.

Preventive **trunk sprays** can reduce new attacks. Coverage of the trunks during periods of adult moth flight activity and egg laying can kill the newly hatching caterpillars. However, coverage of the trunk and branches must be thorough, particularly around active wounds. Two or more treatments per season, repeated over at least two years, may be needed to reduce an infestation. In Colorado State trials, pyrethroids and Dursban have been among the most effective treatments. Experimentally, use of dimethoate injected into the root zone has provided some control in Colorado State University trials. Certain dimethoate formulations are registered for soil injection use.

ZIMMERMAN PINE MOTH

Dioryctria zimmermani (Grote)
Lepidoptera: Pyralidae

Hosts: Pines, particularly Scots and Austrian pines.

Damage and Diagnosis: In recent years, the Zimmerman pine moth has also become introduced and established along the Front Range. Austrian pines have been most commonly infested. Scots and ponderosa pines are also reported as hosts. Branches typically break at the crotch area where they join the trunk. Infestations are commonly marked by dead and dying branches, often in the upper half of the tree. First external symptoms of injury are the production of popcorn-like pitch masses at the wound site. The pitch mass may reach the size of a golf ball and ultimately resemble cluster of small pale-colored grapes.

ABOVE: Typical sap mass at wound site produced by Zimmerman pine moth.

The adults, rarely observed, are mid-sized moths, with gray wings blended with red-brown and marked with zigzag lines. Adults are difficult to distinguish from other members of this genus. Larvae are generally dirty white caterpillars, occasionally with some pink or green coloration. They are found within characteristic popcorn-like masses of sap on the trunks and branches.

Life History and Habits: The Zimmerman pine moth has a one year life cycle. The insect over winters as a very young caterpillar, inside a small cocoon (hibernaculum), underneath scales of bark. In mid-late April and May, they again become active and tunnel into the tree. Tunneling typically occurs at pre-existing wounds or at the junction of the trunk and branch. (Rarely, tips may also be infested.) Initial tunneling may result in sawdust and pitch at the entry site, although these are difficult to detect in early stages. The larvae continue to feed into July and early August, at which time large amounts of pitch are produced. Prior to pupation they gouge out large areas under the bark, leaving a thin bark flap and pupate just underneath this. When adults emerge the pupal skin may be pulled through the emergence opening in a manner similar to that of clearwing borers.

Adult moths are active primarily in late July and August. After mating, female moths lay eggs, often near wounds or previous masses of pitch. Eggs hatch in about a week and the larvae move immediately, without feeding, to protected sites on the bark where they prepared the over wintering hibernacula.

Management: Zimmerman pine moth is most vulnerable to control during the periods when larvae are active and exposed on the bark. Drenching trunk sprays, directed at the central trunk and concentrating the upper half of the tree, are best applied around mid April. August treatments, timed for egg laying and egg hatch periods, may also provide some control. Permethrin (Astro) and chlorpyrifos (Dursban) applied during April and early May have provided a high level of control in Colorado State trials.

ABOVE: Pigeon tremex.

PIGEON TREMEX

Tremex columba (L.)
Hymenoptera: Siricidae

Hosts: Silver maple, elm and many species of hardwoods

Damage and Diagnosis: The larvae create meandering tunnels within the sapwood of dead and dying trees. Densely tunneled trees may be prone to wind breakage. However, damage is usually confined to dead wood or to parts of the tree suffering from severe stress. Pigeon tremex, and other related horntails are not "aggressive" wood borers and can not develop in healthy trees.

The large size of this insect - a thick-waisted cylindrical wasp about one and a half inches long - often attracts attention. Furthermore the females possess a stout "stinger" (ovipositor), although they are harmless. Pigeon tremex (and other horntails) are the only common insects that make a circular exit hole as they emerge from the tree. (Most other wood borer produce exit holes that are oval in cross section).

Life History and Habits: Adults are most commonly present in late summer, searching and probing recently killed and declining trees for egg laying sites. Finding wood that is suitable for their young they insert an egg under the bark. (Sometimes the females are unable to extract their ovipositor and may be found attached to the trunk.) During oviposition they also introduce oidia of white rot fungi which decay the wood around the developing pigeon tremex larva. Larvae feed for the next 8 to 9 months on the wood and fungi, creating tunnels that run through the heartwood. When full grown they create a pupal chamber just under the bark. The emerging adults cut through the bark leaving a perfectly round exit hole. There is one generation per year.

Related Species: Several different horntails (*Sirex* species, *Urocerus* species) are associated with conifers. Most are black or blue-black and somewhat smaller than the pigeon tremex. These horntails attack trees that are in advanced decline or recently killed. They are particularly attracted to fire killed trees. They are not a problem for landscape or forest plant protection, but can degrade lumber by their tunneling and through the introduction of wood rotting fungi.

Management: Large, eye-catching ichneumon wasps with spectacularly long ovipositors often parasitize pigeon tremex and provide natural control of this insect. (See Giant Ichneumon Wasp, *Megarhyssa macrurus*).

Problems with pigeon tremex, and other horntails are limited to wood that is dead, dying, or under severe stress. Attention to cultural conditions that create these stresses can prevent attacks. Chemical controls, if attempted, would involve standard wood borer trunk sprays timed to coincide with the late summer egg laying period.

ROSE STEM-BORING SAWFLY

Hartigia trimaculatus (Say)
Hymenoptera: Tenthredinidae

Hosts: Rose, *Rubus.*

Damage and Diagnosis: A common cane borer infesting rose and raspberry is the stem boring sawfly. Damage is caused by the larvae which tunnel into the stem, often girdling it. The top of the plant, beyond the injury, wilts and dies. Canes break easily at the injury point. The larvae found within the canes are cream-colored and elongate with a distinct head.

Life History and Habits: The stem boring sawfly spends the winter as a full-grown larva within the old canes. They pupate in spring and adults emerge in April and May.

The adult stage is an elongate, 1/2-inch black and yellow wasp. Females insert eggs under the bark at the tip of current season canes, leaving a small puncture wound. Upon hatching the larva enters the stem to feed. As they get older, they tunnel downward under the bark, girdling the tip. They feed in the pith, eventually forming a small chamber in the upper stem during late June and July.

At least some of the stem boring sawflies pupate within the stem and a second generation of adult wasps gnaw their way out and emerge in mid summer. These wasps repeat the cycle, laying eggs which hatch into larvae that cause late summer cane die back. At the end of the summer larvae within the stems tunnel downward in the pith, spending the winter near the base of the plant.

TOP: Stem-boring sawfly larva in raspberry stem.
BOTTOM: Cottonwood borer.

COTTONWOOD BORER

Plectodera scalator (F.)
Coleoptera: Cerambycidae

Hosts: Willow, cottonwood, other poplars, willow. Within the region, this insect is restricted to southeastern Colorado.

Damage and Diagnosis: Larvae tunnel into heartwood of trees. During heavy infestations trees may be severely weakened and break at the base in periods of high wind. Adult feeding on bark of twigs and tender branches may cause some dieback. A fairly uncommon species apparently restricted to the southeastern areas of the state.

The adult is large beetle, from one to one and one half inches long. They are boldly patterned with white and black checkered markings on the wing covers. Larvae are typical roundhead borers, legless elongate grubs with retracted head.

Life History and Habits: Adults are active in late spring or early summer, and feed on tender young shoots. This adult feeding often causes the shoots to break, shrivel and turn black. Eggs are then deposited in pits chewed in the bark at the tree base. Larvae hatch and feed in the phloem, progressing downward into larger roots during their first fall. Larvae spend the second summer feeding in galleries at the tree base. The life cycle requires two years to complete.

POLE BORER (Aberrant Wood Borer)

Parandra brunnea (F.)
Coleoptera: Cerambycidae

Hosts: Willow, maple, elm, poplars and many other hardwoods are known hosts. Pole borer has also been known to attack conifers.

Damage and Diagnosis: The pole borer develops as a typical roundheaded borer, tunneling wood. Attacks of living trees are almost always restricted to the base of the plant, usually initiated at wounds where the sapwood has been exposed. Repeated attacks of this insect, in association with wood rotting fungi that are almost always associated with it, result in a honeycombing of the wood after a few years. Larval tunnels can extensively riddle the wood inviting collapse or storm breakage. Telephone poles and timbers in contact with the soil have also been sometimes damaged by this insect, leading to the name "pole borer".

The adult insect is about 3/4 inches long, but can vary greatly in size. They are shiny, reddish-brown, with prominent jaws. It is sometimes known as the "aberrant wood borer" because, unlike other members of this family, known as longhorned beetles, the antennae of the pole borer are not unusually long.

Life History and Habits: Adults are active in midsummer. During initial attacks the females insert eggs into areas of dead wood, sometimes in groups of up to a dozen eggs at a site. The larvae, a type of round headed borer may take up to three years to develop, producing a tunnel that may extend for several feet. They pupate in cells at the end of the larval tunnels and usually emerge to mate and initiate new attacks. However, as infestations progress, some adults may not leave the tree, mating and laying eggs within the galleries of existing tunnels. Previous exit holes are also sites used by migrant adults visiting a previously infested tree.

In addition to dying trees, any situation where wood contacts the soil may be attractive. On living trees, areas such as wounds or pruning scars are occasionally used. Most larval feeding is in sound wood and the tunneling, along with the subsequent decay fungi continue to expand the damaged areas. Larval tunnels ultimately can honeycomb the wood.

Management: Attacks of pole borer are initiated at recently bruised or wounded sites with high moisture conditions around the base of the tree favoring the insect. Eliminating these risk factors should limit pole borer problems.

Insecticide trunk sprays, standard for many borers, are unlikely to be highly successful for pole borer, particularly after infestations have initiated. The extended larval life span, ability to breed entirely under the bark, and tendency to lay eggs in previously produced larval tunnels all would reduce effectiveness of such an approach.

TOP: Pole borer.
BOTTOM: Bark sculpturing typical of *Callidium* larval feeding.

BLACKHORNED PINE BORER

Callidium antennatum hesperum (Casey)
Coleoptera: Cerambycidae

Hosts: Austrian and ponderosa pines primarily. Some other conifers are infrequent hosts and a closely related species, *Callidium texanum* Schaeffer, occurs in junipers.

Damage and Diagnosis: Blackhorned pine borer is a common insect and commonly attacks dying trees or trees under severe stress, such as recent transplants. As they feed they deeply score the wood under the bark and can kill declining trees. *Callidium* borer tunnels are packed with a distinctive, granular frass.

This borer also causes much concern to home owners by emerging from firewood or unseasoned lumber. In firewood and fresh houselogs, large amounts of sawdust can be produced by larvae, creating some confusion between these borers and the genuinely serious termites or powderpost beetles.

The adult insect is a bluish-black beetle, about 1/2 inch long with antennae the same length as the body. The wing covers are more leathery than those of other beetles. Larvae, found within wood resemble most roundheaded wood borers, being off-white, elongate legless grubs with brown heads.

Life History and Habits: Winter is spent within the trees, as a full-grown larva. Adults can emerge throughout the warm months of the year and have been observed in early May along the Front Range. Adult beetles feed on tender bark of twigs and shoots, and females lay eggs in pits in the bark. The larvae tunnel under the bark making very wide, wavy tracks that characteristically score the outer wood deeply. These tracks resemble those made by a router. Older larvae then excavate oval tunnels deep in the wood, where they over winter. There is one generation per year.

PINE SAWYERS

Whitespotted sawyer [*Monochamus scutellatus* (Say)], **Spotted pine sawyer** [*Monochamus clamator* (LeConte)], and possibly other *Monochamus spp.*
Coleoptera: Cerambycidae

Hosts: Pine, spruce, fir and Douglas-fir.

Damage and Diagnosis: Larvae bore extensively in sapwood and heartwood of dying and recently killed trees. They cause very little damage to living trees, but are commonly associated with firewood and house logs. A typical roundheaded borer, the larvae are elongate, segmented, legless grubs with brownish heads.

Adults cause minor injury by feeding on needles and shoot bark. In the Midwest and other parts of the world this genus includes the primary vectors of the pine wilt nematode, the cause of pine wilt disease, but this is not known from the region. Adults are large beetles (about one inch long), black to brownish-gray with white speckling, and have with extremely long antennae one to three times the body length.

Life History and Habits: Adults are present during the warm months but are most common in late summer. These big, harmless beetles often land on people working or recreating in the forest. They feed in tree crowns and lay eggs in little craters cut in the bark of branches and particularly trunks. The larvae construct wide galleries just under the bark, which are filled with fibrous frass. Large piles of sawdust and fiber often accumulate below infested trees or longs. Larvae enter the wood and tunnel for several months. There is one generation per year.

Related Species: The Asian longhorned beetle (*Anoplophora glabripennis*) is similarly a dark-colored longhorned beetle with very long antennae. Asian longhorned beetle has recently become established in some areas of the United States where it has generated considerable concern due to its potential to serious damage to maple and some other hardwoods. It has not been identified from anywhere in the Central Rockies.

Management: To avoid excessive tunneling to house logs or other recently cut wood, this material should not be produced during the warm months, or should be removed from the forest as quickly as possible. Borers which emerge from firewood or house logs do not reinfest the wood they came out of and are not a threat to wooden furniture, house plants or healthy conifers in the area.

TOP: Blackhorned pine borer.
BOTTOM: Pine sawyer resting on log.

CACTUS LONGHORN

Moneilema armatum LeConte
Coleoptera: Cerambycidae

Hosts: Cholla cacti (*Cylindropuntia* spp.) primarily, occasionally *Opuntia* spp..

Damage and Diagnosis: Larvae tunnel into the pads to feed and damaged pads often collapse from the effects of this injury plus invasion of secondary organisms. Conspicuous amounts of frass are often expelled through openings, sometimes mixed with ooze.

Related and Associated Species: Several species of *Moneilema* occur in the southwestern U.S., all restricted to various cacti. Species that are found in the region include *M. annulatum* Say, *M. appressum* LeConte, and *M. semipunctatum* LeConte. These generally resemble the cactus longhorn, but are somewhat smaller.

TOP: Cactus longhorn.
CENTER: Redheaded ash borer adult.
BOTTOM: Banded ash borer tunneling in log.

Life History and Habits: The cactus longhorn develops by feeding on various cacti in the genera *Opuntia* and *Cylindropuntia*. They survive winter within a pupal cell they construct during late summer and early fall around the base of the cactus. They transform to the pupal stage in spring and adults emerge in late spring and early summer. Adult beetles feed at night, typically eating young cactus pads or oozing sap. After mating, the females glue eggs to the cactus pad. The young larvae attempt to tunnel into the cactus, causing the plant to ooze sap at the wound. The young larvae first feed in this ooze, later entering the plant. They feed throughout the summer and early fall. One generation is usually produced per year, but some of the later larvae may not emerge until the second season.

REDHEADED ASH BORER

Neoclytus acuminatus (F.)
Coleoptera: Cerambycidae

Hosts: Ash, hackberry, several fruit trees, and other hardwoods

Damage and Diagnosis: The redheaded ash borer develops as a roundheaded borer, usually only developing in dead and dying wood. During feeding they can reduce the sapwood to a fine powder and produce meandering oval tunnels. Occasionally they also will bore into twigs and branches.

The adult beetles have a narrow brown body with reddish head and thorax. They are about 5/8 inch long and have wing covers marked with four yellow transverse bands. They have long spindly legs and move rapidly. Adults of the redheaded ash borer, and the closely related banded ash borer, very commonly emerge during early spring from firewood stored indoors.

Life History and Habits: Adults become active in late spring and females deposit eggs beneath bark of dead or dying hardwoods. Newly hatched larvae first feed beneath the bark, then tunnel into sapwood. Pupation occurs in fall. There is one generation per year.

Related Insects: The banded ash borer, *Neoclytus caprea* (Say), is another roundheaded borer that is generally similar in appearance and habits to the redheaded ash borer. It is marked with yellow stripes on the wing covers and is slightly larger than the redheaded ash borer. Biology is similar. Another *Neoclytus* species common in the state is *N. muricatulus* (Kirby), associated with conifers.

Management: Controls would be typical for other wood boring insects involving efforts to increase tree vigor through cultural practices, removal of infested wood prior to insect emergence, and treatment of trunks with protective insecticide sprays to coincide with egg laying.

POPLAR BORER

Saperda calcarata (Say)
Coleoptera: Cerambycidae

Hosts: Primarily aspen, but cottonwood and poplar are also hosts.

Damage and Diagnosis: Larvae develop under the bark and tunnel the sapwood, girdling trees. Early stages of attack are indicated by moist areas on the bark, often with some associated sawdust. Chronically infested trees exhibit a black varnish-like stain on the bark below points of borer attack. Stringy sawdust is pushed out of holes in the bark by the developing larvae and may pile around the base of trees. Branches and trunks may break off in high winds and trees may be invaded by wood rot. Wound sites also develop as rough growths that may split the bark.

The poplar borer is the most destructive insect to aspen in many areas of the state. Attacks are generally restricted to large diameter trees. Trees most affected are unshaded and generally suffering from some growing stresses related to site.

Life History and Habits: The poplar borer has an extended life cycle that likely requires two to three years to complete. Adults are generally grey beetles, about 1 and 1/4 inches long, with a central yellow stripe on the thorax and some black and yellow stippling marks on the wing covers. They emerge beginning in late June and may be found through August. After emerging from the tree they first feed on the leaves and young shoots of their host for about two weeks. The females then begin to lay eggs, which they deposit in niches that they chew from the bark of the trunk. Larvae first feed in the bark and later move to the sapwood where they make tunnel upward making galleries that typically extend about a foot. Larvae are large, yellowish, round-headed grubs, about 1 3/8 inches long when mature, and are found only within their tunnels. Unlike most borers, throughout their period of feeding they maintain an opening to the outside, through which they push the boring dust. Pupation occurs in late spring, within a chamber cut just beneath the bark.

Related Species: The poplar-gall saperda, *S. inornata* (Say), forms galls on the twigs of aspen, poplar and willow. The elm borer, *S. tridentata* Olivier, develops in dead or weakened American elms sometimes causing dieback of individual limbs. Both species are uncommon in Colorado, but are reported to seriously damage trees in other areas of the United States.

Management: Several biological controls, including parasites of eggs and larvae, predators, and fungal diseases have been reported to affect this insect. Siting aspen so that it is shaded and allows optimal growing conditions will reduce risk of attack.

Because of the long (probably three year) life cycle, poplar borer is particularly difficult to control. Standard preventive borer treatments of insecticides applied to the trunk should provide coverage throughout the period of adult activity (primarily July and August). Application should be most thorough to existing areas of attack, in the middle of the tree, where egg laying is concentrated.

Insertion of 'borer crystals' (paradichlorobenzene) or injections of insect parasitic nematodes (*Steinernema* species) into active borer tunnels have given partial control of larvae. Because this species does maintain an external opening, sprays of nematodes may also be effective, as they have for wood-boring species of similar habit.

TOP: External symptoms of poplar borer infestation.
BOTTOM: Poplar borer.

LOCUST BORER

Megacyllene robiniae (Forster)
Coleoptera: Cerambycidae

Host: Black locust (*Robinia pseudoaccacia*).

Damage and Diagnosis: Locust borer is very commonly associated with black locust stands. Larvae develop within trunks, causing deep tunneling that may riddle the plant and produce serious structural weakening.

The adult beetle is a colorful insect, generally black beetles but marked with yellow cross bands on thorax and "W" shaped bands on the wing covers. They are about 3/4 inch in length and are most commonly observed visiting flowers in late summer and early fall. Larvae are robust, cream colored, legless grubs with brown heads, about one inch in length when full grown.

Life History and Habits: Adults are active in late summer and early fall, and are commonly seen feeding on various flowers. Also at this time eggs are deposited in cracks and crevices in bark of host trees. Larvae hatch in late fall, bore into bark and construct small hibernation cells for over wintering. They resume activity in the spring and tunnel

TOP: Locust borer larvae in wood. (S. Krieg)
CENTER: Locust borer adult.
BOTTOM: Raised areas of callous on birch trunk resulting from bronze birch borer tunneling.

extensively through heartwood. The larvae mature in the latter part of July. There is one generation per year.

Management: Preventive trunk sprays of insecticides can be applied in mid to late summer to control the insect during the egg laying/egg hatch period.

BRONZE BIRCH BORER

Agrilus anxius Gory
Coleoptera: Buprestidae

Hosts: Birch

Damage and Diagnosis: Larvae develop by tunneling the cambium of birch. Girdling injuries first causes limb dieback, with thinning of the crown an early symptom of infestation. Raised ridges on the bark often are evident, a response of the plant to injury. Small D-shaped holes are made in the bark by emerging adult beetles.

The adults are elongate, metallic-copper beetles, about 1/2 inch long. Larvae are typical flatheaded borers - white, wormlike grubs with an enlarged flattened area just behind the head and no legs.

Life History and Habits: Adults are metallic, copper-colored beetles that emerge in late May or early June, cutting a D-shape opening through the bark. They feed some on leaves for a week or two before eggs mature. Females lay eggs in bark crevices, around curls of bark or in other protected sites, primarily on the unshaded sides of trunks and branches. During initial phases of attack most egg laying is concentrated in the upper crown on branches less than an inch in diameter; larger branches and, ultimately the trunk are attacked as infestations progress. Eggs hatch in about two weeks and the larvae tunnel into the cambium and spend most of there life in this area, rarely moving into the xylem. Larval galleries often have a zigzag pattern and are packed with fine sawdust frass. Trees that overgrow these wounds may show evidence of tunneling as a raised trail when examining the outer bark. Mature larvae over winter and pupate early in the spring. There is one generation per year.

Related Species: Several species of *Agrilus* attack shade trees and shrubs. Among the more common is *A. difficilis* Gory, which is found in many plants, and commonly is associated with honeylocust wounds. It is a metallic purple color and similar in size to bronze birch borer. Other common species include the bronze cane girdler, *Agrilus aurichalceus* Redtenbacher, that attacks currant, rose, and raspberry and *Agrilus politus* (Say), which develops in many woody plants including willow and poplars. Life cycle of other members of this genus is similar to the bronze birch borer.

Management: Proper siting and maintenance of birch is very important in maintaining vigor which can help resist attacks. In typical landscape plantings, birch is easily stressed sufficiently to become susceptible to bronze birch borer. Perhaps the best cultural practice is providing a large area of mulch over the root system of the tree to help conserve moisture around roots and provide for more moderate fluctuations in soil conditions.

Individual limbs that show evidence of dieback should be trimmed and disposed before beetles emerge in spring. Pruning should be done below the point where there is evidence of infestation. Raised areas on the bark, resulting from callus growth around borer tunnels, is a means for easily identifying much borer activity. If raised areas, or semicircular exit holes, are found on larger branches or the trunk, pruning will have little chance of rescuing the tree. Pruning between May and early July should be avoided as fresh wounds may be attractive to egg laying adults.

Preventive insecticide applications (chlorpyrifos, permethrin) need to cover the trunk throughout the adult activity period when new eggs are laid. This can occur from late May through early July. Under heavy pressure by this insect, two to three applications may be needed.

There has been some success with injections of insecticide *when applied shortly prior to egg laying*. However, these treatments are not very effective against older borers that have already damaged extensive areas of the cambium. Furthermore, wounds caused by injections can further stress the tree leading to later decline.

Some species and varieties of birch have been observed to be resistant to bronze birch borer. In the Midwest the following categorization of susceptibility to this insect has been proposed (Minnesota Extension Bulletin FS-1417-A).

	Highly Susceptible
Betula jacquemontii	Jacquemonti Birch, Whitebarked Himalayan Birch
Betula pendula	European white birch, Silver birch
Betula pendula 'Youngii'	European weeping birch, Young's weeping birch
	Moderately Susceptible
Betula alleghaniensis	Yellow birch
Betula lenta	Sweet birch, Black birch, Cherry birch
Betula papyrifera	Paper birch, White birch, Canoe birch
Betula platyphylla japonica 'Whitespire'	Whitespire birch
Betula populifolia	Gray birch
	Very Low Susceptibility
Betula nigra	River birch, Red birch
Betula nigra 'Heritage'	Heritage birch, Heritage river birch

TOP: Bronze birch borer.
BOTTOM: Larva of bronze birch borer.
(Whitney Cranshaw)

ROSE STEM GIRDLER/BRONZE CANE BORER

Agrilus aurichalceus Redtenbacher
Coleoptera: Buprestidae

Hosts: Currant, gooseberry, raspberry, rose

Damage and Diagnosis: The flathead borer stage (larva) makes meandering tunnels under the bark of rose, cane berries (raspberry, blackberry, etc.), currants and gooseberries. Characteristically a swollen area develops around the wounded area of the stem. Canes often die back or break at these wounded sites.

Related and Associated Insects: Several other *Agrilus* species occur as pests of trees and shrubs, notably the bronze birch borer and honeylocust borer. Another common borer attacking currant is the currant borer, *Synanthedon tipuliformis*. The currant borer concentrates feeding in the crown of the plant and can kill several canes.

Life History and Habits: The bronze cane borer over winters as a nearly full grown larva under the bark of the canes or within the pith. In spring it resumes feeding and pupates within the plant. Adult stage is a bronze colored beetle which emerges in May and is present through early June. Eggs are laid during this period, usually near the base of leaves. The larvae, which emerge from eggs, tunnel under the bark and move upwards in the plant. These early tunnels are often close to the surface and may be externally visible on some plants, such as rose. Later they tunnel more deeply into the canes, producing the damaging girdling wounds. Pupation occurs inside a chamber produced within the cane the following spring. There is one generation per year.

TOP: Rose stem girdler larva in currant stem.
BOTTOM: Rose stem girdler/bronze cane borer.

Management: Canes that show evidence of injury should be removed before midspring to destroy the insects before they emerge.

Effective chemical controls have not been demonstrated. If attempted, they are best applied during midspring to kill adult beetles before they have laid eggs. Larvae within stems can not be controlled with insecticides. Insecticide options are very limited on edible fruit-bearing crops.

HONEYLOCUST BORER

Agrilus difficilis Gory
Coleoptera: Buprestidae

Hosts: Honeylocust

Damage and Diagnosis: The larvae tunnel under the bark of honeylocust, making meandering tunnels that are packed with fine frass. Tunneling is usually restricted to larger branches and the trunk and is concentrated in areas of the tree affected by cankers or wounds. However, this insect is capable of infesting non-diseased bark areas of trees under environmental stress, where there have been injuries through poor trimming or storm damage. During conditions favorable to outbreaks the insect can move into relatively healthy wood and contribute to decline.

Small D-shaped exit holes are made through the bark by emerging adults. The larvae, which are most readily found in early spring, are pale colored, very elongated with a slight widening behind the head.

Associated Insects: A longhorned beetle, *Styleiopus variegatus*, is sometimes found developing in dead branches and canker-killed areas of the trunk. Smaller branches are sometimes tunneled by the apple twig borer (*Amphicerus bicaudatus*).

Life History and Habits: Winter is spent as a larva within tunnels under the bark. When mature they cut into the xylem to form a pupation chamber. Adult beetles emerge over a very extended time and have been observed from May through September in Colorado. After feeding of foliage for a couple of weeks the females begin to lay eggs. Bark scars, recent wounds, around the edge of cankers and trunk crotches are favored sites for egg laying. The newly hatched larvae tunnel under the bark and small damp areas, sometimes with associated gumming, indicate sites of initial attack. There is one generation produced per year.

Management: Honeylocust borer is primarily a secondary pest. Severe growing stresses, trunk wounds and/or canker growths are essential for its development; management should focus on correcting these factors.

The very extended period of adult activity and egg laying suggest, combined with the smooth bark of honeylocust, that chemical control via trunk sprays would be very difficult.

ABOVE: Flatheaded appletree borer larval tunneling.

FLATHEADED APPLETREE BORER

Chrysobothris femorata (Olivier)
Coleoptera: Buprestidae

Hosts: Flatheaded appletree borer has a very wide range of food plants including most deciduous fruit, forest and shade trees. Maples and apple are among the more common hosts.

Damage and Diagnosis: The immature stage (a flatheaded borer) tunnels under the bark of trunks and larger branches, producing broad galleries that are tightly packed with fine sawdust frass. Areas of bark where injury has occurred often appear darkened, somewhat sunken and later may split above injuries. On young trees tunneling may girdle and kill the plant; tunnels are more restricted in area on established trees. Injuries are concentrated on the sunny side and most commonly occur on trees suffering sunscald, wounds or that suffer from drought stress.

Related Species: Two closely related species occur within the region. The Pacific flathead borer (*Chrysobothris mali* Horn) predominates west of the Rockies; the flatheaded apple tree borer in the east. However, their ranges may overlap in the region. Biologies of the two species are similar.

Life History and Habits: Larvae are a typical "flathead borer" - pale yellow, legless with a very enlarged thorax. Winter is spent as a larva under the bark. They complete development the following spring, cut a chamber into the sapwood and pupate. Adults may begin emerging by mid spring but peak activity is from late May through June. Adults are dark olive-gray to brown metallic wood borers about 1/2 inch long. The females may be observed searching the sun-exposed sides of trunks of host trees. Eggs are laid singly in bark cracks or near existing injuries and over the course of a month may lay about 100 eggs. Eggs hatch within 8 to 16 days and the larvae chew through the bottom of the egg and begin to tunnel into the tree. Larval development can be rapid and gallery formation extensive in low vigor trees. Development is retarded and tunneling more restricted in trees of high vigor. They continue to feed for several months, becoming dormant during the cold season. There is one generation per year.

Management: Attacks by flatheaded apple tree borer are concentrated around wounds, cankered areas, and on trees in generally poor health. Healthy trees are less attractive to the egg laying females and the tunneling larvae often are killed by tree defenses, such as oozing sap. By maintaining trees in a healthy, vigorous condition and preventing injuries, problems with these insects can be avoided.

Where sunscald injuries are likely, shade the lower trunk of young trees or use tree wrap to prevent this injury. Whitewashing the trunks can also reduce attacks.

Dying trees and newly cut wood should not be kept near susceptible trees, since large numbers of borers can develop in these materials.

Once the borers are in the trunk, digging them out in late summer or early fall is the only control. This is difficult to do without causing additional tree injury.

Preventive use of insecticides, applied to trunks and branches can limit new attacks. Treatments should be timed to coincide with periods of egg laying and egg hatch.

BELOW: *Buprestis* sp. adult.

BUPRESTIS BEETLES

Buprestis spp.
Coleoptera: Buprestidae

Hosts: Most *Buprestis* beetles develop in conifers such as ponderosa pine, lodgepole pine, grand fir and Douglas fir. However, one common metallic green species, *B. confluenta* Say, develops in cottonwood and aspen.

Damage and Diagnosis: Larvae of Buprestis beetles make girdling tunnels under the bark in a manner similar to that of most flathead borers. These tunnels tend to be larger than those produced by most flathead borers, reflecting their relatively larger size.

Life History and Habits: Adult beetles in the genus *Buprestis* are often a striking green or bronze with a metallic and reflective body. One of the more attractive species is the golden buprestid, *B. aurulenta* Linnaeus, which has metallic green or blue-green wing covers with gold flecks and copper edges. Approximately a dozen species of *Buprestis* occur in the region.

Females lay eggs in masses wedged under the bark of dead or dying trees. The larvae of these insects (flatheaded borers) are legless, elongate, pale colored insects with a flattened area behind the head. They develop under the bark of dead and dying trees, feeding primarily in the cambium area just under the bark producing characteristic meandering tunnels. Most do not damage healthy trees but a few species, notably the golden buprestid, may help kill trees suffering from wounds or stress. The tunneling by the larvae occasionally causes damage to buildings and finished wood products. Adults feed on needles and tender bark. In the forest, the life cycle typically requires two or three years to complete. If development occurs in finished wood, egg to adult development may take up to fifty years.

TOP: Apple twig borer.
CENTER: Poplar and willow borer adult.
BOTTOM: Borings produced by poplar and willow borer.

APPLE TWIG BORER

Amphicerus bicaudatus (Say)
Coleoptera: Bostrichidae

Hosts: Honeylocust, apple, grape, ash and many other deciduous trees.

Damage and Diagnosis: The legless, grub-like larvae tunnel into twigs, girdling them and causing dieback. This insect is considered to be a secondary pest in most situations as larval attacks are restricted to parts of trees previously wounded or diseased and in decline. Adults that emerge during late summer may cut into healthy wood to feed and produce winter shelters, but this is unlikely to occur within the region.

Life History and Habits: Adults emerge in late spring from old larval tunnels or wintering chambers cut into bark. They are cylindrical-shaped brown beetles about 1/4-inch long and in spring cut into the bark of twigs to lay eggs. The grub-like larvae tunnel the twig as they develop, primarily restricting feeding to the pith and packing their tunnels with sawdust frass. Larvae feed continues throughout the summer and most pupate during late fall within the larval tunnel. There is one generation produced per season.

Management: Apple twig borer is primarily a secondary pest limiting attacks to plants and plant parts in decline. By avoiding wounding and other practices that stress plants problems can be largely eliminated. Infested twigs should be pruned and destroyed before adult insects emerge. Chemical controls are not recommended.

POPLAR AND WILLOW BORER

Cryptorhynchus lapathe (L.)
Coleoptera: Curculionidae

Hosts: Willow and, rarely, poplar

Damage and Diagnosis: The larvae tunnel into the lower trunk of trees. Infested trees may become malformed because of excessive sucker growth. Young willows may have bulb-like swellings from borer attack and may break easily due to borer weakening. The poplar and willow borer is found frequently in moist sites such as along streams and ponds.

Larvae are cream-colored, legless, "C"-shaped grubs about 1/4 inch long. At points of feeding large amounts of moist sawdust are pushed from the entry holes. The adults are chunky snout weevils, about 3/8 inches long, and rough-surfaced. They are primarily black in color except for the hind 1/3 of the wing covers, which are gray, somewhat resembling a bird dropping.

Life History and Habits: The poplar and willow borer spends the winter as a partially grown larva in the sapwood. In the spring, the larvae grow and continue boring, pushing large amounts of fibrous frass through exit holes. Larvae pupate under the bark, beginning in May. Adults may be present from late May through mid-July. Eggs are deposited in small slits in the bark. There is one generation per year.

Management: Preventive trunk sprays applied in late May or early June can reduce most injury. For specific recommendations see **Borers, General** in the supplement.

Mistletoes and Other Parasitic Plants

Parasitic flowering plants that live on woody plants range from minor water robbing plants to serious parasites that derive all their nutrients and water from their host. Parasitic plants may attach to roots or the upper stems of trees. In the region we have two important groups of plants called the mistletoes. One type - leafy mistletoes (*Phorodendron* spp) or the true mistletoes grow on both hardwoods and conifers and are only water parasites. Birds eating the seeds spread these mistletoes. In this region we only have one *Phorodendron* sp. and it occurs on junipers in the southwest part of Colorado and down into New Mexico and Arizona. The other group of mistletoes is the dwarf mistletoes (*Arceuthobium* spp.) that derive all their food from the host tree. These parasites grow only on conifers and are some of the most widespread and damaging agents in the region. Dwarf mistletoes spread by means of explosive fruit that shoot out seeds. Dwarf mistletoes cause growth reduction and mortality and predispose the trees to attack by other insects and diseases. Most dwarf mistletoes are host specific. They also may produce witches' broom symptoms.

DWARF MISTLETOES

Arceuthobium americanum (lodgepole pine), *A. cyanocarpum* (limber pine), *A. divaricatum* (pinyon), *A. douglasii* (Douglas-fir), *A. vaginatum* subsp. *cryptopodum* (ponderosa pine)

Hosts: Ponderosa, lodgepole, limber, pinyon and Douglas-fir are the most common hosts.

Diagnosis and Damage: The major symptoms caused by dwarf mistletoes are witches'-brooms, loss of vigor, dieback, and death. The first symptom of dwarf mistletoe infection is a slight swelling of the bark at the infection site. As the parasite's growth within the tree increases, a distorted branching habit or witches'-broom may form. Yellow foliage, reduced foliage and mortality of branches or the entire top of the trees may indicate mistletoe infections are present. The parasite is identifiable when the yellow to green or brownish-green segmented shoots protrude from the infected part of the tree. These woody shoots are one inch to 6 inches long and 1/8 inch to 1/4 inch in diameter. Shoots form two to three years after infection. Mistletoes are only found in native forests unless infected trees are moved to urban landscapes

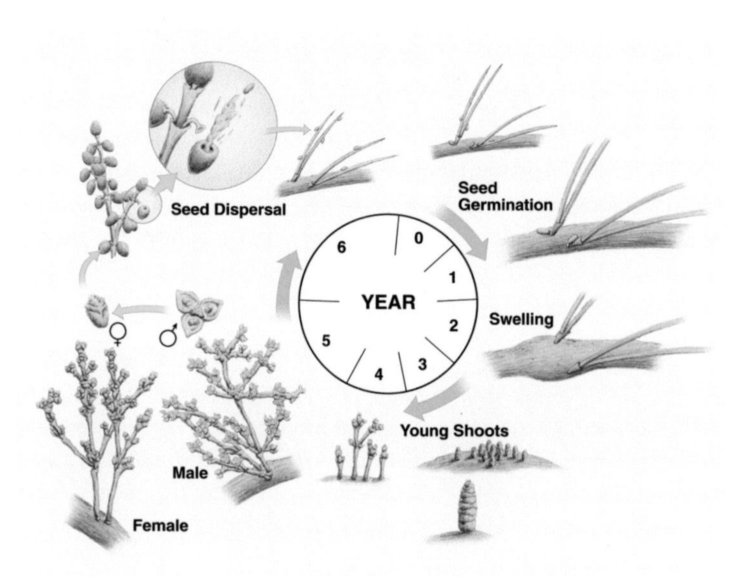

TOP: Dwarf mistletoe on limber pine. (W. Jacobi)
SECOND: Mistletoe broom on limber pine. (M. Schomaker)
THIRD: *Arceuthobium divaricatum* on pinyon pine. (W. Jacobi)
BOTTOM: Dwarf mistletoe infected overstory of lodgepole pine. (W. Jacobi)
BOTTOM LEFT: Lodgepole pine dwarf mistletoe life cycle.

TOP: *Arceuthobium americanum* male flowers on lodgepole pine. (M. Schomaker)
SECOND: *Arceuthobium vaginatum* on ponderosa pine. (M. Schomaker)
THIRD: *Arceuthobium vaginatum* female plants on ponderosa. (USDAFS)
BOTTOM: *Phoradendron juniperunum* brooms on juniper. (USDAFS)

The effects of dwarf mistletoe include growth reduction, loss of wood quality, poor tree form, predisposition to twig and bark beetles and diseases, premature death and reduction in seed crops. Dwarf mistletoe kills by slowly robbing the tree of food and water. Diseased trees decline and die from the top down as the lower infected branches take more food and water. Mortality of infected trees does not occur rapidly in most cases and depends on the severity of infection, and the vigor and size of tree.

Biology and Disease Cycle: Dwarf mistletoe is a small, leafless, parasitic flowering plant. The seeds, explosively discharged from the fruit at almost 60mph, are sticky and adhere to any surface they strike. Seeds that adhere to young needles and branches of susceptible trees germinate and the mistletoe rootlet penetrates the bark.

Dwarf mistletoe grows into the bark and phloem of the tree. The parasite produces root-like structures called "sinkers" that form each year. Older ones become imbedded deep in the wood as twigs grow. These sinkers provide the parasite with water and nutrients obtained from the host. Most dwarf mistletoes are specific to a particular type of tree and do not infect other tree species.

Conditions Favoring and Management of Disease: No specific conditions are necessary for infection development. The worst situation is when older/tall infected trees can shower seeds down on young trees and cause numerous infections. Stressful sites allow more damage to occur to the tree infected. Adequately watered trees can withstand the added stress of mistletoe better than trees in stressed sites.

Pruning infected branches and removing trees is the best management measure available to reduce or eliminate dwarf mistletoe infestations in ornamental trees or urban forests. First, remove trees severely infected or those with only a few live branches. The parasite can be removed from lightly infected trees by pruning infected branches. Do not remove more than 60 percent of the crown when pruning. Examine trees every two or three years and remove any newly infected branches. Buffer zones (50 feet wide) can be cut or formed with resistant trees between infected trees and healthy trees if space allows. In heavily infected areas, plant resistant trees to replace infected and removed trees. Other conifers and hardwoods can be planted in heavily infested areas since the dwarf mistletoes are relatively host specific.

JUNIPER MISTLETOE (Leafy or True Mistletoe)

Phoradendron juniperinum subsp. *juniperinum*

Hosts: Leafy mistletoes or *Phoradendron* spp. are host specific parasites. Juniper mistletoe is the only leafy mistletoe found in the region and its principal host is juniper.

Diagnosis and Damage: Juniper mistletoe plants grow on their host to about 40 to 80 cm in diameter. The mature stems are woody and their surfaces smooth. The nodes are closely spaced and somewhat swollen and plants may appear jointed. The stem contains chlorophyll and is the principal site of photosynthesis. The leaves, simple and oppositely arranged, are scale-like, and only about 1 mm long so the mistletoe plants look like dwarf mistletoes of other conifers. Juniper mistletoe does not usually cause mortality, although it can cause dieback of branch ends and decline of infected Utah juniper, especially in the southwest.

Biology and Disease Cycle: Juniper mistletoe is a shrubby, photosynthetic plant that is parasitic on the stems of juniper species from which it acquires water. Most of its organic nutrients are derived from the products of its own photosynthesis. Seeds are carried from tree to tree by birds. Initial infection usually occurs on a small branch and is followed by multiple infections on the same tree after the initial plant produces fruit. Seeds can germinate anywhere if temperature and moisture are suitable, but only seeds that lodge on thin bark of twigs and small branches of a juniper cause infection. Upon germination the young plant penetrates the bark with a haustorium, which then produces a radiating system of branches. Sinkers are produced from the branches and they grow into the outer surface of xylem. These sinkers become deeply imbedded in the wood as the tree grows larger, elongating as the host stem increases in radius. The life of mistletoe is limited only by that of the host and may extend to hundreds of years.

Conditions Favoring and Management of Disease: Trees growing in the open or in disturbed forests with open canopies are infected more frequently than those in undisturbed forests with closed canopies. This reflects the roosting preference of birds that eat mistletoe berries and disperse the seeds. Mistletoe may increase dramatically within a single tree where birds feed on berries, roost, and deposit seeds on twigs and branches.

Management options include removing infected branches, cutting 30 cm beyond the mistletoe shoots. Shoots on large limbs or trunks may be cut or broken off. This will reduce the parasite's demand for water but must be done repeatedly because new sprouts will replace those removed.

TOP: *Phoradendron juniperunum* brooms on juniper. (W. Jacobi)

CENTER: Bacterial wetwood /slime flux developed in crotch of American elm. (M. Schomaker)

BOTTOM: Bacterial wetwood /slime flux on cottonwood. (W. Jacobi)

Miscellaneous Pathogen-Associated Problems

BACTERIAL WETWOOD AND SLIME FLUX

Hosts: Bacterial wetwood is a disease most frequently seen in poplars (aspen and cottonwood), willow (especially globe), and elm. The disease also affects apple, ash, beech, birch, cherry, fir, honeylocust, linden, maple, mountain ash, mulberry, oak, sycamore, pines, and plum.

Diagnosis and Damage: The causal agents of bacterial wetwood have not been conclusively identified. However, several species of bacteria in the genera *Enterobacter*, *Klebsiella*, and *Pseudomonas*, which are often associated with wetwood, are thought to be directly involved. Research using affected elms suggests that an association of bacterial species, and possibly yeasts, act together to produce the complex of symptoms observed.

Wetwood typically develops in the center of the tree but may be found in the cambial area of stressed cottonwoods and globe willows. Wetwood is most easily recognized by the presence of a liquid that oozes from wounds, crotches, branch stubs, frost cracks, or other weak points in the wood or bark. As the liquid flows down the bark, vertical dark or light streaks remain. Oozing sap is initially colorless. After colonization by various bacteria and yeasts, the liquid becomes slimy and is often called slime flux. Foliage of severely affected trees sometimes wilts, and branches, sections of the trunk, or the entire tree may prematurely die. Symptoms of nutrient deficiency may appear due to poor water movement within affected trees.

Wetwood is important in forest trees harvested for lumber. Abnormal color and moisture in wetwood cause lumber to be devalued. Warping, increased drying time, and some cracking and splitting problems occur in boards cut form wetwood. The slime is toxic to cambial tissue so wounds can not close since the callus is killed. The wetwood also occurs in the cambium or just under the bark killing branches and stems of stressed cottonwoods and globe willows.

Biology and Disease Cycle: Very little is known about the bacteria and other microorganisms associated with wetwood and their transmission to healthy trees. Bacteria are assumed to enter trees through roots and thus it is hard to prevent. Insect vectors do not appear to be involved.

Conditions Favoring and Management of Disease: There are no chemical treatments effective for curing or preventing wetwood. Plant stress especially drought appears to favor disease development and appearance of symptoms. Adequate water during the summer months decreases wetwood symptoms. Recently transplanted trees may show symptoms of wetwood, especially in the cambial area, if not adequately established. In trees where oozing liquid has damaged the bark, the dead bark can be cut away to allow for drying and better wound closure. Use a smooth, continuous cut, to expose an elliptical area of bare wood. If this area exceeds 30 to 50 percent of the trunk circumference, the tree may not close the wound. The use of drain tubes is no longer recommended because the wet wood bacteria can move out into healthy tissue and decay fungi and insects may move into the tree.

TOP: Gummosis on Russian olive. (USDAFS)
CENTER: Sooty mold on American elm. (M. Schomaker)
BOTTOM: Sooty mold on piece of elm wood. (W. Jacobi)

RUSSIAN OLIVE DECLINE AND GUMMOSIS

Stress related

Hosts: Gummosis is a description of a symptom. Russian olive trees suffer from gummosis of unknown cause. Honeylocust and stone fruit species exhibit gummosis for various known reasons.

Diagnosis and Damage: Symptoms of gummosis include exuding of amber color gum from the trunk or branches of a tree that hardens with time. The tree will usually have other symptoms, including death of one or more branches and yellow leaves. Death of the entire tree may take one to seven years.

Biology and Disease Cycle: The causes of gummosis on Russian olive are unknown. It is believed that stress is a component, and that gummosis is an indication of root dysfunction. This root related stress may come from many sources including disease or adverse environmental conditions such as drought and sometimes signs of canker diseases can be found in trees exuding gum, but this is not always the case.

Conditions Favoring and Management of Disease: Conditions that stress the tree, environmental stresses such as too much or not enough water, extreme temperature fluctuations, or other stresses such as heavy pruning, or wounding will favor disease development. Management options include attempts to increase vigor of susceptible trees and decrease stresses to the tree.

SOOTY MOLD

Various genera and species of fungi

Hosts: All species of trees that are hosts to large populations of certain insects in the order homoptera which produce a liquid secretion called honeydew on which sooty mold fungus grows.

Diagnosis and Damage: Sooty molds vary in appearance from thin dark patches to irregular, blackish masses covering large areas. They may be found on any of the above ground surfaces of host plant such as trunks, leaves and upper surfaces of branches. Although widespread and conspicuous, these groups of fungi cause little damage to plants other than affecting, to a limited degree, the function of affected leaves. They can be a nuisance if growing on honeydew deposited on automobiles, outdoor furniture, etc.

Biology and Disease Cycle: Sooty molds are entirely superficial saprophytes that derive nourishment from insect and plant secretions. These fungi reproduce by ascospores, conidia, and sometimes by fragmentation of hyphae. The most common nutritional substrate for these fungi is honeydew, a liquid secretion released from the anus of aphids, soft scales, mealybugs, and some species of leafhoppers. Honeydews are complex mixtures of sugars, amino acids, proteins, other organic substances, and minerals. Droplets of honeydew often fall from insect-infested leaves or branches and stick to other plant parts or to objects such as automobiles, sidewalks, or outdoor furniture. Both honeydew and spores of sooty mold fungi are dispersed in water during rain. Thus dark fungal deposits often occur on plant parts and on other surfaces not infested by insects.

Conditions Favoring and Management of Disease: Plants vulnerable to high population of insects known to produce honeydew often exhibit a layer of sooty mold. Infrequent, minimal amounts of rainfall allow honeydew to build up on plants creating perennial cases of sooty mold on many trees. Management options include controlling levels of honeydew producing insects on host plants, and spraying or washing down plants to remove honeydew. Both of these options are not particularly feasible with large, street trees.

conditions generally associated *with roots*

Root Diseases

Root diseases can be caused by fungi that only attack the fine feeder roots and cause death and decline of the tree or by fungi that decay and kill the larger roots. Root diseases are doubly important in forest situations since many of the fungi can attack and kill living trees but remain living on the dead roots and stump to attack future trees.

Diagnosis Hints: Diagnosis of root diseases is difficult and can seldom be accomplished without excavating and looking for symptoms and signs on roots. Above ground symptoms include yellowing of foliage, reduced tree growth, reduced leaf size, thinning of crown and death. The same symptoms are found with over watering, drought, bark beetles, vascular wilts, and phytoplasma diseases, to name a few. Examining the root crown and larger roots for stain, decay, mycelial fans, rhizomorphs, resin, and fruiting bodies characteristic of each disease is necessary for diagnosis. Look for these symptoms and signs on recently killed or dying trees since secondary insects and fungi destroy the signs and symptoms a year or so after death. It is not always necessary to figure out exactly what fungus is involved. It is necessary to determine if it is an abiotic or fungal caused problem so that the appropriate management options are chosen.

ANNOSUM ROOT DISEASE

Heterobasidion annosum (= *Fomes annosus*)

Hosts: Most conifers. The most vulnerable conifers in the region are true firs, but Annosum may be found on pines and Douglas-fir.

Diagnosis and Damage: Symptoms of annosum root rot are common to other root pathogens and are not diagnostic. Common symptoms include general decline of trees alone or in groups and windthown trees with decayed roots. Infected trees have unusually short, often yellowish-color needles in tufts at twig tips, and heavy crops of cones. Infected roots are initially resin soaked and brownish-red color. They eventually becoming a white stringy mass of decayed tissue. Shelf-like fruiting structures (conks) are white when grown in darkness and have a light brown upper surface if exposed to light. They can be found under duff at tree bases, and under upturned tree stumps. The bottom of the fruiting body is white and has a narrow band at the edge with no pores.

All hosts are subject to root death and decay by *H. annosum*, but the nature of further damage varies with the host. Pines may be killed young, or if older, death may occur within several years. Young fir and cedar trees may also be killed within a few years, and older trees are usually not killed outright, but develop decay in heartwood that increases susceptibility to insect attack and wind damage. Root and butt decay of trees in recreational areas creates hazard and sometimes causes damage to people and property when diseased trees fall unexpectedly.

Biology and Disease Cycle: The fungus can survive for several decades in the roots of dead hosts, so trees are often killed in a gradually expanding clump as nearby plants are infected through root contact or grafts. The most common means of initial entry

TOP: Annosum basidiocarp. (W. Jacobi)
CENTER: Annosum stain in wood cross section. (W. Jacobi)
BOTTOM: Root decay symptoms. (W. Jacobi)

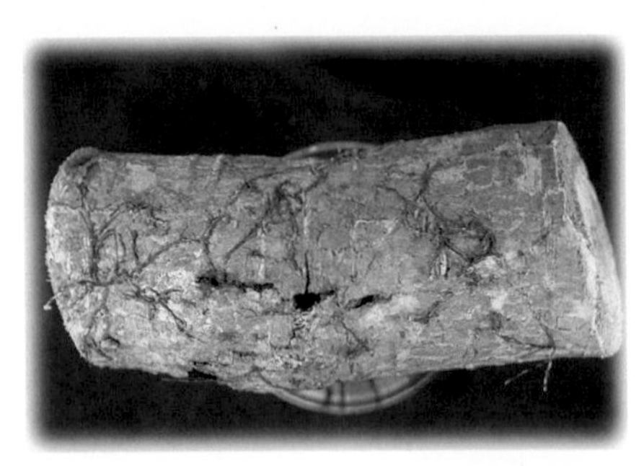

TOP: Armillaria mushroom basidiocarp. (M. Schomaker)
SECOND: Armillaria rhizomorphs on aspen root piece. (W. Jacobi)
THIRD: *Armillaria* on subalpine fir. (W. Jacobi)
BOTTOM: *Armillaria* mycelial fans on xylem of conifer. (W. Jacobi)
BOTTOM RIGHT: *Armillaria* infection with rhizomorphs producing resin area on root impregnated with soil.

into a new area is via airborne spores that germinate on freshly cut stumps or other fresh wounds that expose sapwood.

Conditions Favoring and Management of Disease: The disease is common in forest situations where fire management and selective harvesting have allowed susceptible species to dominate. Soil conditions that allow drought stress are related to high incidence of the disease. Landscape and ornamental plantings are seldomly affected.

Management options include planting less susceptible species, immediate treatment of freshly cut stumps with products to prevent infection, and removal of hazard trees in recreation sites. Attempts to increase vigor of susceptible trees is advised as trees weakened by adverse environment or suppressed by dominant neighbors are most likely to be attacked.

ARMILLARIA ROOT DISEASE

Armillaria spp.

Hosts: Nearly all conifers and many hardwoods

Diagnosis and Damage: Loss of needles creates a thin crown; infected conifers may maintain only two to three years of leaf growth, rather than the five to seven years held by healthy trees. Reduced shoot growth in conifers and hardwoods are apparent in older trees; younger trees are usually killed too quickly by the fungus for noticeable loss of growth on shoots. In advanced stages, chlorotic leaves occur on terminal branches. Often, the disease will weaken the host to the point where it becomes susceptible to attack by bark beetles, turning the foliage red. As the tree approaches death, it is common to see an increased crop of undersized cones.

Some conifers, e.g. Douglas-fir, produce resin at the tree base when attacked by Armillaria. These areas, exuding resin, usually become evident when the fungus has moved up the roots to the root collar. Decayed wood initially is gray- to brown-stained and appears water-soaked. Advanced decayed wood is a yellow-brown color and a white stringy rot. The most prominent diagnostic tool in differentiating between Armillaria and other root diseases is the presence of white mycelial fans in the cambium and bark of the roots and stems. These fans will be present initially only in the roots, but are very common in the stems of dead trees. Fruiting bodies, in the form of honey-colored mushrooms, may be found at the base of the tree during wet periods in the fall. *Black rhizomorphs*, root-like structures that the fungus uses to penetrate the soil in order to colonize other hosts, can be an additional diagnostic tool. Damage is from loss of growth, mortality, and structural weakness leading to hazard.

Biology and Disease Cycle: Armillaria spreads from a host tree or stump to an uninfected live tree in one of two ways. Root systems of the two trees come into contact and knit together (root grafting), or *rhizomorphs* (root-like fungal structures) grow through the soil to susceptible tree roots and establish new infections. Once contact has been made, the fungus spreads along the root system of the uninfected tree, penetrating

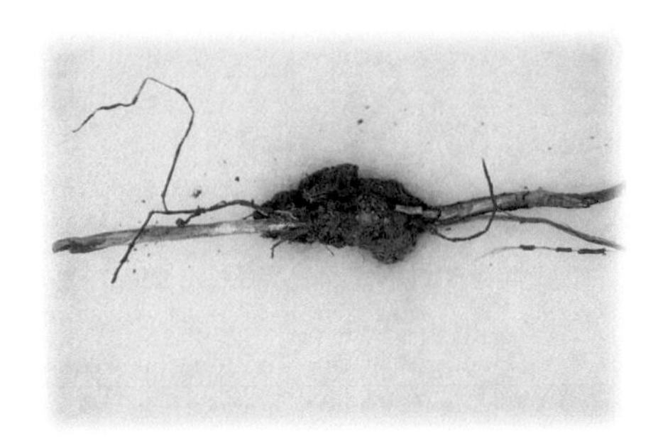

its bark and entering the cambium. The fungus continues to spread along the root until it reaches the root collar, where it spreads to other primary roots. Death of the host occurs when the tree is girdled at the root collar or when bark beetles attack or windthrow occurs. *Armillaria* survives as a saprophyte on dead roots and stumps for 20 to 30 years. Coming in contact with the old roots and stump can infect new trees.

Conditions Favoring and Disease Management: Reducing the stress experienced by individual trees through reducing density and improving growing conditions reduces susceptibility to attack by Armillaria. A long-term control option is shifting the composition of species in a given area to represent more tolerant and resistant species.

BLACK STAIN ROOT DISEASE

Sexual stage: *Ophiostoma wageneri*
Asexual stage: *Leptographium wageneri* (various subspecies) (formerly *Verticicladiella wageneri)*

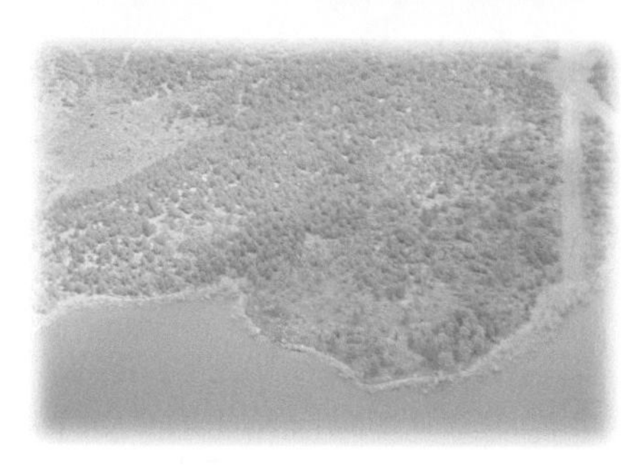

Hosts: In this region, black stain is primarily found on pinyon (*L. wagneneri* var. *wagneneri*) and occasionally on ponderosa pines (*L. wagneneri* var. *ponderosum*). It also occurs on Douglas-fir (*L. wagneneri* var. *pseudotsugae*) and lodgepole (*L. wagneneri* var. *ponderosum*) in other areas of the west.

Diagnosis and Damage: External symptoms are typical of a vascular wilt, and may become visible several years before death. These include a thinning crown and reduced needle length. In some cases, affected trees will produce a large crop of small cones. As the symptoms worsen, the foliage will become chlorotic and fade to brown. The fungus itself will be visible in the wood and/or roots as a dark stain that noticeably proceeds along (but not across) the wood's annual rings. Other blue stain fungi grow in ray cells giving a radial stain pattern. The stain will be present in one or more of the roots before it becomes visible in the trunk. It is common for trees weakened by black stain to become infested with bark beetles, in which case small holes will be visible on the trunk, often exuding resin. The disease is common in the pinyon forests of the southern west slope of Colorado.

Black stain generally kills affected trees within several years. Over time, the disease can produce patches of mortality several acres in size. Additionally, secondary infestation by bark beetles may allow the beetle population to grow to the point that healthy trees are attacked as well as those affected by root disease.

Biology and Disease Cycle: Transmission of the black stain fungus is generally via root to root contact and direct growth through the soil over short (inches) distances. Insects are known to carry the fungus from host to host in both ponderosa pine and Douglas-fir, but vectors have not been identified yet for the fungus on pinyon pine. Once the fungus enters a host root, the fungus begins to grow inside the xylem. The growth of the fungus eventually clogs these cells so that water transport to the leaves is diminished. Eventually, the tree dies from water stress and/or attack by beetles.

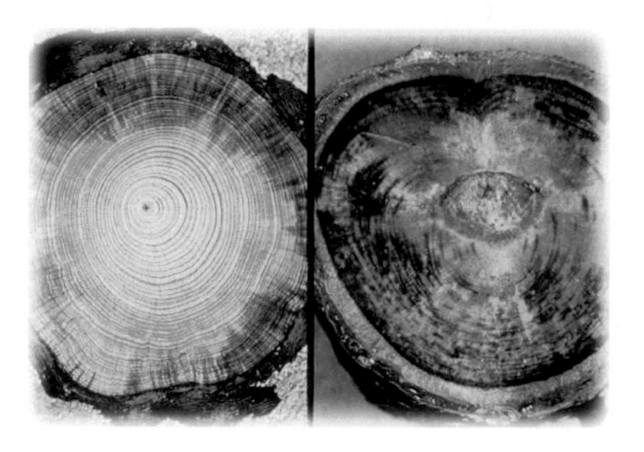

TOP: Black stain in pinyon pine. (W. Jacobi)
SECOND: Black stain disease centers (gray trees). (W. Jacobi)
THIRD: Black stain in pinyon pine. (W. Jacobi)
BOTTOM: Black stain in cross section of pinyon pine root. (W. Jacobi)
BOTTOM LEFT: Blue stain on left and Black stain on right. (W. Jacobi)

Conditions Favoring and Management of Disease: Increased infection of various conifers by the pathogen has been connected to various environmental conditions, including high soil moisture, low temperature, disturbances caused by harvest and road construction, and air pollution. No cultural practices have proven useful in reducing the spread of the disease although trenching to cut root connections would seem reasonable. Planting non-susceptible species is the best tactic in high disease incidence areas.

TOP: Crown rot symptoms on apple tree.
CENTER: Root collar rot on honeylocust. (W. Jacobi)
BOTTOM: Root collar rot, healthy honeylocust collar tissue is creamy white color, diseased is yellow to tannish-brown. (W. Jacobi)

PHYTOPHTHORA ROOT AND CROWN ROTS

Phytophthora spp.

Hosts: Common hosts of root rots caused by *Phytophthora* are honeylocust, apple and crabapple, stone fruits and junipers. However, many other species are infrequently infected.

Diagnosis and Damage: The most obvious symptoms of root and crown rot are stunted, sparse and chlorotic leaves, premature coloration and drop of leaves in fall, and twigs and branch dieback. By the time foliar symptoms develop, the rot canker may extend halfway or more around the stem of the plant. In early stages, the diseased bark is firm and intact while the inner bark is slimy and may produce a moist, gummy exudate. Later the affected area becomes shrunken and cracked.

Collar rots are lethal to many of the plants that they affect, especially single stemmed trees with root collar cankers. Affects on multistemmed plants are limited to the stems and roots infected.

Biology and Disease Cycle: Phytophthora fungal species exist as spores in soil or killed plant material, but most species do not grow to a significant extent without a host plant. Species that cause root and crown rots enter host tissue near the root collar via wounds or the succulent parts of small roots. Fungal spores are motile in water and are attracted to the exudates from roots and other plant parts. Wounded root collars and the surfaces of tender, succulent root tips are suitable for infection. There they cause lesions at and on the adjacent parts of major roots, or smaller roots of trees or shrubs. The pathogen destroys roots or the physiological connection between roots and stem, thereby causing the failure of host.

Conditions Favoring and Management of Disease: Flooded or water-saturated soil and low temperatures favor root rots caused by Phytophthora. Such conditions not only promote reproduction and dispersal of the causal fungi but also promote the susceptibility of plant roots. Roots stressed by reduced oxygen supply in waterlogged soil chemically attract zoospores of Phytophthora, and root resistance to pathogens is impaired. Management options include rapid reversal of conditions that are promoting the disease. Roots that have been damaged are not repairable, but those that have not been damaged do not need to be removed from the predispositioning conditions. Avoid overly saturating root zones with poor drainage or wounding plants at the crown or below.

ROOT COLLAR ROT

Various soil organisms

Hosts: Honeylocust

Diagnosis and Damage: Root collar rot symptoms include premature fall coloration of a portion of the tree, and discolored bark and wood at the ground line. Early fall coloration of a portion of the tree may indicate a large amount of damage. Small drops of gum on the stem near the ground or farther up the stem usually indicate that girdling by collar rot has occurred below that point. Loose bark and discolored (yellow to brown instead of white) wood, just below the bark, indicates initial collar rot and is the most indicative symptoms. Extensive death and discoloration of bark and wood can occur over several months. Thyronectria or Tubercularia cankers at the tree base usually indicate collar rot is or was active in the past. Symptoms of root collar rot need to be recognized promptly because the disease can kill trees rapidly. Root collar rot is quite common in urban areas and is responsible for the death of many honeylocust.

Biology and Disease Cycle: The biology and disease cycle of this disease involves the interaction of many soil organisms and the exact process is unknown. Frequent irrigation that keeps the soil too wet at the tree base, allowing soil microorganisms to kill the bark and cambium just below ground-line, appear to be the main cause. Thyronectria or Tubercularia may infect the weakened tree above the area previously killed by collar rot.

Conditions Favoring and Management of Disease: Frequent irrigation of turf is a factor that creates conditions favoring this disease. Root collar rot can be prevented by keeping the soil at the base of trees as dry as possible. Placing sprinklers and sprinkler heads far away from trees will keep the least amount of water from falling on the tree stem and at the tree base. Remove flowers, turf or other vegetation that requires frequent irrigation from around the tree's base and replace with small gravel or mulch. When planting honeylocust trees, place the root ball above grade a number of inches to allow for drainage away from the tree trunk.

Abiotic Disorders

OXYGEN STARVATION OF ROOTS/OVERWATERING

Hosts: All plants

Diagnosis and Damage: Leaves may wilt, be distorted and smaller than normal. They may also drop prematurely, yellow from the base to the top of the plant and from the inside to the outside. Drooping and other drought stress symptoms as well as chlorosis may develop. Petioles may droop while the leaves remain firm and full of fluid. The entire plant may wilt or exhibit shoot dieback. New roots may form on stems. Growth is slowed and plants can be killed by this condition.

Oxygen starvation also damages or kills roots, inhibiting their function. Depending on the soil, the cause of oxygen starvation, the health and age of a plant and the care it has received, symptoms of oxygen starvation may appear quickly or slowly. Plants may be more susceptible to secondary disease and insect problems as a result of stress from oxygen starvation.

Biology and Disease Cycle: Normally, a certain percent of soil pore space, the space between the solid particles, is occupied by oxygen. The amount of oxygen found in a soil depends on the soil texture, its structure, permeability, amount of water and the degree of compaction.

Oxygen starvation occurs when soil oxygen is displaced by water or another gas, is unable to penetrate soils because of an impermeable barrier, like black plastic, or soil is added above the roots. Soil compaction, the pressing together of soil particles, reduces the amount of air pores present and reduces permeability so that water and air infiltration is reduced.

The first response to decreased oxygen levels occurs in the roots. Their function is either greatly reduced or death occurs. Poorly functioning or non-existent roots cannot absorb water or nutrients. Plant pathogens usually attack the stressed roots and cause extensive damage.

Conditions Favoring the Disease and Management: Conditions that favor development of oxygen starvation are heavy, compacted, poorly drained clay soils, periods of high rainfall, snowmelt, excessive irrigation, or flooding during the growing season. Plastic or other non-breathable material laid over the root zone, and natural gas leaks in

TOP: Overwatered oak with yellow leaves.
SECOND: Overwatered junipers.
THIRD: Tree mortality from raising the soil depth ten inches.
BOTTOM: Asphalt caused oxygen and water starvation.

the vicinity will decrease oxygen levels and cause symptoms to develop. Increasing the soil depth even a few inches over the tree's root zone will restrict oxygen.

If faced with a poorly drained site, choose plants that tolerate reduced oxygen levels. Prepare heavy clay soils before planting by incorporating adequate organic matter (approx. 3 cubic yards per 1000 square feet). Don't work wet soils, this damages soil structure, and avoid foot and vehicular traffic over root zones when possible. Mulch to reduce compaction from raindrop and irrigation water impact. Some sites can be improved by installation of drainage tile, or by construction of berms or raised beds above poorly drained sites. Avoid the use of plastic barriers; use weed barrier fabrics instead. Remove existing plastic barriers, and replace with weed barrier and/or bark mulch. Removal of plastic barriers may be impractical for various reasons. In this case, punch holes at regular 4 to 6 inch intervals through the material so oxygen can move into the soil. If overwatering occurs, gradually decrease the amount applied and/or the intervals in between applications.

Root Feeding Insects

BLACK VINE WEEVIL

Otiorhynchus sulcatus (F.)

Coleoptera: Curculionidae

TOP: Black vine weevil injury to euonymus.
BOTTOM: Black vine weevil adult. (J. Capinera)

Hosts: Several shrubs, particularly euonymus, lilac, *Taxus*, and rhododendron; occasionally lilac

Damage and Diagnosis: Adult weevils feed on leaves at night, producing characteristic notching wounds along the leaf margin. When weevils are abundant, plants may be heavily defoliated.

Larval stages, pale-colored legless grubs, feed on plant roots causing dieback problems. These injuries may be more destructive than those caused by adult feeding. In areas of the Pacific Northwest and Midwest this is often considered to be the most important nursery insect pest; problems have been minor in central Rockies.

The adults are dark-gray or black snout beetles of about 1/3 inch length with wing covers marked with gold flecking. They occasionally are minor nuisance invaders of homes in late summer and fall.

Life History and Habits: The black vine weevil spends the winter usually as a larva, in the soil around the root zone of plants on which it feeds. Occasionally some adults may survive winters if they find the shelter of homes. Black vine weevil occasionally occurs as a nuisance invader of homes in summer and fall.

Larvae resume feeding in spring and can extensively damage roots during May and June. When full-grown, they pupate in the soil and adult weevils start to emerge in mid-June. Black vine weevil adults feed on the leaves of various plants during the night, and cause characteristic notching wounds that sometimes resemble grasshopper injury. After about two weeks, the females begin to lay eggs around the base of plants. Eggs begin to hatch in midsummer and the legless larvae feed on plant roots until cold weather temporarily stops development. There is one generation produced per year.

Related Species: The strawberry root weevil, *Otiorhynchus ovatus* (L.), and the rough strawberry root weevil, *Otiorhynchus rugosostriatus* (Goeze), are two related root weevils that also are common. Both are smaller in size, but generally the same shape as the black vine weevil. Although they have similar habits neither is as damaging to plants. However, the strawberry root weevil invades homes more frequently, particularly at higher elevations and during midsummer periods when hot, dry conditions persist.

Management: Black vine weevil has proved to be quite resistant to insecticides and difficult to control. Best control of adult stages has been with certain pyrethroids, notably Talstar and Samurai. Bendiocarb (Turcam, Ficam) has also been widely used in some parts of the country.

Larval stages are even more difficult to control. However, research indicates that certain species of insect-parasitic nematodes can be very effective, at least in container-stock. Nematodes in the genus *Heterorhabditis* (e.g., Oti-Nem) have been particularly effective, although these are not commercially available at present. Nematodes in the genus *Steinernema* (e.g., Exhibit, BioSafe) have also been quite promising and are now being sold for this purpose. Talstar (bifenthrin), incorporated into potting media, is effective and limits movement of this insect in nursery stock.

Specific recommendations for black vine weevils appear in the supplement on insect control recommendations marked **Black vine weevil**.

WHITE GRUBS/SCARAB BEETLES
Tenlined June Beetle *Polyphylla decemlineata* (Say)
May/June Beetles *Phyllophaga* spp.

Diplotaxis obscura LeConte
Dichelonyx spp. and others
Coleoptera: Scarabaeidae

ABOVE: White grub.

Hosts: Larvae and adults may feed on roots and foliage, respectively, including both conifers and deciduous plants.

Damage and Diagnosis: Both adult and larval stages of some scarab beetles can damage trees and shrubs. Adults chew on foliage, although serious defoliation is very rare. They are moderate to large beetles of generally oval form and also known by names such as "May", "June" beetles or "chafers".

Larvae are root feeders and occasionally cause serious root pruning. They are c-shaped "white grubs", some of which may reach the size of a half dollar.

Life History and Habits: Among the many scarab beetles associated with trees there is a range of life histories. All lay their eggs in soil and larvae develop feeding on plant roots. Some species, notably those in the genera *Phyllophaga* and *Cyclocephala*, are particularly damaging to grasses and can be important turf and rangeland pests. Other white grubs, including *Phyllophaga* and *Polyphylla* species, have a more general feeding habit and may damage roots of shrubs and trees. Larval habits of *Diplotaxis* and *Dichelonyx* are poorly understood.

Adults fly at dusk and rest on trees and shrubs to feed. Many are easily disturbed during feeding, dropping readily. Many scarab beetles are also attracted to lights.

Length of life cycles vary but annual life cycles predominate. However, the larger species (*Phyllophaga, Polyphylla*) require two or three years to complete. Adults of early emerging scarabs (*Phyllophaga, Dichelonyx*) emerge in late spring (May, June); others have greatest activity in July and early August (*Polyphylla*).

Related Insects: The family of scarab beetles (Scarabaeidae) is very large, containing approximately 1400 species in North America, few of which are injurious to plants. Adults of most species are dark-colored, oval and heavy-bodied. The larvae of all are known as "white grubs" and most develop as scavengers of animal manure, decaying plant matter and carrion. In this role, as macrodecomposers, white grubs include some of the most important insects involved in the nutrient cycling.

A scarab beetle that commonly visits tree wounds in summer is the bumble flower beetle, discussed later in this publication.

One of the most notorious of all scarabs is the Japanese beetle, *Popillia japonica* Newman. In eastern North America, adults feed on a wide variety of woody plants, particularly favoring roses, crabapple and other roseceous hosts. Larvae are also serious white grub pests of turfgrass in these areas.

The status of the Japanese beetle in the region is uncertain. Some trapping has indicated that it is possibly widespread, but at very low populations. It is also still unclear whether regional infestations can sustain themselves or result from repeated introductions on midwestern nursery stock. However, the detected presence of some has resulted in regulatory concerns regarding shipments of plant material to states where Japanese beetle has not yet been detected.

Management: Most white grubs, including plant damaging species, will feed on decaying organic matter and may increase in areas where there is heavy use of compost or animal manures. Species that normally damage grasses may move to adjacent roots of woody plants if the grasses are killed after egg laying. This has sometimes caused problems in nursery plantings.

Insect parasitic nematodes in the genus *Heterorhabditis* can be effective biological controls, applied to moist soil as a drench. Imidacloprid (Merit, Marathon) and halofenzimide (MACH-2) are among the more effective insecticides for white grubs. Adult beetles rarely, if ever, warrant control on foliage.

conditions generally associated with *fruiting structures*

Insects that Feed on Flowers, Fruit, and Seeds

CHERRY CURCULIO

Anthonomus consors (Dietz)
Coleoptera: Curculionidae

Hosts: Sour cherry, occasionally chokecherry

Damage: The adults chew small holes in the base of flowers and cause abortion of developing fruit. Larger fruit are pitted by this injury. Eggs are inserted into fruit and larvae tunnel the fruit, ultimately feeding on the pit.

Life History and Habits: The cherry curculio spends the winter as an adult snout beetle ('weevil') around trees infested the previous season. These small (1/8-inch) brown "snout beetles" fly to trees in spring as flower buds form. They feed on the buds and flowers, making small chewed punctures. After several weeks, the developing fruits appear and females insert eggs into them. The young larvae, pale legless grubs, feed and develop primarily within the pit. Larvae then pupate within the pit and emerge from the cherry fruit as they ripen. There is one generation per year.

Related Insects: Related species of small weevils attack plum [the plum gouger, *Coccotorus scutellaris* Leconte)]) and apple [the apple curculio, *Anthonomus quadrigibbus* (Say)], but rarely cause serious injury to these plants. The plum curculio, *Conotrachelus nenuphar* (Herbst), a fruit infesting weevil that seriously damages plums, apples, and apricots in eastern states, does not occur in the region.

The apple curculio, feeds on *Amelanchier* (serviceberry, shadbush) and apple. It is difficult to detect since they remain motionless, with beak raised up, when disturbed and closely resemble bud scales or dried petals). The apple curculio makes small holes in leaves and feeding punctures in fruit. Fruit injuries made in late spring heal over by late season feeding results in dry, sunken areas on the fruit. Similar to the cherry gouger, the females excavate holes in which eggs are laid. Larvae and pupae within the apple develop only within the shed apples (growing apples kill them.) Only some varieties of apples are susceptible, including Delicious. Cleaning up June drop apples can control the insect, although not eliminate spring feeding damage during the first year.

The plum gouger feeds on hard types of plums. Adult weevils make numerous feeding punctures in the developing fruit, many of which result in a flow of clear ooze from the wound. Eggs are laid in some of the punctures and the larvae develop within the pit. This species apparently spends the winter as an adult within the pit. Removal of damaged fruit should help control this species.

Management: Several types of parasitic wasps commonly attack the cherry curculio.

The adult weevils can be shaken from trees and collected on sheets. However, they are difficult to see and 'play dead' when disturbed. To prevent the insect from developing, damaged fruit should be picked and destroyed as it is observed.

TOP: Cherry curculio on fruit.
BOTTOM: Plum gouger adult on plum.

Little work has been done on the chemical control of this insect. Effective insecticides, applied before flowering, should provide control.

ASH SEED WEEVIL

Lignyoodes (= Thysanocnemis) helvolus (LeConte)
Coleoptera: Curculionidae

Host: Ash

Damage and Diagnosis: The grub-like larvae develop in the seeds of ash, largely consuming them. In Fall, as the larvae leave the tree to pupate they may attract attention and concern as they drop from the seeds.

Adults are small, yellow-gold weevils active in late spring and early summer. The larvae are typical legless weevil grubs, pale yellow and only about 1/10 inch in length when full grown.

Life History and Habits: Ash seed larvae leave seeds during Fall and early Winter, moving into the soil where they pupate. Adults emerge in late spring and early summer. Eggs are laid in the developing seeds during summer and the larvae feed on the seed. One or two larvae will be found in an individual seed, which may be completely consumed. They then drop to the ground to pupate. There is one generation produced per season.

TOP: Larva of the ash seed weevil.
BOTTOM: Rose curculio.

ACORN WEEVILS

Curculio spp.
Coleoptera: Curculionidae

Host: Oak

Damage and Diagnosis: Larvae develop within acorns of oak, largely destorying the seeds. Oozing fermented sap from infested acorns also is sometimes observed. The grub-like larvae sometimes attract attention and concern as they migrate out of acorns to pupate. Adults of *Curculio iowensis* (Casey), known from Colorado, are about 1/4-in, mottled tan, and have a pronounced snout.

Life History and Habits: The adult insects are active in summer and females chew small holes in developing acorns to lay eggs. Larvae then develop within the acorn feeding on the kernel for about three months. They then cut their way out of the acorn at about the time of nut fall, drop to the ground and burrow into the soil to pupate. There is one generation per year.

ROSE CURCULIO

Merhynchites bicolor (F.)
Coleoptera: Curculionidae

Hosts: Rose, particularly wild rose

Damage and Diagnosis: The rose curculio damages roses by making feeding punctures into flower buds, resulting in ragged flowers. During periods when the buds are not common, the feeding occurs on the tips of shoots, killing or distorting the shoot. A "bent neck" condition of rose appears to often be caused by feeding rose curculio feeding wounds in developing stems.

Life History and Habits: The adult stage of the rose curculio is a red snout beetle (weevil) with a black "beak". They become active in late spring and lay eggs in developing flowers. The larval (grub) stage feeds on the reproductive parts of the flower. Blossoms on the plant, including those clipped off by a gardener, are suitable for the insect to develop. When full grown, the grubs fall to the soil and form an underground cell; pupating the following spring. There is one generation per year.

Management: Regular hand picking, and removal, of spent blossoms will prevent populations from developing. However, other brambles are hosts and can serve as additional sources of the insect.

Adult weevils drop readily from plants and feign death. Where plantings allow, shake plants over a collecting container to speed hand collection of rose curculio. Rose curculio can be controlled with most insecticides, applied during late May and June.

CHOKECHERRY GALL MIDGE

Contarinia virginianae (Felt)
Diptera: Cecidomyiidae

Hosts: Chokecherry

Damage and Diagnosis: The larvae distort developing chokecherry fruit causing them to enlarge and become hollow. Within this galled fruit the bright orange-red maggots develop and feed.

Life History and Habits: Winter is spent in the pupal stage, around the base of previously infested chokecherries. Adults emerge in early spring, around the time of blossoming, and females lay eggs in the flowers. After eggs hatch the young maggots tunnel into developing chokecherry fruit, causing it to become enlarged and hollow. The maggots feed within the fruit until midsummer. After they have finished feeding, they drop to the ground and pupate. There is one generation per year.

Management: Damage by this insect is negligible. No insecticides are registered for use on chokecherry grown for edible fruit. Thoroughly hand-picking infested fruit before the maggots emerge should reduce populations the subsequent season.

ABOVE: Chokecherry gall.

ROSE MIDGE

Dasineura rhodophaga (Coquillet)
Diptera: Cecidomyiidae

Hosts: Rose, particularly hybrid tea types

Damage and Diagnosis: The small maggot stage of the rose midge feeds by making small slashes in developing plant tissues to suck the sap. Developing flower buds are usually killed or distorted by this injury. "Blind" shoots, where no flower buds appear to form may also be the result of rose midge damage.

Life History and Habits: The rose midge overwinters in the pupal stage in the soil. The adult stage, an inconspicuous, small fly, emerges in late spring, sometimes after the first crop of blossoms. The adult stage lives only 1 or 2 days, but during this time the females lay numerous eggs under the sepals, in opening buds and in elongating shoots. Hatching larvae slash plant tissues and feed on sap. Eggs hatch in a few days, and the larvae feed for about a week before dropping to the soil to pupate. The complete life cycle can take about two weeks, with numerous generations occurring during a growing season.

Management: To avoid introducing rose midge on soil, purchase bare root roses or roses that were recently potted by local nurseries in the spring from bare root stock. Infested plantings should be examined every few days and all damaged buds trimmed and removed.

The rose midge is difficult to control with insecticides. (Permethrin and Cygon have been effective against the related honeylocust podgall midge but have not been tested against rose midge.) Treatments of soil at the base of the plant with insect parasitic nematodes or soil insecticides such as Diazinon may reduce populations by killing the insect as it attempts to pupate.

WALNUT HUSK FLY

Rhagoletis completa Cresson
Diptera: Tephritidae

Hosts: Walnut, occasionally late maturing varieties of peach

Damage and Diagnosis: The larvae develop under the skin of walnut, making meandering tunnels. Injured fruit turns dark and rots readily causing staining of the nut. Rarely this insect may also be found tunneling the flesh of late varieties of peach.

Life History and Habits: Walnut husk fly is a typical "picture-winged fly", housefly sized with dark patterned markings on the wings. Winter is spent in the pupal stage, buried shallowly in soil near previously infested trees. Adults emerge in mid to late July and females first feed on honeydew and other foods for about two weeks as eggs mature. Eggs are inserted under the skin of the fruit and hatch within a week. Larvae tunnel the husk of walnut for about one month after which time they drop to the ground to pupate. There is one generation produced per year.

Related Species: The apple maggot, *Rhagoletis pomonella* (Walsh), is commonly found in native hawthorn fruit and more recently apple infesting strains have become established in parts of the Front Range. In southwestern Colorado counties the western cherry fruit fly, *Rhagoletis indifferens* Curran is a pest of sweet cherry fruit.

Management: Activity of adult walnut husk fly can be easily monitored by the use of yellow sticky traps. The use of several such traps may be sufficient to provide adequate control. Insecticides applied in late July, when trap captures indicate activity by this insect, can provide control.

APPLE MAGGOT

Rhagoletis pomonella (Walsh)
Diptera: Tephritidae

Hosts: Apple and large crabapples are the primary hosts of introduced strains found in eastern Colorado. European plum and cherries are infrequently infested. West Slope strains are native and develop on hawthorn.

Damage and Diagnosis: Most damage results when the young maggots tunnel fruit, producing meandering brown trails that hasten rots. Egg laying by the adults involves small puncture wounds to the fruit surface that cause dimple-like distortions. Apple maggot is of consider concern as a pest of regulatory concern as movement is restricted of apples produced from known-infested areas to non-infested areas.

Life History and Habits: Apple maggot is a "picture-winged fly", housefly sized with dark patterned markings on the wings. During winter the apple maggot is in the pupal stage, buried shallowly in soil near previously infested trees. Adults emerge in early summer and females first feed on honeydew and other foods for about two weeks as eggs mature. Eggs are inserted under the skin of the fruit and hatch within a week. Larvae feed within the fruit for 3 to 4 weeks before dropping to the soil to pupate. Peak egg laying tends to occur during mid to late July. There is one generation produced per year.

Management: Apple maggot is one of the few insects that can be well controlled by trapping. Newly emerged adults are highly attracted to yellow sticky traps. Females with matured eggs can be trapped by sticky red spheres - "super apples" - that are highly attractive to them. In many cases the use of only a couple such traps per tree can adequately control this insect.

Apple maggot is most damaging to soft, early maturing varieties. Survival is reduced in firm-fleshed late varieties. A cover of vegetation under trees has been shown to reduce survival of the insects moving from trees to pupate in soil.

Apple maggot can be well controlled by cover sprays of several insecticides that are labelled for use on apple. As with codling moth, timing is critical and should coincide with peak periods of egg laying. The use of yellow sticky cards is very useful for making this determination, the dark bands on the wings of the flies being distinctive.

Related Species: The walnut husk fly, *Rhagoletis completa* Cresson, is widespread in the West, where it feeds on walnuts. It also rarely develops in certain late varieties of peaches. West of the Continental Divide the western cherry fruit fly, *Rhagoletis indifferens* Curran, can be a serious pest of sweet cherries. As with apple maggot, the adult female flies "sting" the fruit producing small puncture wounds. The developing maggots chew through the flesh of the fruit and infested berries are misshapen, undersized and mature rapidly. In some areas (e.g., parts of Utah) strains have developed in some regions that will attack and tunnel apples.

TOP: Western cherry fruit fly. (Oregon State University)
BOTTOM: Apple maggot injury to apple.

WESTERN CHERRY FRUIT FLY

Rhagoletis indifferens Curran
Diptera: Tephritidae

Hosts: Cherries, rarely apple

Damage and Diagnosis: The adult female flies 'sting' the cherry fruit with her ovipositor, producing small puncture wounds. Eggs are often laid in the punctures and the immature maggots chew through the flesh of the fruit. Infested berries are misshapen, undersize, and mature rapidly. Western cherry fruit fly is only known from southwestern Colorado.

In most areas of North America, the western cherry fruit fly limits damage to cherry, including wild types. However, strains have developed in some regions (e.g., Utah) that attack and tunnel into apple.

Related Species: The western cherry fruit fly is closely related to the apple maggot (*Rhagoletis pomonella* Walsh), a serious apple pest in the eastern United States that has become established in parts of Colorado. Controls are similar to western cherry fruit fly. Another related species, the walnut husk fly (*Rhagoletis completa* Cresson), is widespread in western North America, primarily feeding on the husk of walnuts. Rarely it may develop in late ripening varieties of peach.

Life History and Habits: The western cherry fruit fly spends the winter in the pupal stage around the base of previously infested trees. In late June, the flies emerge and feed on aphid honeydew and other fluids, including oozing sap from wounds made by fruit punctures. The females then insert eggs into the fruit.

The immature maggots feed on the fruit, particularly the area around the pit. They become full-grown in two to three weeks and drop to the ground to pupate. In most areas there is one generation per year. However, a few flies may emerge and produce a second generation if susceptible fruit is available.

Management: Adult flies are easily trapped by yellow sticky cards or sticky red spheres. However, it has not been shown that trapping can adequately control the western cherry fruit fly.

Prematurely ripening cherries, often infested with developing fruit flies, should be picked and destroyed before the insects drop to the soil. Control of aphids, that produce honeydew fed on by the flies, can reduce egg laying.

The western cherry fruit fly can be controlled by several insecticides applied during periods when the adults are mating and beginning to lay eggs. Yellow cards or red spheres with a sticky covering can be used to determine when adult flies are present.

TOP: Boxelder bug adult.
SECOND: Boxelder bug egg mass.
THIRD: Boxelder bug adults and nymphs in mass.
BOTTOM: Golden raintree bug. (Whitney Cranshaw)

BOXELDER BUG

Boisea (= Leptocoris) trivittata (Say)
Hemiptera: Rhopalidae

Hosts: Boxelder, sometimes silver maple

Damage and Diagnosis: Nymphs and adults usually feed on sap from seeds, flowers and leaves, but cause little damage to trees. Occasionally they feed on developing fruits, such as apples, and can produce puckered 'catface' injuries to these plants.

The major "damage" from these insects is their appearance in nuisance numbers on the windows and porches of homes. This is occurs most often in the spring and fall. Boxelder bugs are brownish-black, about 1/2 inch long, with three red lines on the head, and a bright red abdomen beneath the wings. The nymphs, which do not survive winter, are sometimes called "red bugs" because their bright body coloration is not covered by the wings.

Life History and Habits: Boxelder bugs overwinter in the adult stage in protected sites, often including homes. They emerge in midspring and lay eggs near fallen seeds of boxelder, ash, and silver maple. The first generation nymphs feed on these seeds, as well fruit trees, various low growing plants and, occasionally, dead insects. They become full-grown in early summer.

A second generation occurs during late summer. Eggs are laid almost entirely on boxelder, particularly seeds produced by female trees. The nymphs develop on these seeds often into October, if weather permits. After frosts, boxelder bugs move to winter shelter, during which time homes are invaded. Boxelder bugs can move hundreds of yards from boxelder trees.

It is commonly noted that some buildings are repeatedly infested by large numbers of boxelder bugs, while others are much less commonly infested. It is unclear why there can be such differences. Surface texture of buildings and, in particular, direct sun exposure seem to be important. Also there is a noticeable odor that the insects produce so that there may be an aggregation pheromone produced.

Similar Species: Another common black and red bug that may be confused with the boxelder bug is the small milkweed bug, *Lygaeus kalmii* Stal. *L. kalmii* belongs to the seed bug family (Lygaeidae).

Management: Only adult stages survive the winter and many nymphs which fail to complete development are killed by early frosts or shorter growing seasons.

Boxelder bugs do not require controls for protection of trees. However, treatments are often made to reduce nuisance movements into homes. Boxelder bugs are quite resistant to most insecticides with the pyrethroids being among the more effective.

CONIFER SEED BUGS

Leptoglossus occidentalis Heidemann
Leptoglossus phyllopus (L.)
Hemiptera: Coreidae

Hosts: Developing seeds of pines, Douglas-fir, dogwood and other plants

Damage and Diagnosis: Conifer seed bugs primarily feed and develop upon developing seeds. Conifer seeds are preferred but developing seeds and fruits of a wide variety of plants may be fed upon. Occasionally they do minor damage to fruits, causing pitting, and can be serious pests where pine is grown for seed.

Conifer seed bugs are a common nuisance invader of homes in Colorado during fall and winter. Also, as they may fly readily and can produce a odd, somewhat piney odor when provoked, their behaviors can produce further concern. However, they are harmless to humans.

Conifer seed bugs are fairly large insects (ca. 5/8 to 3/4-inch) and of rather bizarre appearance. They are reddish-brown to a dark gray in color and light marking on abdominal margins (3/8 to 1/2 inches long). These bugs possess an enlarged hind tibia that is broad and flat; thus the name leaffooted bugs is often given to them.

Life History and Habits: Conifer seed bugs primarily feed and develop upon seeds of various trees and shrubs. Seeds of pines, Douglas-fir and other conifers are preferred, but developing seeds and fruits of a wide variety of plants may be fed upon, including dogwood and sumac.

The insects overwinter in the adult stage under protective debris and other sheltering sites. Frequently, they move into nearby homes where they may cause concern. However, during the cool season the insects are in a semidormant state neither reproducing nor feeding, but living off fat reserves developed during the summer.

In spring the conifer seed bugs move to trees and feed on the male flowers and one year-old cones. Beginning in late May the females begin to lay eggs, which are glued in small groups to needles and leaves. The immature stages (nymphs), which somewhat resemble the adults but lack developed wings, feed on the seeds through the summer, becoming mature in August and September. Adults continue to feed on cones until moving to winter shelter. There is only one generation produced per year.

Management: There is probably little that will effectively control the occasional nuisance movement of conifer seed bugs into homes. The best action is to ensure that homes are well sealed during the September/October period when most are migrating from trees to buildings and other winter shelter. It is probable that some use of insecticides on building exteriors, particularly directed at cracks and openings where the insects may use to enter, should also assist in reducing numbers found indoors. However, insecticides have not been tested against these insects. Insecticides have no place for control of insects that are already present in a home; swatting, vacuuming or otherwise disposing of individual insects is the appropriate response.

However, conifer seed bugs are harmless. Members of the "leaffooted bug" family (Coreidae), they are seed feeders and do not bite humans. Their presence in a home is related to their habit of seeking warm protected sites to overwinter, a habit shared by many other insects such as boxelder bugs and elm leaf beetles. The insects in the home do not reproduce nor damage any household items.

TOP: Conifer seed bug.
CENTER: Conifer seed bug. (USDA)
BOTTOM: Pale western legume bug.

PLANT BUGS

Tarnished Plant Bug: *Lygus lineolaris* (Palisot de Beauvois)
Pale Western Legume Bug: *Lygus elisus* Van Duzee
Mullein Bug: *Campylomma verbasci (Meyer)*
Hemiptera: Miridae

Hosts: *Lygus* sp. plant bugs have a very wide range of hosts, particularly legumes. Economic damage in Colorado is most common to tree fruits.

Damage and Diagnosis: Plant bugs feed on developing leaves, fruits and flowers, killing the areas around the feeding site. This can cause abortion of young flowers, seeds, or buds. Older tissues may continue to grow but be deformed. Leaf curling and corky 'catface' injuries to fruit are common distortions due to lygus bug feeding injury. Peach, apricot, strawberry, and beans are among the garden plants most commonly damaged.

Some flower abortion of most plants is normal and plant bug feeding injuries have little effect on plant yields unless the insects are abundant.

Adults plant bugs are generally oval in shape, being about twice as long as wide, and 1/4 inch in length. Most common species of *Lygus* in Colorado are pale green, but brownish and mottled forms occur. The nymphs are more rounded in general form and usually dark green, often with dark spotting.

Life History and Habits: Lygus bugs spend the winter in the adult stage, under the cover of piled leaves, bark cracks or other sheltered sites. They emerge in early in spring and feed on emerging buds of trees and shrubs. Most then move to various weeds and other plants, and females insert eggs into the stems, leaves, and buds of these plants. The young hatch, feed and develop on these plants becoming full-grown in about a month. There are several generations produced during the year.

Lygus bugs also occasionally feed on insects, and can contribute to the biological control of aphids and other small, soft-bodied species. This habit is particularly well documented with the mullein (campylomma) bug, which can damage fruit during spring, but later is an important predator of orchard insects. Furthermore, there is one group of plant bugs (*Deraeocorus* spp.) that are strictly predators of insects and mites.

Management: Legumes, particularly alfalfa, are important host plants for lygus bugs. If these plants occur around a garden, they should not be cut during times when fruits and vegetables are in susceptible stages, such as fruit set. Cutting can force migrations of lygus bugs.

Lygus bugs, and most other 'true bugs' are fairly difficult to control with insecticides. Since most injury occurs during early fruit development, insecticide sprays are best timed either immediately before flowering and/or immediately after petal fall. Insecticides should never be sprayed during flowering to avoid killing beneficial pollinating insects.

TOP: Codling moth larva in apple.
BOTTOM: Codling moth. (Clemson University)

CODLING MOTH

Cydia pomonella (L.)
Lepidoptera: Tortricidae

Hosts: Apple, pear, some large fruited crabapples

Damage and Diagnosis: Larvae tunnel into the fruit of apples, pears, and crabapple. (It is almost always the "worm" in a wormy apple.) Less commonly it also may damage other fruits, including apricot and peach. It is the single most important insect pest of tree fruits in the western US.

The larvae are pale colored, with a dark head and are found associated with the fruit. Adult moths are small (1/2-in long) grey moths with a coppery tip to the forewings.

Life History and Habits: Codling moth larvae spend the winter inside a silken cocoon attached to rough bark or other protection locations around the tree. With warm spring weather, they pupate and later begin to emerge around blossom time as small (about 1/2-inch), grey moths. The spring appearance of this adult stage may primarily occur over the course of one or two weeks, but can be much more prolonged if weather is cool.

During periods when early evening temperatures are warm (above 60 F) and not windy, the moths lay small, white eggs on the leaves. The larvae hatching from the eggs may first feed on the leaves but then migrate to the fruit, usually entering the calyx (flower) end. They tunnel the fruit, feeding primarily on the developing seeds. After about three to four weeks, the larvae become full grown, leave the fruit, and crawl or drop down the tree to spin a cocoon and prepare to pupate.

After about two weeks most, but not all, of the pupae develop to produce a second generation of moths. The remaining moths remain dormant, emerging the following season. (For example, in western Colorado only about 2/3 go on to produce a second generation and fewer than 50 percent of their progeny go on to produce a third generation.) These moths lay eggs directly on the fruit and damage by the larvae to fruit is greatest at this time. Becoming full-grown, the larvae emerge from the fruit and seek protected areas to pupate.

In the warmer, southern parts of the region, a small third generation is produced in late summer. This generally causes much less damage to apples and pears than the earlier generations.

Management: Codling moth has many natural enemies, although these biological controls often are not adequate to provide control. Birds will sometimes feed on the larvae and pupae in cocoon. Perhaps most important, codling moth larvae and pupae are often killed by several types of parasitic wasps. The activity of these wasps has been improved in some areas by the presence of nearby flowering plants which provide alternative foods. Codling moth larvae also are attacked by several general predators such as ground beetles and earwigs. Keeping loose bark scraped from trees and removing debris from around the tree can eliminate shelter used by the insects when they pupate. This will cause them to be more exposed to birds and other natural controls.

Since caterpillars often need some leverage to help them cut into the fruit, many larvae enter where two fruits are touching. Thinning apples to prevent this can reduce the survival of the delicate young larvae. Pupal stages of the moth can be concentrated by placing bands of corrugated cardboard or burlap around the trunk. They can then be easily collected and destroyed from such sites.

Adult moths of both sexes can be attracted to certain baits and trapped. A typical design might be a gallon jar baited with a pool of fermenting molasses and water (1:10 to 1:15 dilution is suggested). This attracts both male and female moths. (Pheromone traps only capture male moths.)

Insecticides should be applied during periods of peak egg laying by the adult moths. Two damaging generations of codling moth are common, one in late spring (after petal fall) and the other in midsummer. Feeding by the larvae of the latter causes most fruit damage.

Use of pheromone traps containing the sex attractant of the female moth can improve treatment timing. Using this technique, insecticides are most efficiently used 10 to 14 days after peak flights are detected in pheromone traps.

In many large block orchard plantings pheromones have been effectively used in the 'male confusion' method for control of codling moth. This involves permeating the air with the sex attractant used by the female to attract males, reducing successful mating. In backyard settings, this technique is thought not feasible because of difficulties in disrupting mating over the large area needed for this mating disruption to be successful.

PLUM POCKETS

Taphrina communis (*T. pruni*)

Hosts: All plum species and varieties, most common on American, less common on European or Japanese plum

Diagnosis and Damage: Symptoms appear as small white blisters on the young fruit. The blisters enlarge as the fruit develops and eventually involve the entire fruit. The infected fruit increases in size and becomes spongy and distorted. The seed does not develop, and a hollow cavity forms in the infected fruit. The fruit is reddish initially, then becomes covered with grayish powder consisting of asci. Young shots and leaves also may become infected and malformed.

BELOW: Plum pocket on plums. (USDAFS)

ABOVE: Plum pocket on wild plums. (M. Schomaker)

Fifty percent or more of the fruit may be lost in years when the disease is severe. Buds and twigs may be affected, thus stressing the tree. Once the disease is established in a tree, it will appear each year unless controlled.

Biology and Disease Cycle: The fungus overwinters as conidia or ascospores on twigs and buds on the tree. In the spring these spores are blown to young, succulent tissues. The spores germinate and penetrate developing leaves or other tissues through stomata or directly through cell walls. The mycelium grows between cells, causing abnormal cell enlargement and division. Infected fruit or tissue is distorted and much larger than normal. The fungus produces asci below the epidermis of the fruit which eventually enlarge and break through the surface of the fruit. Spores are released and windblown to new tissue where infection takes place.

Conditions Favoring and Management of Disease: Low temperatures and high humidity favor infection of susceptible tissue. All tissues become resistant as they mature. The pathogen can be controlled by a single fungicide application in late fall or early spring before leaf buds swell. All fungicides that are labeled for use on plum trees would be acceptable.

vertebrate damage to *woody plants*

Birds

SAPSUCKERS

Yellow-bellied Sapsucker [*Sphyrapicus varius (L.)]; Red-naped Sapsucker* [*Sphyrapicus nuchalis* Baird]; Williamson's Sapsucker [*Sphyrapicus thyroideus* Cassin)]

Hosts: Many, but particularly pines, junipers, alder, willow and aspen.

Damage: All peck holes (called "sap wells") in live tree trunk and branch bark and drink the resulting sap flow. Sap wells and other pecking injuries, which might be confused with boring insect exit holes, are usually just physical wounds which will heal over in time. Sap wells tend to occur in patterned rows, whereas insect emergence holes and holes made by woodpeckers preying on insects tend to occur at random. Occasionally, various stain and decay-causing fungi are introduced via sap wells and, in coniferous hosts, the wounds can attract attack by pitch-mass borer moths in the genus *Dioryctria*.

Many other species of birds, mammals, and insects, the majority considered desirable (for example, hummingbirds), utilize sap wells as a food source.

All of the sapsuckers found in the region are medium-sized woodpeckers with a basic red, black and white color scheme. They have long, chisel-like bills and cling on the sides of trees, using their stiff tails as a brace. In both adult and immature plumages, sapsuckers are distinguished from other Colorado woodpeckers by the presence of a large white shoulder patch.

Life History and Habits: All sapsuckers are usually migratory in the region. Red-naped and Williamson's sapsuckers are present in the mountains from May through October. The yellow-bellied sapsucker is a winter species in the region, usually restricted to lower elevations.

All three sapsuckers utilize trees in the same manner. They drill characteristic 1/4 inch x 1/4 inch rows of holes to encourage sap flow. These are revisited on a daily basis and particularly productive wells are reworked along their top edge to allow sustained flow. Such reworking also results in enlargement of the wound. Long-term activity of this sort produces distinctive rectangles in the bark that reveal the outer wood, and in severe cases can effectively girdle branches or trunks, causing foliage discoloration and dieback.

The Williamson's sapsucker is fond of ponderosa pine and Douglas-fir. The red-naped sapsucker particularly utilizes willow and aspen. The yellow-bellied shows a preference for ornamental Scots, Austrian, and ponderosa pines. All can drill many other tree species on occasion (e.g., Siberian elm, black walnut, Russian olive, juniper, eastern redcedar, maples, American linden, fruit trees, etc).

Management: Only rarely is sapsucker damage to live trees seriously damaging. In the great majority of cases their activities should be tolerated and appreciated as mostly beneficial. These birds are protected by law and as such cannot be shot or trapped. Most legal "controls" involve discouraging their presence or mechanical exclusion. Such

TOP: Yellow-bellied sapsucker injury to aspen.
BOTTOM: Sapsucker damage to lodgepole pine.

ABOVE: Downy woodpecker injury while feeding on insect.

things as owl decoys, flutters and mesh coverings over favored feeding areas have met with mixed success.

OTHER WOODPECKERS

Downy Woodpecker [*Picoides pubescens* (L.)]; Hairy Woodpecker [*Picoides villosus*]; Ladder-backed Woodpecker [*Picoides scalaris*]; Three-toed Woodpecker [*Picoides tridactylus*]; Black-backed Woodpecker [*Picoides arcticus*]; Red-bellied Woodpecker [*Melanerpes carolinus*]; Red-headed Woodpecker [*Melanerpes erythrocephalus*]; Northern Flicker [*Colaptes auratus*]; Lewis's Woodpecker [*Melanerpes lewis*]; Acorn Woodpecker [*Melanerpes formicivorus*]

Hosts: Virtually all types of trees at one time or another, especially mature ones.

Damage and Diagnosis: Occasionally peck the bark of living branches and trunks to cause sap flow, similar to the habits of sapsuckers. Such wounds created by woodpeckers other than sapsuckers are rare, and are found primarily on trees with sweet sap (such as boxelder and other maples). These woodpeckers also often visit sap wells made by the sapsuckers to "steal" a drink and can be mistakenly accused as the makers of these wounds.

These woodpeckers are medium-sized birds, 6 (downy)-13 (flicker) inches long. All have the basic woodpecker shape, long chisel-shaped beak, tend to occur on the sides of trees and use their tails as a brace. For detailed descriptions, see one of the standard western bird field guides.

Life History and Habits: Unlike the highly migratory sapsuckers, most other woodpeckers are present all year. Some of the high mountain species, such as the three-toed, may move to lower elevations for the winter. By far the most common tree activities of woodpeckers are: 1) excavating and nesting in trunk holes, and 2) feeding on bark beetles and wood boring insects. The latter is highly beneficial and almost always requires the chipping of bark and wood to get at prey. Hairy, black-backed and three-toed woodpeckers are particularly helpful as natural controls of conifer bark beetle outbreaks. The wounds that result from woodpeckers extracting beetle larvae from under bark may be confused with the emergence holes of these same insects. In some situations, woodpecker-caused bark flakes accumulating on snow at the base of insect-infested trees (for example, spruce beetle), can be noted during aerial surveys and prove helpful in detecting infestations. The random appearance of woodpecker predation holes usually separates it from the patterned rows made by sapsuckers. Also, woodpeckers are particularly fond of tree fruits in fall. Apples and other large fruits may show peck marks that resemble insect holes. Yet another interesting habit of some woodpeckers is that of caching food. Hard-bodied insects like beetles, seeds and other food items are stored in trunk crevices, telephone poles and other places for later consumption. The Lewis's and red-headed woodpeckers are noted for this. The acorn woodpecker, which just gets as far north as southern Colorado, is famous for its habit of pounding acorns and other mast into the bark of large, heavily-used trees called "granaries".

Mammals

TREE SQUIRRELS

Fox Squirrel [*Sciurus niger (L.)]*
Other arboreal mammals such as Pine (="Red") Squirrel [*Tamiasciurus hudsonicus* (Erxleben)], Abert's (= "Tassel-eared") Squirrel [*Sciurus aberti* Woodhouse], Chipmunk species (*Eutamias* spp.), Rock Squirrel [*Citellus variegatus* (Erxleben)] and Golden-mantled Ground Squirrel [*Citellus lateralis* (Say)]

Hosts: The fox squirrel debarks many species of deciduous trees, with elms, hackberry, honey locust, and Russian olive being common hosts. Twig clipping follows a similar pattern. Both deciduous and coniferous fruits and cones are eaten, particularly oak acorns, walnuts, elm seeds, fruit-tree fruits, and honey locust pods, plus pine and spruce cones. The red squirrel strongly favors conifers in Colorado and mostly feeds on fleshy fungi and spruce and pine cones. The Abert's squirrel relies heavily on ponderosa pine for shelter and food. Rock squirrels spend a surprising amount of time in trees and occasionally feed on the bark of elms, Gambel oak and other trees. Chipmunks and the other ground squirrels spend most of their time on the ground but often can be seen feeding on various types of tree bark or fruits and cones.

Damage and Diagnosis: The feeding habits of these mammals, particularly the fox squirrel, are quite variable according to location, season and food availability. Squirrel damage to trees could be confused with that of insect in three cases: when they debark branches or trunks, clip twigs, or feed on fruits/cones. With debarking, incisor marks on the outer wood are characteristic. Squirrel clipped twigs littering the ground usually show missing parts, such as chewed buds. Squirrel damage to fruit and cones is usually indicated by teeth marks and the season in which it occurs (fall/winter). Prior to periods of brood-rearing (early spring and early summer), some fox and rock squirrels strip bark from Russian olive, honeysuckle and other live, woody plants for use as nest bedding.

The fox squirrel has a ten to 15-inch long body with a bushy tail about equally as long. Its body is rusty yellowish with a yellow to orange belly and a rusty brown tail. (For a description of the others, refer to a mammal book such as the *Peterson Field Guide to the Mammals* or *Mammals of Colorado*.)

Life History and Habits: The fox squirrel generally has two broods per year. They spend most of their time in trees, traveling between trees or burying food items gathered from trees. Being very opportunistic, their feeding is strongly tied to food available at the time. Thus, in spring they commonly feed on swelling flower and foliage buds. Later in spring, seeds and fresh foliage are commonly eaten. In summer all manner of food items are taken including insects, eggs, and plant parts including bark/phloem tissue. In fall, fruits and cones are heavily relied on. A common food item in early fall is hackberry nipplegall psyllid nymphs, taken by biting off the tops of the galls. In winter it is back to buds, bark and stored items from fall foraging activities.

Management: This can be a controversial issue. Squirrels are loved and even imported by some tree owners, despised by others. Live-trapping and relocation can be a lifelong operation, as new squirrels often fill in as fast as problem animals are removed. Seamless metal skirts can be wrapped around trunks to prevent squirrel access to tree crowns, provided the target trees do not have branches within jumping distance of adjacent trees.

With bird feeders often being a source of attraction to rodents, converting to squirrel-proof designs may be helpful (although "squirrel-proof" has proven to be an elusive concept). Squirrel-killed branches should be pruned to prevent human hazard and breed-

TOP: Fox squirrel.
BOTTOM: Bark stripping by fox squirrel.

ing by the smaller European elm bark beetle (insect vector of Dutch elm disease). Most squirrel damage is of a minor nuisance or cosmetic nature and should be tolerated.

VOLES and DEER MICE

Microtus spp. ("voles")
Peromyscus spp. ("deer mice")]

Hosts: Most damage occurs to small trees. Both conifers and deciduous trees are affected. Fruit trees in orchards are among the more valuable species with recurrent small rodent problems. In certain areas, mature aspen in the mountains show considerable superficial bark injury from voles feeding under winter snow.

Damage and Diagnosis: Voles (also called "meadow mice") and deer mice injure fruit and ornamental trees and shrubs by gnawing on the bark. Extensive injury produces girdling wounds that can kill smaller plants. The small size (about 1/8 inch wide) of the chewing wounds differs from those of other gnawing animals, such as rabbits. Small rodent damage to tree bark is most frequent during the winter. They will eat almost anything green, including grasses, tubers, bulbs and garden plants. They also commonly damage lawns by clipping grass while making surface runways.

ABOVE: Vole damage to aspen.

Voles are very similar in shape and size to the familiar house mouse. The various species are separated by such features as fur color, tail length, and tooth arrangement, in addition to their distribution and typical habitats in which they occur.

Life History and Habits: In this region, there are at least five species of voles in the genus *Microtus* and six species of "deer mice" in the genus *Peromyscus*. The pine vole has been particularly damaging to orchards, and the meadow and prairie voles have done damage to bark of seedlings in nursery settings. The deer mouse (*P. maniculatus*) can also be a pest of trees. Their food base during the warm months contains a high complement of insects and succulent plant parts, with their winter diets being more reliant on seeds and bark. These two genera actively feed and reproduce throughout the year. Peak breeding occurs during spring and summer. Several young are produced in each litter, which develop rapidly and become full-grown in about six weeks. Their life span is fairly short, averaging less than one year. Population numbers fluctuate greatly, with peak injury to trees and shrubs associated with periodic outbreaks, particularly during the winter. They produce and maintain runways and shallow tunnels that cross the soil surface. Of necessity, in winter these runways are used and maintained beneath the snow.

Management: Natural predators of voles and mice include large snakes, shrikes, hawks, owls, weasels and predaceous mammals, including domestic cats.

To occur in large numbers, these rodents require cover and ready access to food. Mowing and other types of weed control in, around, and near tree-growing areas can reduce favorable cover. Mulches also provide cover for small rodents. To reduce bark feeding, mulch should be kept at least a few feet from the base of susceptible trees and shrubs. The depth of mulch should also be kept to a few inches.

Individual trees and shrubs can be protected from rodent injury by tree guards or wire mesh screening, 1/4 inch or less in diameter. Because these little animals can tunnel, the barriers should be buried 6 inches. Placing small, sharp pebbles in planting holes for bulbs can also deter them. During the winter, compacting the snow around trees and shrubs can act as a barrier to vole tunneling. Mouse traps, placed with the trigger along the runways, can be used to kill small rodents. Baiting is not necessary if traps are properly placed in the runway.

Hot pepper sauce and thiram-based repellents have been registered for control of meadow voles. Their effectiveness is considered questionable.

COTTONTAIL RABBITS

(*Sylvilagus* spp.)

Hosts: Many types of small coniferous and deciduous trees and shrubs. Willows, poplars and fruit trees are often favored.

Damage and Diagnosis: Cottontail rabbits feed on and can destroy many different garden plants and trees. During winter, cottontails will eat buds, twigs and bark of fruit trees, willows, junipers and many other woody plants. Succulent shoots of young trees are also browsed by rabbits, clipped at snow height. Rabbit damage is characterized by a sharp, clean 45° angle feeding cut. When there are lots of rabbits, other signs, including round droppings and footprints, can be used to separate rabbit injury from that of other gnawing mammals, such as voles and squirrels. At least three species of cottontails occur in the region. The eastern cottontail (*Sylvilagus floridanus*) is generally most abundant east of the Rockies, although other species occur in the West. Cottontail rabbits prefer to nest in areas with brush or other cover, and landscaping of developed neighborhoods is ideal for this activity. Cottontails are rarely found in dense forests or open rangeland. In open areas of the plains, jackrabbits (species of *Lepus*) may predominate. Jackrabbits are not particularly damaging to trees but may cause other damage when drought or overgrazing force them into yards and gardens. At elevations above 8000 feet, the snowshoe hare can be common. Trees and other woody plants make up a high percentage of their diet, particularly in winter.

Among the woody plants most frequently damaged by rabbits are most rose family plants (including apple, raspberries, cherry, plum, mountain ash), basswood, red maple, honey locust, oak, willow, sumac, and dogwood. Planting favored garden plants such as tulips, peas, beets, carrots and beans near valuable woody plants may invite damage to the latter. Garden plants least favored by rabbits include corn, squash, cucumbers, tomatoes, potatoes, and peppers.

Cottontail rabbits are cat-sized mammals with long, generally erect, ears. Their hind legs are enlarged for jumping. They have large eyes and short, thick coats of hair.

Life History and Habits: Despite the well-known exploits of Peter Cottontail (actually a European hare), cottontail rabbits do not dig underground nests. During warmer months, they form shallow hollows in dense vegetation for cover; for winter protection, natural cavities or burrows are used. Cottontails typically produce two to three litters per year and each litter contains about three to five young. The young rabbits can leave the nest about three weeks after birth and are sexually mature within a few months. Populations can increase rapidly when food is abundant. Cottontails do not hibernate and are active throughout the winter. They can walk on snow and feed on plants at the level the snow cover allows.

Management: In most of the region, cottontail rabbits are classified as a game animal. As such, they fall under regulations of state wildlife agencies, which generally restrict hunting and trapping to specific seasons, requiring a license. However, exemptions may be granted by these state agencies for rabbits that are damaging property.

Biological Controls: Rabbits are preyed upon by many animals in the wild, including owls, foxes, snakes and hawks. Rabbits also succumb to various diseases. Cottontail rabbits rarely live for more than fifteen months under natural conditions. In areas of dense human habitation, house cats are important predators, feeding on young rabbits in nests. Dogs will also deter rabbits from roaming in yards, although they rarely kill them.

Cultural Controls: Rabbits tend to avoid open areas where they are particularly vulnerable to predators. Keeping areas mowed and landscaping plans that provide open areas will deter rabbits. Removal of brush piles and other dense, protective cover will eliminate sites where rabbits hide and nest.

TOP: Cottontail rabbit.
BOTTOM: Twig damage by rabbits.

Mechanical Controls: Rabbits are easily excluded from gardens by fencing. A 2-foot-high chicken wire fence, buried shallowly in the soil, can prevent most rabbits from entering gardens during the summer. Plastic tree wraps are commonly available and will generally provide control for 3 to 5 years. Also, flexible polypropylene netting (many styles and generally called "rabbit guards"), galvanized poultry wire and even aluminum foil provide a barrier to rabbit feeding on trees and shrubs. They are most commonly used for seedling plantings. Experience has shown rabbit guards to also reduce tree damage from deer, grasshoppers and blowing sand. Holes in any type of barrier should not be larger than 1/2 inch to exclude small rabbits.

Repellents: Various types of odor repellents marketed as animal repellents, such as naphthalene moth balls, are not effective for rabbit control outdoors. Taste repellents, usually involving the fungicide repellent thiram, are effective for preventing rabbit feeding. However, thiram is a toxic material to other mammals, such as humans, and cannot be applied to plants that are to be eaten.

Traps: Rabbits can be trapped easily in box live traps baited with apples, carrots, ears of corn or cabbage. Traps should be placed close to areas of cover used by rabbits.

TOP: Eskers produced by pocket gophers.
CENTER: Pocket gopher damage to roots.
BOTTOM: Pocket gopher damage to seedling pine.

POCKET GOPHERS

(*Thomomys talpoides* (Northern Pocket Gopher), *Geomys bursarius* (Plains Pocket Gopher) and others)

Hosts: All parts of many herbaceous and woody plants are fed on by pocket gophers. Pines, spruce, cottonwoods, shrubs and fruit trees are among the most-damaged woody plants.

Damage and Diagnosis: The primary damage to woody plants is feeding on the roots. Pocket gophers are among the most damaging organisms to seedling tree plantings in many parts of the region, particularly on the plains. However, trees over 20 feet tall have been killed. The crowns of trees with gopher-caused root damage often first show foliage thinning, then foliage discoloration and finally, death. Root systems can be reduced to nubs.

The name "gopher" is at times incorrectly applied to most small mammals that live underground. The animals themselves, symptoms of their presence, and damage are often confused with moles, various voles and mice, prairie dogs, ground squirrels and kangaroo rats. In general, true pocket gophers are medium-sized (8 to 12 inches long), brown, plump rodents with short fur, short tails, large heads, small eyes, small ears and short, stout, front legs. Their name comes from external cheek pouches that open on either side of the mouth, used to transport soil. They live a burrowing existence and rarely are seen above ground. Their subterranean life style usually results in above-ground mounds of dirt that appear to have no entrance. This is because openings are typically plugged with soil. (Thus, mounds with conspicuous holes in them were made by something other than a pocket gopher.)

Life History and Habits: Gophers construct extensive burrow systems, with some of the tunnels being shallow and others more than 15 inches deep. A tunnel system may be linear or highly branched and can consist of up to 200 yards of tunnels. The tunnels are 2-3.5 inches in diameter. Entrances to the systems are marked by rather large mounds with plugged openings. Interestingly, soil castings (called "eskers") from tunnels are also brought to the surface by means of the cheek pouches and stuffed into the snow. Following snow melt, the connected trails of soil making up an esker create a very characteristic pattern that roughly mirrors that of the underground tunnels. Gophers are ecologically important and influence soil formation, soil mixing, soil erosion, plant succession, rainfall infiltration, and the distribution of other animals. Each gopher brings at least one ton of soil to the surface each year. A single litter of 1-10 (usually three to four) young is produced annually, usually in early summer. Roots and tubers make up most of the winter diet, while leaves and stems are the primary summer foods.

Management: Pocket gophers are not protected by federal or state laws but it should be recognized they can be environmentally valuable and control is difficult. Given these points, where tree damage is excessive, the following methods could be employed:

Mechanical Controls: Barriers have some use but to be effective must extend well below the soil line. Screen or plastic-mesh cylinders should be placed around trees to a depth of 18 inches. To avoid root damage on larger trees, this barrier should be placed 2 feet from the stem. Obviously, this is not always a practical method. Trapping is another method often used in larger infestation areas to reduce populations prior to poison baiting. Body-gripping traps (several types available) are placed within main or side tunnels by finding and excavating mound soil plugs. Traps should be checked frequently and moved if unsuccessful after a few days. Trapping works best in spring or fall when most tunnel construction occurs.

Repellents: While some experimental materials based on predator odors show promise, none of the highly-touted plant-based repellents (such as those utilizing castor-oil plant) as yet have any supporting data.

Chemical Control: Approved baits (strychnine alkaloid, zinc phosphide, chlorophacinone and diphacinone) are mixed with grain or formulated into paraffin blocks. [NOTE: some uses of these materials are "Restricted-Use"]. Baits are placed in tunnels by hand, by probe and spoon, or special bait-dispensing probe. Another larger machine called a *burrow builder* constructs artificial burrows and automatically baits them. The system works best if burrows are constructed in soil with moderate moisture and at depths typical of gophers using the area. The gophers intercept the artificial burrows and contact the bait. This is a tractor-pulled device and would be best-suited for economical treatment of large areas.

Detailed gopher management information is contained in Fact Sheet 6.515.

TOP: Porcupine.
BOTTOM: Porcupine damage.

PORCUPINE
(*Erethizon dorsatum*)

Hosts: Mostly coniferous trees such as ponderosa pine, but many deciduous trees and shrubs such as willow, aspen, cottonwood and serviceberry are also utilized as food and for roosts. Trees with smooth bark are preferred over rough-barked species.

Damage and Diagnosis: Porcupines are rather large (2 to 3 feet in total length), waddling rodent with short legs, a short, thick tail with many hairs on the back and tail modified into distinctive barbed quills. They have shiny black eyes and ever-growing incisors.

Porcupines feed on foliage, buds and the inner bark. Large patches of bark are consumed. On pines and other conifers, these areas are typically golden yellow, covered with pitch and marked by wide incisor scrapes. While "barking" of the lower trunk and higher areas of large-diameter is the norm, branch barking, twig-clipping, and consumption of buds and leaves also occurs. Wooden tools, siding and other objects may also be damaged.

Life History and Habits: Porcupines occur mostly in conifer forest habitats, including pinyon-juniper areas, but can be found in mountain parks and on prairies far from trees during movements between feeding areas. They are active in all months, night and day. Breeding usually occurs in the fall and after a 200+ day gestation period, the single young is born. Softer tree parts are eaten during the spring and summer, with a higher proportion of bark and needles making up the winter diet. Porcupines are highly attracted to salt and may chew on wooden objects containing various resins or exposed to human sweat and urine (as with outhouse seats). For reasons which may be related

TOP: Mule deer.
CENTER: Deer rub on cottonwood.
BOTTOM: Deer guards.

to proximity to den sites and tannin content, certain trees are favored for heavy feeding, while other trees escape feeding injury altogether. Usually solitary, porcupines may be found roosting in groups in winter. When closely approached, porcupines usually turn their back to danger and swing their quill-laden tails in an unpredictable, lurching motion. Contrary to myth, they can not eject or "shoot" quills.

Management: Mechanical and electric fences can be used to exclude porcupines from small tree plantings or orchards. Individual trees can be protected with wire-mesh cylinders or with aluminum flashing skirts that extend to a height of 30 inches. Seedlings or very small saplings could be protected by complete enclosure within wire baskets. While not specifically labeled for use against porcupines, thiram applied for squirrel or rabbit protection may incidentally repel them. Wood treated with normal preservatives can be protected, although care should be taken not to use materials made up of metal-salt solutions (these are attractive to porcupines). There are no legal poisons for porcupines. Control of porcupines by trapping or shooting should be rare and limited to particularly troublesome individuals. Eradication attempts will likely meet with failure due to reinvasion. In summary, exclusion techniques, promotion of non-attractive plants and tolerance should be the norm in most situations.

MULE DEER

Odocoileus hemionus

Damage and Diagnosis: Throughout much of the region, no pest is more destructive and difficult to manage than the mule deer. (The white-tailed deer, *Odocoileus virginianus*, is also found in much of the region and causes similar plant injury.) Deer feed on a very wide variety of plants, depending on the season and the availability of alternative foods. Many vegetable crops may be browsed by deer feeding; the chewing of sweet corn ear tips and nipping of the center from heads of broccoli and cauliflower are examples. Deer also chew on twigs of fruit and ornamental trees, sometimes causing severe damage to younger trees. Feeding damage by deer is characterized by a ragged wound. Deer (and elk) lack upper incisor teeth, which produce the clean cuts of other gnawing animals, such as rabbits.

Male deer may also damage trees by scraping them with their antlers during the late summer and fall rutting season. Young trees are a particularly common target for this practice.

Life History and Habits: Deer prefer areas of mixed vegetation and have been described as creatures of the forest edge. Ideal habitat for deer includes areas near shrubby plants that provide year-round food of leaves, twigs and buds and where denser growth is available for cover. This mix is sometimes produced by landscaping activities, which can attract deer into neighborhoods. Where deer have lost their fear of humans, they may prefer the plantings found in yards and become serious pests. Most deer feed during dusk and dawn. They may wander a half mile or more from resting areas to search for food. In the more northern areas, deer often gather in areas of dense cover for winter protection. Deer breed during late fall and bear young about 202 days later; peak numbers of births occur during May and June. Twin births are most common. The young fawns develop rapidly and some does may be able to reproduce when only six months old. Reproduction of deer is very dependent on the amount of available food. Deer are long-lived, with some individuals living almost twenty years. Hunting is the most important factor in deer life expectancy in most areas.

Legal Status: Deer are protected animals and are only to be killed during specified hunting periods by licensed hunters. Deer hunting within many communities is further restricted by local ordinances. In some cases, special permits may be issued by state wildlife agencies to farmers with specific deer problems.

Mechanical Controls (Exclusion): In areas where deer damage is severe, fencing provides the only satisfactory control. However, deer are excellent jumpers that can easily get over a typical 6-foot yard fence. They can also climb through or under fence openings less than a foot wide. Construction of a deer-proof fence is a substantial project. Among the various designs are the following.

Basic deer fence. A properly constructed fence built at least 8 feet tall will exclude deer. A thick wire mesh is best for this purpose and highly durable. Lighter mesh materials (chicken wire, plastic netting) can also be used, but these degrade rapidly. Unless well marked, the light mesh may not be seen by the deer, which may then accidentally push through. The fence should reach to the soil line to prevent deer from crawling under it. Solid wood or brick fences, through which the deer cannot see, may be somewhat shorter (6 to 8 feet) and still deter deer from jumping into the yard.

Slant fence. An outward-slanting fence can be more simply and cheaply constructed than an 8-foot vertical fence. For example, a 7-foot fence angled at 45° can usually deter deer. Deer will tend to approach and travel under the fence, then be unable to jump through it. However, the area under the fence should be kept mowed to encourage the deer to move under it.

ABOVE: Elk damage.

Electric fence. A variety of effective electric fence designs have been proposed for excluding deer. Although not generally appropriate for most yards, these are much cheaper and easier to construct than fences built to physically exclude the deer. Deer will generally try to go under or through a fence, even if they could easily jump over it. Most electric fence designs encourage this habit so that the deer will touch the fence. After this "shocking experience," deer will often learn to avoid the fence and stay several feet away. Since they usually jump from a point very close to the fence, their avoidance of the fence vicinity reduces their inclination to jump relatively low fences. A typical electric fence design would be about 5 feet in height with wires placed at 8- to 12-inch spacings. Usually wire spacing is shorter at the lower end of the fence. Alternatively, two shorter electrified fences, spaced 3 feet apart, can be used to exclude deer.

Repellents: Many different repellents have been attempted for deterring deer. Some of these are contact repellents, which act by taste. Contact deer repellents include hot pepper sauce (commercially available as well as home brews) and thiram (a fungicide and animal repellent). An important limitation of these contact repellents is their inability to protect the favored new growth that emerges after treatment. To make some of the less residual sprays, such as hot pepper, last longer during rainy periods, it is suggested to mix them with additives that reduce evaporation (e.g., Vapor-Guard, Wilt-Pruf). The commercially available contact repellents cannot be applied directly to food crops.

Area repellents protect plants by odor, and dozens of materials have been suggested. For example, human hair or blood meal are often suggested as repellents, although the general consensus of the seriously deer-plagued is that these are marginally effective at best. Bags of some brands of deodorant soaps are repellent. Also the manure and urine of large cats (provided by your obliging neighborhood cougar or zoo) or coyotes are more widely recognized as having fairly good repellent activity against deer.

Some commercial repellents appear to be the most promising. These include fermented egg solids (Deer-Away, MGK-BGR, Big Game Repellent) or ammonium soaps produced from certain fatty acids (Hinder). The latter is one of the few repellents that can be applied directly to plants that are to be eaten. A highly effective homemade repellent is a mixture of eggs blended with water. Crack the eggs and remove the small white membranous allantois, which tends to clog the sprayer. Blend in a 1:4 egg/water ratio and use this for a plant spray.

Dogs kept around garden areas will repel deer. However, unrestricted dogs are often not allowed in many residential areas, since they also may threaten livestock and other wildlife.

Some Trees and Shrubs that Are Least Likely to be Browsed by Deer

Barberry	Red-osier dogwood
Forsythia	Honeylocust
Beautybush	Norway spruce
White spruce	Colorado spruce
Mugho pine	Austrian pine
Scotch pine	Common lilac
Southern sage	Four-winged saltbush
Quince	Rabbitbrush
Fernbush	Winged euonymus
Rose of Sharon	Vanhoutte spirea
Tatarian honeysuckle	Mountain ninebark
Potentilla	Wild plum
Nanking cherry	Gambel oak
Fragrant sumac	Common hackberry
Hancock coralberry	Silver buffaloberry
Snowberry	Concolor fir
Common juniper	Mahonia
Pyracantha	Alpine currant

OTHER LARGE HOOFED ANIMALS

In addition to deer, certain other large hoofed animals such as elk (*Cervus elaphus*), moose (*Alces alces*) and pronghorn (*Antilocapra americana*) are known to also feed on woody trees and shrubs.

Elk are famous for "barking" aspen trunks up to a height of about six feet. If excessive or followed by fungal infection, this can be a serious injury. In addition, over half of their winter diet in some parts of the region consists of browsed shrub and tree buds and small branches.

Moose do not typically do as much bark damage to trees as elk, but consume large quantities of small stems, branches, buds and leaves. They are particularly fond of willow, so much so that game regulations pertaining to moose can be based on local consumption of this resource. Other favored plants include aspen and birch, with other habitat requirements being spruce, fir or lodgepole pine cover.

ABOVE: Old damage from elk. (Bill Jacobi)

Pronghorn seldom cause problems with trees, but do include some woody shrubs in their winter diet, notably sagebrush and bitterbrush.

As with deer, perhaps the only reliable way to prevent woody plant injury with other large hoofed animals is isolation of valuable plantings with fences (see deer section for types). This can be quite expensive and impractical in certain areas. Conflicts sometimes arise between human neighbors of mountain communities, when one person attracts these animals with supplemental feeding stations (salt blocks, hay, etc.) and another person wishes to dissuade them because of plant damage. This becomes a sociological problem, not an animal damage problem. Continuing this theme, another approach which has met with mixed success, is the establishment of food plots. To have any chance of success, the idea is to provide a food at a location somewhat removed from valuable plants that is more desirable to the animals than those being protected. Alfalfa and wheat are sometimes used for food plots. Putting desirable food items within or near valuable plants invites trouble.

ABOVE: Bark injury by moose. (David Leatherman)

DOMESTIC ANIMALS

Horses, mules and donkeys, bison, cows, pigs, sheep, goats, rabbits, dogs, house cats and many other domesticated species can cause considerable damage to trees. The nature of their damage can be surmised by observing the types of problems their nearest wild relatives cause.

Horses and other domestic hoofed animals often chew bark from trees in corrals and other fenced areas. In addition, soil compaction and toxicity resulting from concentrated animal waste can harm trees.These animals should not be placed in confined areas with trees unless tree injury is acceptable. Where unavoidable, covering the lower trunks with chicken wire or hardware cloth can reduce injury. Over time and with normal trunk expansion, wire-mesh barriers may need maintenance or reattachment.

Domestic rabbits do the same things as wild cottontails and hares. Consequently, management of them would be the same.

The main problem with cats is their habit of clawing lower trunk bark. When heavy, this alone may kill small trees, but often subsequent fungal infection (for example, cytospora canker) is the real long-term problem. Cat-clawing can be easily prevented by attaching a window screen cylinder to the lower three feet of the trunk. This is relatively invisible and a simple office stapler can be used. Take care not to put the staples into the bark, as this, too, could lead to infection.

miscellaneous insects associated with *trees and shrubs*

DOG-DAY CICADAS

Tibicen dorsatus (Say)
Tibicen dealbatus Davis
Homoptera: Cicadidae

Life History and Habits: Immature stages (nymphs) of the dog-day cicadas develop by feeding upon sap from the roots of various trees. Boxelder and cottonwood trees in fairly loose, moist soils are common hosts. Development takes a few years to complete.

When full-grown the immature nymphs emerge from the soil and climb onto a nearby upright object. This typically occurs form mid-July to mid-August. There they molt and change to the adult form. After a brief period, during which the newly emerged adults darken and harden, they fly and feed upon the sap of leaves and twigs. Mating occurs in the trees and the females insert their eggs into twigs.

Males of the dog-day cicadas produce a shrill buzzing song used to attract the female cicada. Females do not produce noise. Cicadas are harmless to humans, although they often startle people by their buzzing and flight when disturbed.

T. dealbatus is the most common species associated with shade trees and found in landscape plantings. *T. dorsatus* is a species of native prairie shrubs.

Associated Species: A very large hunting wasp, the cicada killer, *Sphecius speciosa* (Drury), hunts dog-day cicadas and carries them back to nests they build in holes in the ground. The cicada killer appears similar to a giant yellowjacket wasp. Cedar beetles, *Sandalus niger*, develop as parasites of cicada nymphs. Mississippi kites commonly feed on dog-day cicadas in parts of the Arkansas Valley.

Related Species: Approximately 26 species of cicadas occur in Colorado. None are considered seriously damaging to trees, although some egg laying wounding has been associated with Putnam's cicada.

TOP: Dog-day cicada.
BOTTOM: Dog-day cicada nymph crawling on trunk.

ROUGH STINK BUG

Brochymena sulcatus Van Duzee
Hemiptera: Pentatomidae

Life History and Habits: Although a common species in Colorado, little is known of the biology of this insect. Winter is spent in the adult stage, and it often enters homes for shelter during the cold months. The insect has been collected from several kinds of plants. Although it feeds primarily on insects, it also sometimes feeds on leaves of many deciduous trees.

Other members of this genus have one generation per year. Eggs are laid in late spring and nymphs develop on their host plants through much of the summer. First generation adults are present by late August.

Related species: Other stink bugs are associated with trees and shrubs, most notably those that are predators of insects such as elm leaf beetles. These are discussed under the section on beneficial insects.

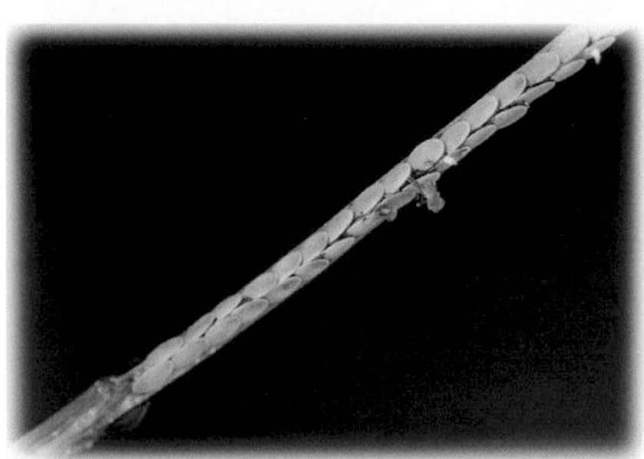

ARID-LAND SUBTERRANEAN TERMITE

Reticulitermes tibialis Banks
Isoptera: Rhinotermitidae

Life History and Habits: The arid-land subterranean termite is a social insect that produces a permanent colony underground. Separate castes occur in the colony including workers, soldiers, reproductive females (queens), and reproductive males (kings).

Colonies are initiated by winged reproductives. These emerge during late winter or early spring during mating flights and the paired queen and king attempt to initiate a colony. Very few are successful in the effort to create a colony and most die out within a few weeks after the mating flights. However, if successful the queen produces eggs and pale, flightless worker termites are produced. The colony slowly increases in size over a period of years and ultimately may later produce reproductive stages that disperse.

The workers are the stage that forage for cellulose-containing products. They avoid light and are very sensitive to drying so they remain underground or within the food (wood, cattle manure, etc.). Workers can not reproduce but may survive for months outside the colony. (Note: Firewood is not a means of acquiring a termite infestation.)

As with other termite species, the arid-land subterranean termite feeds on cellulose materials, including wood. It is a common scavenger of brush, wood, and animal manures in natural areas of the state and may be found in dead parts of living trees. It can feed on wooden structures but is considerably less damaging than are other termites, such as the eastern subterranean termite.

Related Species: The eastern subterranean termite, *Reticulitermes flavipes* (Kollar), has moved into eastern Colorado in recent decades and has become a serious pest of structures in parts of the state. The eastern subterranean termite has a higher moisture requirement than does the arid-land subterranean termite and its spread has likely been favored by irrigated landscapes.

BROADWINGED KATYDID

Microcentrum rhombifolium (Saussure)
Orthoptera: Tettigoniidae

Life History and Habits: The overwintering stage of katydids are eggs, which are very distinctive and attract attention. Females lay eggs by roughening the bark and gluing the eggs to the plant, which are laid on twigs and appear as overlapping rows of pale scales. The eggs hatch the following spring and the young katydids develop during the season, becoming full grown in late summer. Although katydids feed on leaves, this damage is insignificant. In addition, katydids may feed on small insects.

Singing to attract mates occurs in late summer, lasting several weeks. Songs may be a rustling noises or a loud 'lisps' and 'ticks'. Only the males produce the noises which are used to attract mates.

These are large insects, with some at least two inches from the head to the tip of the wings. They are pale-green color and blend in well with foliage. There is one generation per year.

Related Species: Recently, the true katydid, *Pterophylla carnellifolia* F., has become established in parts of the Front Range area of Colorado. The males of this large insect,

TOP: Rough stink bug.
SECOND: Subterranean termite workers.
THIRD: Broadwinged katydid.
BOTTOM: Eggs of the broadwinged katydid.

present in late summer and early fall, are night singers that make loud mating calls, described as "lisps" and "ticks". The females lay flat brown eggs on smaller twigs in overlapping double rows.

EUROPEAN EARWIG

Forficula auricularia L.
Dermaptera: Forficulidae

Damage and Diagnosis: Earwigs feed on a wide variety of plant and animal materials. Although they can damage soft plant parts, such as flowers and seedlings, on trees and shrubs they primarily feed on insects. They are important predators of many pest species including leafcurling aphids and elm leaf beetles.

Adults reach a size of about 3/4 inches and are dark brown with short wing covers. The most distinguishing characteristic are the forceps, or pincers, at the tip of the abdomen. Those of the male are more bowed than the female.

Life History and Habits: The adult earwigs spend the winter under rocks or similar protected sites and may become active during warm periods in winter. The females live within a small chamber they create and lay their eggs during these winter. Eggs are laid in groups, numbering around 40 to 60, and the mother carefully guards them.

Eggs hatch in early spring, around the time that fruit trees start to blossom. During the early stage of their life the mother continues to care for the young, periodically leaving the nest to collect food. Only after the nymphs have molted do they leave the nest, although they often continue to return to it for several weeks. The females often then produce a second, smaller group of eggs.

Earwigs are active at night and feed on a wide variety of foods. Soft parts of plants, such as corn silks and flower petals, are common foods and garden plants may be damaged. However, earwigs primarily feed on other insects and can be important natural controls of many garden pests. During the day they seek cover, preferring tight, dark locations. This includes existing holes in bark or fruit or curled leaves and galls.

Most of the overwintered earwigs die by late spring. The young earwigs become full grown in about two months and continue to be active throughout the summer. There is one generation a year, although nymphs may be found for much of the season since many overwintered females rear a second brood.

GIANT ICHNEUMON WASP

Megarhyssa macrurus (L.)
Hymenoptera: Ichneumonidae

Life History and Habits: The giant ichneumon wasp is the largest of the parasitic wasps with a body often reaching two inches and a much longer ovipositor (stinger). However, despite their appearance they are harmless to humans and do not damage trees. Instead they develop as a parasitoid of the pigeon tremex horntail, a minor pest species of wood borer associated with many species of hardwood trees.

TOP: European earwig, male and female on leaf.
BOTTOM: Giant ichneumon wasp.

The giant ichneumon wasp overwinters as a full grown larvae on its devoured host, within the tree. They pupate in spring and the adults emerge in mid-summer. The wasps are most common from late June through early August.

The females are capable of detecting the presence of developing larvae of the pigeon tremex within wood. They then drill through the wood and lay an egg on or near the young horntail larva. The emerging wasp larva soon kills and devours the horntail.

Adults are very large wasps, up to two inches long from head to tip of abdomen, females have extremely long ovipositors, making them look even larger. Larvae are cream colored, grub-like parasites of the pigeon tremex, found associated with their host in tunnels.

The giant ichneumon wasp causes no damage to trees or people. However, they can attract considerable interest because of their rather frightening appearance.

Related Insects: Two other large parasitic wasps are *Megarhyssa nortoni* Cresson and *Rhyssa persuasori* L. Both develop as parasites of horntail larvae (*Sirex* spp., *Urocerus* spp.) that develop in conifers.

CICADA KILLER

Sphecius speciosus (Drury)
Hymenoptera: Sphecidae

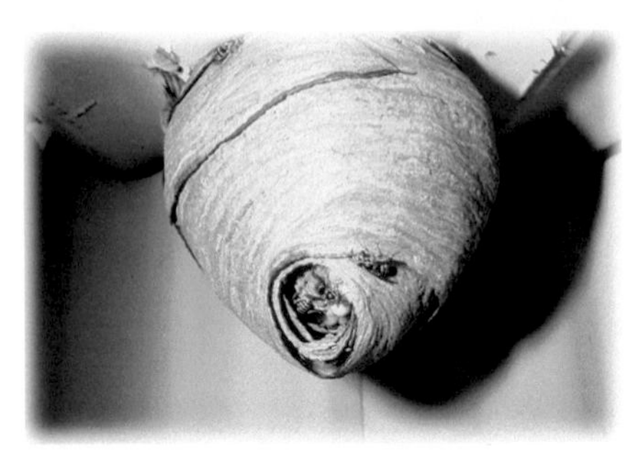

TOP: Cicada killer with cicada prey. (H. Evans)
BOTTOM: Large nest of the aerial yellowjacket. (H. Evans)

Life History and Habits: Adult cicada killers are most abundant in late July and August at which time they search for adult cicadas and prepare nests. Nests are dug out of the ground in sandy soils and typically to a depth of about of a foot or more. Nests are located in loose sandy soils that are sun exposed. Several chambers (an average of around 16) are excavated in each nest. After provisioning with the cicada prey the female lays an egg, seals the chamber and prepares a new cell. The young wasps develop on the cicadas, completing development in early summer of the following year.

Cicadas are captured by the wasps and immobilized with a paralyzing sting, then carried back to the nest in a series of short flights. Usually only the larger, egg-filled female cicadas are used by the cicada killer. Because the cicada prey weighs more than the wasp, often it is dragged up trees or buildings to allow the adult wasp to make short gliding flights to the nest. The cicada is then pulled into the nest hole and an egg laid upon it.

Two cicadas are used when the wasp rears a female; one cicada is sufficient for the food needs of the smaller male wasps. The young wasps feed on the paralyzed cicada becoming full-grown in a few weeks. The overwintering stage of the wasp is as a diapausing larva within a cocoon. There is one generation of the insect per year.

Female cicada killers will sting if handled. The sting is painful but the pain passes more quickly than a bee sting.

The cicada killer is primarily found in southern and southeast Colorado, along with its prey the dog day cicadas. It is also present in the eastern plains areas.

Related Insects: The family Sphecidae contains a large number and wide variety of hunting wasps and members occur throughout Colorado. One group are the wasp in the genus Pemphredon that excavate pith of various plants for nesting. They then pack the nest cells with paralyzed aphids or other small insects on which their young develop.

BALDFACED HORNET

Dolichovespula maculata (L.)
Hymenoptera: Vespidae

Life History and Habits: Life history is similar to that of other social wasps such as yellowjackets (*Vespula* species) and *Polistes* paper wasps. Overwintering stage is a fertilized female wasp that seeks protected locations for shelter. In spring the queens become active and seek sites to initiate colony nests. Trees and shrubs are commonly used nesting areas. Nest construction is made of paper-like materials produced from ground bark and wood mixed with saliva. Nests take the coloration of the wood used for construction and often are strikingly banded in colors ranging from light gray to reddish brown.

The overwintered queen produces a few cells and rears her young on a diet of various insects such as caterpillars that she collects. The first wasps produced are small and infertile females. They assist with further colony development and food collection. Subsequently reared wasps are feed improved diets due to better food collection and wasps produced late in the season are full-sized and fertile. Colony and nest size grow continually becoming football-sized or larger by the end of summer. At the end of the season some male wasps are also produced. Mating occurs at this time. The old queen, males, and early worker wasps die at the end of the season. Fertilized queens disperse for overwintering sites. Colonies are abandoned and are not reused.

Baldfaced hornets do not harm trees and are usually nonaggressive. However, they can produce a painful sting and will defend nests vigorously. Accidental stings can occur when work is done in trees where nests occur.

Hornets and many species of yellow jackets (*Vespula* species) frequently visit trees in late summer to collect honeydew produced by aphids and other insects. The large numbers of these stinging wasps can be a significant nuisance.

Related Species: A slightly smaller hornet, known as the aerial yellowjacket, *Dolichovespula arenaria* (L.), also occurs in Colorado. Habits are similar to the baldfaced hornet although nests are somewhat smaller. This species has the yellow and black coloration more commonly associated with yellowjackets (*Vespula* species), wasps which similarly construct paper nests but are located in wall voids, underground in abandoned rodent burrows, or similar protected sites. Yellowjackets most commonly visit trees in late summer and fall to feed on honeydew produced by aphids and soft scales.

CARPENTER ANTS

Camponotus species
Hymenoptera: Formicidae

Life History and Habits: Most species of carpenter ants in Colorado form nests inside rotting wood, although the species most common in the eastern Plains areas will nest in the soil. Unlike termites, which consume cellulose, the wood is not eaten and sawdust piles will be dumped at colony entrances. Worker ants (incompletely developed females) collect food (insects, honeydew, etc.) and return it to the colony. Several hundred ants may occur within a carpenter ant colony.

Fully developed and fertile female "queens" are produced. These queens are winged and emerge and periodically fly from colonies, usually in late June or July. The smaller winged males, also produced by the colony, also fly at this time and mate with the females. New colonies are then individually started by these mated female queens. Shortly after finding a suitable site for beginning a colony, the females shed their wings.

Carpenter ants are very common insects in forested areas of the state and serve an important ecological role in the decomposition of wood. Infrequently they also will invade structures, initiating colonies at areas softened due to decay. Various species of carpenter ants can be found throughout Colorado. However, the most common and largest species are typically found above 6500 feet elevation, in mountainous forested areas of the state.

TWOTAILED SWALLOWTAIL/WESTERN TIGER SWALLOWTAIL

Papilio multicaudatus Kirby
Papilio rutulus L.
Lepidoptera: Papilionidae

Life History and Habits: Swallowtails are large, strikingly colored butterflies marked with yellow and black. The larvae are variable in coloration and markings. Younger stages mimic bird droppings, are black splotched with white. Older larvae are lime green or generally brown, marked with conspicuous "eye-spots" on the back. When disturbed, swallowtail larvae can evert a pair of fleshy yellow "horns" from behind the head.

Overwintering stages of swallowtails are as pupae which form a grayish chrysalis that colors and camouflages with the background color on which it is formed. Adults emerge in May and June, mate, and lay eggs on the plants fed upon by the caterpillar

TOP: Carpenter ant larvae.
CENTER: Carpenter ant winged male and female.
BOTTOM: Western tiger swallowtail adult on zinnia flower.

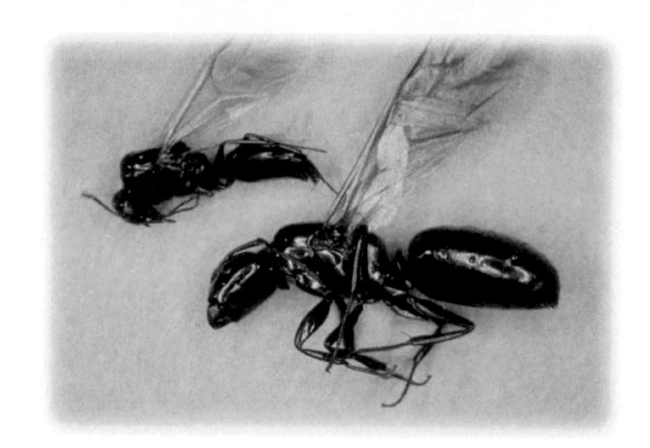

stage. Green ash and chokecherry are host plants of the caterpillar stage of the twotailed swallowtail; willow, cottonwood, and chokecherry are fed upon by the western tiger swallowtail.

Although the caterpillars do feed on the leaves of certain trees and shrubs, they never occur in numbers sufficient to cause plant injury.

PLEASING FUNGUS BEETLE

Cypherotylus californicus Lacordaire
Coleoptera: Erotylidae

Life History and Habits: The pleasing fungus beetle develops on soft conk fungi on aspen, ponderosa pine and other logs in forested areas. The biology of the insect is largely unknown; some apparently spend the winter in the adult stage laying eggs in spring, others survive as larvae within the fungus. The larvae feed on the fungus during late spring and early summer, consuming large quantities. When full-grown the larvae hang from the underside of the logs and transform to a pupa, often in groups of several dozen. With this habit, the pupal stages may appear somewhat like a miniature bat roost. After about a week the adults emerge and are present through the end of summer and early fall. There is one generation per year.

BUMBLE FLOWER BEETLE

Euphoria inda (L.)
Coleoptera: Scarabeaidae

Life History and Habits: The adult beetles feed on a wide variety of sweet or fermenting liquids. They are commonly attracted in late summer to the bacterial ooze produced by infection of many trees. They also may occasionally damage ripening corn, ripe apples, grapes, melons, and peaches. The pollen and nectar of flowers such as sunflower, strawflower, and daylily may also be food plants.

This beetle primarily attracts interest and concern when it appears in great numbers on the wet bacterial ooze produced from infected trees, such as cottonwood, elm, and willow. Despite its occurrence at the site of this ooze, it is not involved in transmission of any diseases to woody plants.

On rare occasions, the bumble flower beetle has damaged day lily and strawflower. There is also some evidence that they can transmit disease organisms which can cause wilting in strawflower.

The overwintering stage is the adult beetles. These are broadly oval, about 1/2 to 5/8 inches long and densely cover with yellowish-brown hairs. In spring the beetles usually lay eggs in fresh manure (particularly horse manure), rotten wood or compost, and the C-shaped grubs develop in the decaying organic matter. As they feed, they form small, packed chambers in which they later pupate. The grubs are commonly found in gardens fertilized with manure or compost, but do not feed on roots.

The adults emerge in mid- to late summer and feed on a wide variety of sweet or fermenting liquids. They are commonly attracted in late summer to the bacterial ooze produced by infection of many trees. They may also occasionally feed on certain vegetables and flowers. There is one generation per year.

PONDEROUS BORER

Ergates spiculatus (LeConte)
Coleoptera: Cerambycidae

Hosts: Ponderosa pine and Douglas-fir are the principal hosts

Damage: The ponderous borer only develops on fallen trees and is not a threat to landscape plantings. However, this is the largest species of wood boring beetle found in Colorado and their large size attracts attention.

Life History and Habits: Larvae of the ponderous borer are known to foresters as timber worms. As larvae they feed in the wood of dead and dying pines. Borer weakened snags often fall more quickly than other trees so that it reduces the fire danger. Eggs are laid in the bark crevices of dead trees and stumps. Larvae excavate large tunnels through the sapwood and heartwood. The life cycle takes several years to complete.

Related Species: Several large, dark-colored longhorned beetles in the genus *Prionus* occur in Colorado. Although all are smaller than the ponderous borer, one species, the California prionus (*Prionus californicus* Motschulsky) can reach a length of two to three inches. In Colorado, the California prionus is limited to areas west of the Continental Divide.

ABOVE: Ponderous borer.

biological controls of insects associated with *trees and shrubs*

LADY BEETLES

Coleoptera: Coccinellidae

Often called "ladybugs" or "ladybird beetles", lady beetles (Coccinellidae) are the most familiar insect predator to most people. Although dozens of species occur in Colorado, they are all typically a round-oval shape. Most are also brightly colored and often spotted.

Females periodically lay masses of orange-yellow eggs. The eggs are quite distinctive, although they somewhat resemble those produced by elm leaf beetle. Eggs are usually laid near colonies of insects (aphids, scales, etc.) which will later be fed on by the larvae.

During the summer eggs hatch in about five days. The immature or larval stages look very different from the more familiar adults and often are overlooked or misidentified. Lady beetle larvae are elongated, usually dark colored and flecked with orange or yellow. They can crawl rapidly over plants, searching for food.

Adult and larval lady beetles feed on large numbers of small soft-bodied insects such as aphids. Lady beetles also eat eggs of many insects. Pollen, nectar and honeydew are other common foods.

One group of very small black lady beetles, aptly dubbed the "spider mite destroyers" (*Stethorus*) are also very important in controlling spider mites. Another unusual group are *Coccidophilus* spp. which are important predators of scales. Larvae of some lady beetles, e.g., those which specialize on aphids within leaf curls or feed on mealybugs, produce waxy threads which cover their body.

Lady beetles reproduce rapidly during the summer and can complete a generation in less than four weeks under favorable conditions. As a result, they often overtake a pest outbreak, controlling many potential insect problems.

Unfortunately, lady beetles tend to be 'fair weather' insects that are slow to arrive in the spring and often leave the plants by late summer. (A few kinds along the Front Range even 'head for the hills', spending the cool seasons at high elevations, protected under the snow.) As a result, late season 'blooms of aphids sometimes occur, as they continue to feed and escape their natural enemies.

TOP: Lady beetle with pupal skins. (T. Weissling)
BOTTOM: Lady beetle larva feeding on greenbug. (F. Peairs)

CLERID BEETLES

Coleoptera: Cleridae

Clerid beetles, or checkered beetles, are usually brightly colored insects that are generally elongate in form and somewhat flattened. Several species are important predators of bark beetles and other wood boring insects. The immature clerid beetles are commonly found in tunnels of bark beetles.

TOP: Green lacewing. (F. Peairs)
SECOND: Green lacewing larva.
THIRD: Brown lacewing adult.
BOTTOM: Syrphid fly adult at flower.

GREEN LACEWINGS

Neuroptera: Chrysopidae

Several species of green lacewings commonly are found on trees and shrubs. The adult stage is a pale green insect with large, clear, highly-veined wings that are held over the body when at rest. They are delicate and very attractive insects that primarily feed on nectar. The females lay a distinctive stalked egg, approximately one half inch in height. They may be laid in small groups or singly on leaves of plants throughout the yard.

Lacewing larvae emerge from the egg in about a week. These larvae, sometimes called aphid lions, are voracious predators capable of feeding on small caterpillars and beetles as well as aphids and other insects. In general shape and size, lacewing larvae are superficially similar to lady beetle larvae. However, immature lacewings usually are light brown and have a large pair of viciously hooked jaws projecting from the front of the head. Whereas lady beetles often are limited to smaller insects such as aphids, the green lacewings are capable hunters that can easily kill insects larger than themselves. Several generations of lacewings occur during the summer, and a green-brown cold tolerant species can be found late into fall.

BROWN LACEWINGS

Neuroptera: Hemerobiidae

The brown lacewings are related to the green lacewings with generally similar habits, being predators of insects in both their adult and larval stages. Many are more specialized predators, feeing primarily on woolly aphids and mealybugs. Brown lacewings also are almost always associated with dense vegetation, including trees and shrubs.

Adult brown lacewings are somewhat smaller than green lacewings with light brown wings. Mouthparts of the larvae are designed to pierce and capture prey, but are a bit more pronounced as those of green lacewings. Egg stages are not stalked.

SYRPHID FLIES

Diptera: Syrphidae

Syrphid flies, also called flower flies or hover flies, are common brightly colored flies. Typical markings are yellow or orange with black, and they may look like bees or yellowjacket wasps. However, syrphid flies are harmless to humans. Usually they can be seen feeding on flowers.

It is the larval stage of the syrphid fly that is an insect predator. Variously colored, the tapered "maggots" crawl over foliage and can daily down dozens of aphids. Syrphid flies are particularly important in controlling aphid infestations early in the season when its still too cool for lady beetles and other predators.

A few species of syrphid flies, such as the narcissus bulb fly, develop by feeding on and tunnelling plant tissues. These plant feeding syrphid flies often develop into large, stout-bodied flies that may resemble bumblebees.

LEFT: Syrphid fly larva feeding on aphids. (F Peairs)

LONG-LEGGED FLIES

Diptera: Dolichopodidae

The long-legged flies are moderately small flies noted for their metallic coloration. Adult stages feed on small insects, such as gnats and midges. Larvae also are predaceous and some (*Medetera* spp.) live under bark and are predaceous on bark beetle larvae.

PREDATORY STINK BUGS

Hemiptera: Pentatomidae

Although stink bugs include many species which feed on plants, particularly fruit or seeds, some are predators. Stink bugs, whether plant feeders or predators, are characterized by their distinctive shield-like body and ability to produce an unpleasant 'perfume' when disturbed. Predatory stink bugs feed by piercing the prey with their very narrow mouth parts and sucking out body fluids. Stink bugs also produce very unusual egg masses, which appear as clusters of small barrels fringed with spines at the top. Common Colorado species that feed on insects include *Perillus bioculatus*, *Apateticus bracteatus* and *Podisus placidus* that primarily feed on larvae of beetles and caterpillars.

ASSASSIN BUGS

Hemiptera: Reduviidae

Assassin bugs are equally capable predators, that can subdue large insects such as caterpillars and beetles. Most assassin bugs are elongate in form, have a pronounced 'snout' on the front which is the base for the stylet mouthparts, and are spiny. Despite their prodigious ability to dispatch most garden pests, they rarely become very abundant since they in turn have too many enemies of their own (mostly egg parasites).

The largest assassin bug associated with woody plants is the wheel bug (*Arilus cristatus*) an inch long species found in SE Colorado. Assassin bugs of elongate body form and stick-lie legs, notably those in the genus *Zelus*, are most commonly observed associated with trees and shrubs. Distinctive pods of eggs attached to leaves are produced by many assassin bugs.

PREDATORY PLANT BUGS

Hemiptera: Miridae

The family of plant bugs (Lygaeidae) include many species that can be seriously damaging to plants, such as the honeylocust plant bug. However, many have omnivorous habits, such as the Campylomma bug of western Colorado, and frequently feed on insects as well as plants. A few, such as *Deraeocoris nebulosus*, appear to be primarily predaceous and may be among the most important biological controls of spider mites, gall midges, and other shade tree pests.

TOP: A longlegged fly.
SECOND: Predatory stink bug feeding on elm leaf beetle larva.
THIRD: Assassin bug.
BOTTOM: Minute pirate bug.

MINUTE PIRATE BUGS

Hemiptera: Anthocoridae

The smallest of the "true" bugs commonly associated with trees and shrubs, minute pirate bugs also are among the most effective predators. Spider mites, thrips, aphids and insect eggs are the most common prey of these insects. Their small size (typically about 1/16-in) and the black and white coloration of adults are distinctive.

TOP: Predatory mite feeding on spider mites.
SECOND: Crab spider. (F. Peairs)
THIRD: Hunting wasp stinging caterpillar prey. (H. Evans)
BOTTOM: Bronze birch borer with its primary parasite, a chalcid wasp. (Whitney Cranshaw)

PREDATORY MITES

Acarina: Phytoseiidae, Bdellidae, Camerobiidae and others

Many types of mites are predators of plant-feeding spider mites. Typically, these predatory mites are little larger than spider mites and faster moving than their prey. Predatory mites can often provide good control of spider mites, but most are slowed down by dry weather - they like it hot and humid. The predatory mites are also more susceptible to insecticides than are plant-feeding species.

SPIDERS

Araneida: Thomisiidae, Salticidae, Lycosidae, Araneaidae, Agelenidae and others

Although hardly a favorite of most, spiders are often the most important predators of insects found on trees and shrubs. All spiders feed only on living insects or other small arthropods.

Most people primarily observe the many web-making spiders, banded argiope or "monkey-face" spiders. However there are many other spiders (wolf spiders, crab spiders, jumping spiders) which do not build webs. These spiders move about the plants and hunt their prey. These less conspicuous spiders can be very important in controlling insect pests such as beetles, caterpillars, leafhoppers and aphids.

HUNTING WASPS

Hymenoptera/Families: Sphecidae, Vespidae

Many wasps prey on insect pests and feed them to their young. These hunting wasps build nests out of mud or paper. Other hunting wasps construct nests by tunnelling into soil or pithy plant stems, such as ash, rose or caneberries. The adult wasps then capture insect prey and take them back to the nest, whole or in pieces, to feed to the immature wasps.

Most hunting wasps are solitary wasps. The females construct the entire nests, working alone. The females then search for prey which they immobilize with a paralyzing sting and carry back to the nest. The young wasps develop by eating the food the mother wasp has provided.

The solitary hunting wasps (Sphecidae) have very specialized tastes which cause them to search only for selective types of prey. For example, some develop on leafhoppers; others attack caterpillars; beetles are prey for some hunting wasps. The largest hunting wasp is the cicada killer, which resemble a giant yellowjacket and kills the dog-day cicada. Despite their often fearsome appearance, the solitary hunting wasps rarely sting and do not contain the potent venom of the social wasps.

Other hunting wasps, such as the paper wasps and yellow jackets (Vespidae), are social species where many individuals work together and there are a variety of specialized **castes** (queen, drone, workers) in a colony. These make nests of a papery material that they form by chewing wood, cardboard or other materials. The nests are constructed under eaves, in trees, or underground in holes around building foundations or abandoned rodent burrows. These social wasps create new colonies every year and colonies may be aggressively defended by stinging guard wasps. However, most social wasps rear their young on a diet of caterpillar paste or other insects and some are very useful for control of pests such as fall webworm. At the end of the season, the nests of the social wasps are abandoned.

TACHINID FLIES

Diptera: Tachinidae

Tachinid flies develop as parasites inside other insects. Tachinids are about the size of a house fly, generally gray or brown, and covered with dark bristles. They are rarely seen but often leave their 'calling card', a white egg laid on various caterpillars, beetles and bugs, usually near the head. Douglas-fir tussock moth, tent caterpillars, and fall webworm are among the insects commonly attacked by tachinid flies.

The eggs hatch within the day and the young fly maggots tunnel into their host. There they feed for about a week (carefully avoiding the vital organs until the end), eventually killing the host insect.

BRACONID AND ICHNEUMONID WASPS

Hymenoptera/Families: Braconidae, Ichneumonidae

The parasitic wasps, including the braconid and ichneumonid wasps, are a diverse group of wasps which develop as insect parasites. Some are very small and rarely observed, attacking small insects such as aphids. Others even live in the eggs of various pest insects. Larger parasitic wasps attack caterpillars or larvae of the wood boring horntails.

There is often little external evidence of parasitic wasp activity since the young wasps develop inside the host insect from eggs that were inserted by the mother wasp. However, parasitized insects may be somewhat different in form. For example, aphids that are parasitized by these wasps are typically small and discolored, and called "aphid mummies." Other common braconid wasp species spin conspicuous yellow or white pupal cocoons after emerging from a host.

CHALCID WASPS

Hymenoptera/Superfamily: Chalcidoidea

There are hundreds of species of chalcid wasps which attack and kill other insects. However, most chalcid wasps are very small and are rarely observed. Like the braconid and ichneumonid wasps, chalcid wasps do not sting and are harmless to humans.

Some of the more important chalcid wasps attack aphids and various caterpillars such as the fall webworm.

INSECT DISEASES

TOP: Tachinid fly. (J. Capinera)
SECOND: Tachinid eggs laid near head of hornworms. (J. Capinera)
THIRD: Parasitic wasp.
BOTTOM: Virus-killed gypsy moth larva.

Infrequently observed, insects and mites often suffer from lethal disease. Periodically, epizootics resulting from infection by fungi, bacteria, protozoa, or viruses may sweep through an insect population.

Although the classes of organisms that cause human disease are the same for many other animals and plants (e.g., viruses, bacteria, fungi), it is important to keep in mind that insect diseases are very specific in their effects. Insect diseases do not infect mammals or birds, restricting their effects to the arthropods. Furthermore, most insect diseases are so specific in their effects that they only can infect a few insect species.

Viruses are most commonly found among the caterpillars and sawflies. One particularly gruesome group of these viruses (nuclear polyhedrosis viruses/NPV) cause 'wilt disease'. Caterpillars infected by these viruses are killed rapidly, their virus filled bodies hanging limply by their hind legs. At the slightest touch, the insects rupture, spilling the virus particle on the leaves below them to infect other insects. One type of wilt disease is an important biological control of the Douglas-fir tussock moth. Other types of viruses cause less spectacular infections. Most are slower acting than the wilt viruses. External evidence of these viruses may be a chalky color of the insects and a general listlessness.

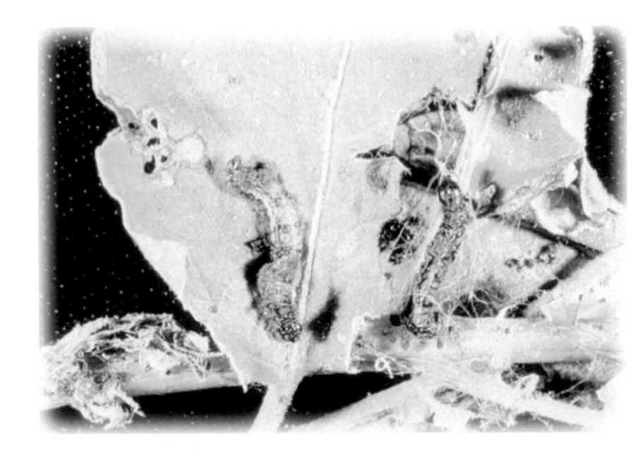

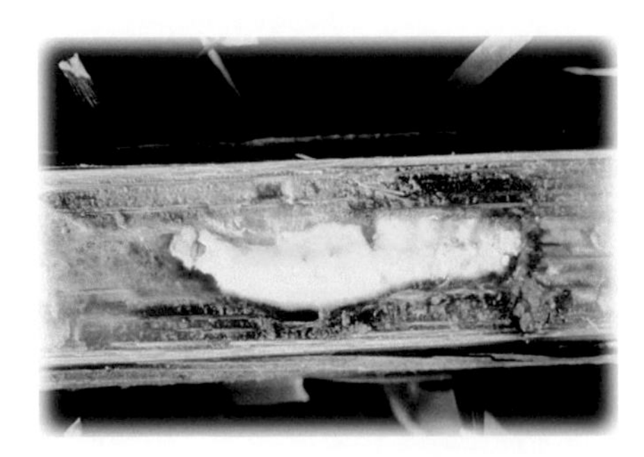

TOP: *Bacillus thuringiensis*-killed alfalfa webworms.
(J. Capinera)
BOTTOM: *Beauvaria*-killed European corn borer larva.
(J. Capinera)

Although virus diseases of insects are widespread in nature, rarely have they been adapted for applied biological control. Much of this is due to problems in registering these as insecticides. Regulatory agencies have had difficulty in deciding how to insure the safety of such mysterious particles as viruses. Manufacturers have also been leery of developing viruses due to problems in production (they need live cells to develop) and because the selectivity of viruses allows them to only be used against a few insects.

Bacteria have received more attention, due almost entirely to the successful adaptation of *Bacillus thuringiensis*. Several manufacturers have produced and marketed various strains of this famous bacteria. *Bacillus thuringiensis*, and most other bacterial diseases of insects, work by disrupting the 'gut' lining of susceptible insects, ultimately killing them by a type of blood poisoning. Infected insects usually shrivel and darken. Strains of this bacteria are effective against caterpillars (*kurstaki, thuringiensis, aizawai* strains), leaf beetles (*tenebrionis/san diego* strain) and larvae of certain flies such as mosquitoes and blackflies (*israelensis* strain).

Fungi produce some of the more spectacular diseases of insects. A wide variety of insects succumb to fungus disease around the yard and garden. Fungus killed insects and mites become stiff and often are tightly attached to a leaf or stem. When conditions are right they become covered with a white, light green or pink 'fuzz', the spores of the fungus. At least one fungus, *Beauveria bassiana* (e.g., Naturalis, BotaniGard), is currently marketed to control insects on ornamental plants.

Protozoa tend to cause debilitating infections among insects. Effects are often subtle, such as reduced feeding, activity or reproduction. Immature stages are usually much more susceptible to protozoan infections and survival can be reduced. Spruce budworms and grasshoppers are among the groups of insects that are common hosts of protozoa.

Additions to 506A

Add as a Related Species, with rabbitbrush beetle, page 32

Coleothorpa dominicana (F.) is occasionally damaging to native sumac (skunkbrush). Adults are dark-colored leaf beetles and larvae are pale brown grubs that skeletonize leaves. It is smaller than the more common leaf beetle of sumac, *Blepharida rhois.*

Add following table on page 33

APPLE FLEA BEETLE

Altica foliacea LeConte

Hosts: Evening primrose (*Oenothera* spp.) is the only known larval host. Adults disperse widely and may damage a wide range of herbaceous and woody ornamentals. Plants that sustained injury during the 2003 outbreak included: *Zauschneria garretti, Epilobium fleischeri,* evening primrose, *Gaura* (whirling butterfly), alfalfa, yellow flax, crabapple, grape, wild rose, *Ribes aureum, Prunus besseyi,* and coyote willow

Damage and Diagnosis: Larvae feed generally on the leaves and flowers of evening primrose (*Oenothera* spp.). They are generally dark and pull themselves along plants with three pairs of legs on the thorax. Adults are shiny green, or blue-green beetles that jump readily when disturbed. Typically they chew small roundish holes in foliage. Heavily infested plants may become very lacy in appearance.

Life History and Habits: Biology in Colorado is poorly understood. Apparently adults are the overwintering stage and these move to evening primrose in spring. Eggs are laid as small masses and larvae feed on leaves and petals. Adults disperse and feed on a wide variety of summer hosts, not laying eggs at this time. A small second generation may occur in late summer, but most adults move to winter shelters following the period of feeding in the summer.

TOP RIGHT: Whitish sawdust characteristic of ambrosia beetle. TOP LEFT: Brownish sawdust produced by Ips beetle in tree killed by Mountain pine beetle. (David Leatherman)
BOTTOM: White pine blister rust, 3-5 year old infections with bark yellowing and aecia. (Bill Jacobi)

Add to elm leafminer, page 38

Scientific name of the elm leafminer has been changed to *Kaliofenusa ulmil.*

Related Insects: The hawthorn leafminer, *Profenusa canadensis* (Marlatt), makes blotch-type leaf mines in hawthorn during late spring.

Note for table on page 45:

Norway maple aphid, *Periphyllus lyropictus* (Kessler), is a very common honeydew producer on Norway maple throughout late spring and early summer.

Page 53, Hawthorn mealybug

Add pyracantha as an occasional host of this species

Correction on 110, Spruce Broom Rust

Spermatia and aeciospores are produced on *spruce* (not true fir); "Infection of *spruce* (not fir) requires moist and temperate weather...."

TOP: Apple flea beetles on coyote willow. (Whitney Cranshaw)
BOTTOM: Galleries produced by banded elm bark beetle. (David Leatherman)

Insert at page 137, before Twig Beetles
MAGDALIS WEEVILS

Magdalis gentilis LeConte, *M. lecontei* Horn

Hosts: Pines, particularly ponderosa and lodgepole

Damage and Diagnosis: Adults are black (*M. gentilis*) or bright blue (*M. lecontei*) and are usually seen on foliage (feeding) or on new shoot growth (laying eggs). Larvae, which are legless, bark beetle-like grubs, feed beneath the bark of twigs and branches. They can cause damage similar to - and often mistaken for - that of small bark beetles in the so-called "twig beetle" group. Feeding injury on needles is very rarely noticeable, but has been reported as serious on lodgepole pine in Montana. Branch dieback is considered minor, although an occasional leader on pines has been killed.

Life History and Habits: Adults are often seen on new candles in late spring, but may be found in the outer periphery of pine crowns throughout the summer. Presumably there is one generation per year, with larvae under bark being the overwintering stage.

Related Insects: The red elm bark weevil, *M. armicollis* (Say), is found in small to mid-size branches of American elm. The adults are dull reddish, and cause petite skeletonizing of leaves in mid-summer. The larvae construct galleries that generally run parallel to the main dimension of the branch. These galleries could be confused with those of the various *Scolytus* bark beetles attacking elm. This weevil is not a vector of Dutch elm disease, and only attacks elm wood that is under serious stress.

Follows smaller European elm bark beetle, page 160
BANDED ELM BARK BEETLE

Scolytus schevyrewi Semenov

Hosts: Records to date in Colorado are from several species of elm. However, this insect is native of China, Mongolia, Korea, and Russia, this insect has a host range that includes the following deciduous tree genera: *Ulmus* (elms), *Salix* (willows), *Prunus* (various stone fruit trees including cherries and apricots), *Caragana* (pea-shrubs), *Persica* (peach), and *Eleagnus* (Russian olive and relatives).

Related Insects: The banded elm bark beetle (BEBB) is a close relative of the smaller European elm bark beetle (SEEBB) and similar can develop under elm bark. On most individuals the elytra (also called "wing covers") are divided into three bands, with the first and third bands being light reddish brown and the middle one being blackish-brown. Even to the unaided eye, the two-toned elytra make BEBB easily distinguished from SEEBB. Also, BEBB is generally slightly larger that SEEBB.

Damage and Diagnosis: So far, beetles have been observed attacking stressed elms of all types in all situations (including urban and rural, conservation-type plantings and including trees broken by storms, suffering from drought, or infested by DED). Trunk attacks on reasonably moist hosts are characterized by sap-soaked areas of wet bark around the entrance holes. Similar to its European relative, BEBB chews a beetle-sized notch in the twig crotches of healthy elms. This habit, widely observed in Hemingford, NE in September 2003, is troublesome if it is proven BEBB vectors *Ophiostoma ulmi*, the primary causal fungus of Dutch elm disease (DED).

Life History and Habits: BEBB larvae develop under the bark and feed on phloem of stressed or unthrifty host plants and adults engage in maturation feeding on tender twig crotch bark of healthy host plants. Attacks are no doubt regulated by a pheromone system, which is being investigated by U.S. researchers. Arborists engaged in elm pruning and removal report being "swarmed" by adult beetles. This apparently indicates host volatiles are involved in the orientation of beetles to elms. BEBB has at least three generations per year at lower elevations. Colorado observations during 2003 showed a flight of over-wintering beetles in March and April, a second generation emergence in late June through July, and a third emergence in late August through early October. Freshly-attacked Siberian and American elm wood collected in Cheyenne, Wyoming and western Nebraska in late September 2003 and promptly brought inside produced a fourth generation in mid-November. Presumably these November adults would have emerged under outdoor conditions in spring of 2004. Thus, it appears the beetles have a generation time of two to three months under summer temperature regimes.

The gallery pattern of BEBB is very similar to that of SEEBB, with the egg gallery being roughly parallel to the long dimension of the wood and the larval galleries radiating off the egg gallery to form a "fan-shaped" pattern typical for the genus. On average the egg galleries of BEBB appear to be shorter than SEEBB, with fewer larval galleries radiating from them, and with the larval galleries being more widely spaced from one another. However, it should be stressed that given current limited experience with BEBB, it is far safer to determine the identity of these two bark beetles in elm, which frequently co-attack hosts and occur side by side, on the basis of adults than galleries. Pupation of BEBB is usually in the corky bark. The exit holes are about 1 mm in diameter, dry, and pepper the bark. Unlike the entrance holes, which tend to be concentrated in bark crevices, exit holes can occur anywhere on the wood surface.

TOP: Banded elm bark beetle. (Whitney Cranshaw)
CENTER: Maple wood wasp adult. (David Leatherman)
BOTTOM: Tunneling produced by Gambel oak borer. (David Leatherman)

Page 141, change in Scientific Name

Two species of fungi are associated with Dutch elm disease in North America, *Ophiostoma ulmi* and *O. nova-ulmi.*

Page 165, add this table between Engraver beetles and spruce ips:

Ips species known to affect pines and spruce in Colorado.

Species	Hosts	Comments
Ips hunteri are	Spruce	This is a common species affecting Colorado blue spruce in landscape settings. Upper portions of the tree typically infested first.
Ips pilifrons	Spruce	A forest species often called the "spruce ips," tends to infest the upper part of fallen trunks.
Ips pini	Ponderosa, lodgepole, other pines	The most common species associated with pines
Ips knausi	Ponderosa pine	Common at base of trunk and in fresh stumps
Ips calligraphus	Ponderosa pine	
Ips confusus	Pinyon, rarely other pines	
Ips latidens	3 and 5 needle pines	
Ips borealis	Engelmann spruce	
Ips integer	Ponderosa pine, primarily	
Ips woodi	Limber pine	
Ips mexicanus	Lodgepole and limber pines	

TOP: Conk produced by Ganoderma infection of aspen. (Bill Jacobi)
CENTER: Gambel oak borer, with one wing extended. (Whitney Cranshaw)
BOTTOM: Red elm bark weevil galleries exposed by downy woodpecker. (David Leatherman)

Insert at bottom of page 165
AMBROSIA BEETLES

Trypodendron spp. including *T. retusum* (LeConte) and *T. rufitarsis* (Kirby)

Hosts: Pines, true firs, Douglas-fir, and spruce; *T. retusum* develops on aspen and poplar.

Damage and Diagnosis: The so-called "pinhole borers" attack dead and dying trees, often those under attack by more aggressive bark beetles. Darkish gray or black stain often develops in the wood immediately surrounding the galleries and their primary impact is degradation of lumber intended for paneling and other high-grade wood products.

Life History and Habits: Several habits of ambrosia beetles differ from other bark beetles, primarily in the use of mutualistic fungi and the pattern of tunneling. Initial attacks are made by the adults, which tunnel through bark and produce a light-colored boring dust. This is distinctive from the darker, bark-colored boring dust of bark beetles or the coarser fibers of wood borers and ants. With the bark removed, the only visible sign of ambrosia beetles are small, round holes going straight into the xylem wood.

The adults cut small chambers for rearing larvae, called "cradles," which radiate at right angles from the parental galleries. The resulting characteristic pattern is found deep within the wood, and usually only seen when the tree is cut or trunkwood is split. These tunnels are colonized by special and specific fungi, and the young develop by eating the fungi, rather than further tunneling. As such, ambrosia beetle may be considered "gardeners" within their gallery systems. Occasionally a purple stain the color of raspberry juice occurs at the base of large-diameter pines killed by bark beetles and subsequently invaded by ambrosia beetles.

Page 168, in table among Metallic Wood Borers, add:

Chysobothris texana
Common name: not established

Common hosts: eastern redcedar
Typical flight periods: mid-June through early August

Agrilus quercicola
Common name: Gambel oak borer

Common hosts: oaks
Typical flight periods: late May through mid-July

Insert at page 177, before Cottonwood Borer

Maple wood wasp
Xiphidria maculata Say

Hosts: In Colorado, has only been found in silver maple. Reported from red and sugar maples in the East.

Damage and Diagnosis: Apparently only attacks dead and dying trees. Usually detected by the premature falling of branches riddled with boring tunnels, or by woodpecker activity. The primary problem caused by this insect is the acceleration of hazard represented by dead limbs and stems. The adult wasps are 7-20mm long, slender with black, white, and yellow markings and narrow necks. The larvae, when full grown, are 18-20 mm long, white, with small dark horn at the upper part of the rear abdomen.

Life History: Adults have been observed emerging in May in Fort Collins, the only place it has been detected so far in Colorado. This corresponds to the reported biology in Indiana. Adults lay eggs on declining trunks or branches. The larvae tunnel beneath the surface, mostly in horizontal galleries in the sapwood that run the long dimension of the wood. Larvae overwinter. There is one generation per year.

Insert this after honeylocust borer, page 184

GAMBEL OAK BORER

Agrilus quercicola Fisher

Hosts: Oak. Native of Gambel oak but can colonize most oaks

Damage and Diagnosis: Larvae tunnel under the bark making girdling wounds. Oozing sometimes, but not always, is present around wound sites in late summer. Decline is usually first expressed as crown thinning. Dieback and tree death can occur from severe and sustained injury.

This insect emerged in 2003 as a major pest of nursery-grown oak over a broad part of the Front Range from Colorado Springs through the Metro Denver area. I had previous been associated with decline of Gambel oak that had suffered sustained stresses from drought, late frosts and other injuries. Much of the 2003 outbreak was associated with migration from native Gambel oak, but this insect may also be established on oaks in landscape settings.

The Gambel oak borer is similar in size and form to other *Agrilus*. The most distinguishing feature is the golden prothorax. Wing covers are very dark gray.

Related and Associated Insects: *Agrilus quercicola* seems to be the only *Agrilus* species associated with borer-related oak decline in Colorado. However, other oak infesting Agrilus occur in the US and could be easily moved into the state. A likely species that may find its way to the state on midwestern-grown oak is the twolined chestnut borer, *Agrilus bilineatus* Weber.

Life History and Habits: Life history studies are underway. It is likely that it will be similar to the other *Agrilus* species that occur in North America, which include:

- One generation produced per year.
- Emergence of adults in late spring, likely beginning by late May and continuing into July.
- A 2 to 3 week period after adults emerge during which they feed on leaves, mate and females mature eggs.
- Eggs laid on the outside of the bark, in crevices, over a 4-6 week period, probably beginning in early to mid June.
- Entry of the larva into the tree following egg hatch and subsequent development in the cambium.
- Overwintering form is a nearly full-grown larva that pupates beneath the bark in spring.

Additions to flatheaded appletree borer, page 184

Hosts: include hackberry

Related Species: Several *Chrysobothris* species occur in conifers. *C. breviloba* Fall, *C. dentipes* (Germ.), and *C. trinervia* (Kirby) are generally associated with overmature or recently felled pines; and *C. ludificata* Horn has been reared from spruce. None of these are damaging to growing trees. However, *C. texana* LeConte has recently been associated with declining eastern redcedar in the eastern plains. All of these species appear to have an approximate flight period of from mid-June through early August.

TOP: Larval tunneling by *Magdalis lecontei*. (David Leatherman)
CENTER: Rough bark caused during infection by white pine blister rust. (Bill Jacobi)
BOTTOM: Blistering resulting from white pine blister rust infection. (Bill Jacobi)

TOP: Cytospora canker, spring symptoms on aspen. (David Leatherman)

Species similar to boxelder bug, page 206

The **goldenrain tree bug,** *Jadera haematoloma* (H.S.), sometimes known as the redshouldered bug can become very abundant on its host goldenrain tree. It may mass on adjacent buildings, but does not winter indoors, as does the boxelder bug.

In the Grand Junction area there have been occasional reports of **sycamore seed head bug,** *Belonochilus numenius* (Say). Very high populations may occur on or in the near vicinity of sycamore.

A

Abdomen - the posterior of the three main body divisions in an insect; the posterior of the two main body regions in a mite, spider or other arachnid

Acervulus - a subepidermal, saucer-shaped, asexual fruiting body producing conidia on short conidiophores.

Adelgid - a family of insects (Adelgidae) closely related to aphids known as "woolly aphids"

Aecium (pl. **aecia**) - cup-like fruiting structure of a rust fungus, often a yellow to orange mass.

Aeciospore - Spore produced in a cup-shaped aecium of a rust fungus.

Alternate host - a plant required to complete the life cycle of an insect or fungus, other than the host plant that is primarily damaged. This habit is very common among certain groups of aphids and the rust fungi.

Annual - a plant or fungal fruiting body in an active, living state for one season

Apothecium - an open cup- or saucer-shaped ascocarp of some Ascomycetes

Ascocarp - the fruiting body of Ascomycetes bearing or containing asci

Ascomycetes - a group of fungi producing their sexual spores, ascospores, within asci.

Ascospore - a sexually produced spore borne in an ascus.

Ascus (pl. **Asci**) - a saclike cell of a hypha in which contains the ascospores (usually eight)

Asexual - any type of reproduction not involving a union in which fertilization and meiosis occurs

B

Basidiomycetes - a group of fungi producing their sexual spores, basidiospores, on basidia

Basidiospore - a sexually produced spore borne on a basidium

Basidium - a club-shaped structure on which basidiospores are borne

Blight - general term for sudden, severe withering and/or killing of leaves, flowers, shoots, fruit, or the entire plant. Usually young growing tissues are attacked.

Bud break - time at which dormant buds begin expanding and opening

C

Callus - a mass of thin-walled undifferentiated cells, developed as the result of wounding or culture on nutrient media.

Cambium - a layer of cells between the bark and wood that divide into phloem outward and xylem (wood) inward. If destroyed or exposed the plant dies.

Canker fungi - disease causing fungi that often cause the death of bark and cambium resulting in sunken areas on trunks, branches and twigs of woody plants

Caterpillar - the larva of a butterfly, moth, or sawfly

Chelated - Chelating agents chemically hold elements in a form usable to the plant.

Cleistothecium (pl. **cleistothecia**) - A spherical, entirely closed fruiting structure that ruptures at maturity to release spores; typical of powdery mildew fungi.

Chlorosis - yellowing of normally green tissue due to chlorophyll destruction or failure of chlorophyll formation

Cocoon - silken case within which the pupal stage of many insects is formed

Complete metamorphosis - a pattern of metamorphosis used by many insects (e.g., beetles, moths/butterflies, sawflies, flies) which involve eggs, followed by immature nymphs, a transition pupal stage, and finally adults. Adult and immature stages often have very different habits and appearance among insect groups with this metamorphosis. Also called holometabolous metamorphosis.

Conidiophore - a specialized hypha on which one or more conidia are produced

Conidium (pl. **conidia**) an asexual fungus spore formed on a conidiophore

Conk - fruiting body usually of a wood- rotting fungus that forms on tree stumps, branches, trunks, or lumber. Usually they are spongy to hard, become large when mature, and persist for one or more years.

Cuticle - waxy protective membrane over plant epidermis, broken only by natural openings

Coremium - an asexual fruiting body consisting of a cluster of erect hyphae bearing conidia

Cornicle - a tubular structure (paired) on the posterior of aphids through which alarm pheromones are released

Crook - a bending of the terminal growth of a plant, often caused by injury

D

Defoliation - the loss of leaves as that occurring in natural shedding, or from the feeding activities of insects and other plant feeders

Diapause - a period of dormancy in which many insects undergo to avoid adverse conditions (e.g., winter cold) that can only be terminated by certain stimuli such as day length or a prescribed length of exposure to cold

Dieback - the decline and dying of the upper or terminal growth of a plant

Disease - any malfunctioning of host cells and tissues that results from continuous irritation by a pathogenic agent or environmental factor and leads to development of symptoms

Disease cycle - the chain of events involved in disease development, including the stages of development of the pathogen and the effect of the disease on the host

Domatia - areas of dense plant hairs at the junction of leaf veins. A speculated function of these sites is to provide protection to predatory mites.

Dorsal - the back or upper side

E

Epidermis - the outermost layer of cells of leaves, young stems, roots, flowers, fruits, and seeds. In insects the epidermis is the single, outermost layer of cells that secretes the cuticle.

Erineum - a change in plant growth, or gall, where plant hairs are produced in great abundance creating a felty patch on a leaf surface. Many eriophyid mites cause such changes in leaf growth.

Exoskeleton - a skeletal structure that is formed on the external surface of an animal, such as occurs with insects, mites, and other arthropods

Eyespots - prominent markings resembling eyes on the wings or body of certain insects

F

Fingergalls - abnormal growth forms in the shape of fingers, such as are produced by eriophyid mites on certain plants (e.g., wild plum, cherry)

Forewings - the pair of wings closest to the head of an insect

Frass - solid insect excrement typically consisting of a mixture of chewed plant fragments

Fruiting body - a complex fungal structure producing spores

Fundatrix - the female aphid emerging from the overwintered egg that initiates colonies in spring

Fungicide - a compound toxic to fungi

Fungistatic - a compound that prevents fungus growth without killing the fungus

Fungitoxic - able to kill or inhibit fungal growth, sporulation or spore germination

G

Gall - the abnormal growth of plant tissues, caused by the stimulus of an animal, microorganism, or wound

Gregarious - living and feeding in groups

H

Haustorium - a projection of a parasitic plant or fungal hyphae into host cells which acts as an absorbing organ

Hibernation - winter dormancy

Hibernaculum - a tiny cocoon spun by first or second instar caterpillars for overwintering shelter (pl. **hibernaculae**)

Honeydew - the sugary, liquid excrement produced by certain aphids, scales and other insects that feed in the phloem of the plant

Host - the plant on which an insect or pathogen feeds; the animal on which a parasite develops

Hyaline - transparent or nearly so; translucent; clear or colorless.

Hypha (pl. **hyphae**) - a single branch of a mycelium

I

Immune - cannot be infected by a given pathogen or damaged by an insect

Infection - the establishment of a parasite within a host plant

Injury - damage of a plant by an animal, physical, or chemical agent

Inoculum - fungal spores, mycelium, bacterial cells, or viral particles that can initiate infections

Internodes - the region between two adjacent nodes on a stem

Instar - the stage of an insect between periods when it molts

L

Larva (pl. **larvae**) - the immature stage, between egg and pupa of an insect with complete metamorphosis; i.e., caterpillars, maggots, grubs

Lateral - directed to the side. Used to described markings on the sides of an insect

Leafminer - an insect which has the habit of developing by feeding on internal leaf tissues which it chews as it mines the leaf. Insects which tunnel needles in a similar manner are called **needleminers**.

Lenticels - a small pore (natural opening) on a stem, tuber, root or fruit through which carbon dioxide and other gasses pass. Pathogens often enter through lenticels.

Lesion - localized, often sunken area of diseased or disordered tissue; a wound.

M

Metamorphosis - changes in form undergone by insects as they grow and develop

Mine - to form a burrow or excavate a tunnel. Used to describe the activities of insects that live or feed within a leaf or needle

Molt - the shedding of the exoskeleton by an insect in the process of development

Mosaic - symptom of certain viral diseases of plants characterized by intermingled patches of normal color with those that are light green or yellowish color

Mycelium - the mass of hyphae that make up the body of a fungus.

Mycorrhizae - a symbiotic association of a fungus with the roots of a plant.

N

Necrotic - dead and discolored

Needlesheath - protective structure at the base of needles

Nematode - generally microscopic, wormlike animals that live saprophytically in water or soil, or as parasites of plants and animals

Node - enlarged joint on a stem that is usually solid; site from which a leafy bud and branch arises

Nymph - an immature stage of an insect with simple metamorphosis; e.g., aphids, bugs, grasshoppers

O

Oviposition - the process of laying an egg by an insect. The verb is **oviposit**.

Ovipositor - the egg laying apparatus of a female insect

P

Parasite - an organism that lives at the expense of another. (The term *parasitoid* is often used to described parasitic insects that kill the host in which they develop.)

Pathogen - an entity that can incite disease

Perithecium - the globular or flask-shaped ascocarp, having an opening or pore (ostiole).

Perennial - a plant or fungal fruiting body that persists in active, living state for more than one growing season.

Petioles - the leaf stem or stalk that attaches to a twig

Pheromone - a chemical used to communicate between members of the same species. For example, many female moths produce sex pheromones to attract mates.

Phloem - food-conducting tissue located in the bark of woody plants that consist of sieve tubes, companion cells, phloem parenchyma, and fibers

Pitch - a resinous material exuded by conifers either naturally or in response to a wound

Predator - an animal that moves and hunts smaller animals (*prey*)

Pupa - the transitional stage, between larva and adult, of insects with complete metamorphosis. In moths and sawflies the pupal stage usually takes place in cocoons

R

Resin - sticky (fresh) to brittle (old) plant exudate of conifers

Resistance - the ability of an organism to exclude or overcome, completely or in some degree, the effect of a pathogen, insect or other damaging factor

Root graft - union of roots from two or more closely situated trees of the same or closely related species; often an avenue for transmission of pathogens.

Rot - the softening, discoloration, and often disintegration of a succulent plant tissue a result of fungal or bacterial infection

Rust - a disease giving a "rusty" appearance to a plant and caused by one of the Uredinales (rust fungi)

S

Sap wells - holes produced in plants by sapsuckers and other birds to produce a flow of sap on which they then feed

Saprophytes - an organism that feeds on dead organic matter commonly causing its decay.

Sapwood - young, physiologically active zone of wood; outermost growth layers of xylem in woody plants that is conducting water.

Sclerotium (pl. **sclerotia**) - vegetative fungal resting structure which may remain dormant in soil, plant refuse, or seed for long periods, surviving unfavorable environmental conditions.

Scorch - "burning" of leaf margins as a result of infection or unfavorable environmental conditions.

Secondary agent - any infection or infestation agent caused as a result of a primary stress agent

Sexual - produced as a result of a union in which fertilization and meiosis occurs.

Serpentine - winding, twisting pattern typically used to describe the shape of certain leaf mines

Simple metamorphosis - a pattern of metamorphosis used by many insects (e.g., true bugs, aphids, grasshoppers, earwigs) which involve eggs, followed by immature nymphs, and finally adults. Adult and immature stages usually feed in the same manner and are primarily differentiated by the adult features of sexual maturity and (usually) wings. Also called gradual metamorphosis or hemimetabolous.

Sign - the pathogen or its parts or products seen on a host plant

Skeletonize - describes the feeding pattern of certain leaf-feeding insects that avoid feeding on the main veins of the leaf, and often a leaf surface leaving a "skeleton" of the leaf surface

Sooty mold - a dark, typically black, fungus growing on insect honeydew

Spiroplasma - a type of pleomorphic bacteria that lacks a cell wall and are found in the phloem of infected plants. The sprioplasmas are related to phytoplasmas but differ in being able to be cultured and, often, having a helical shape.

Spore - the reproductive unit of fungi consisting of one or more cells; it is analogous to the seed of green plants

Sporodochium - a fruiting structure consisting of a cluster of conidiophores woven together on a mass of hyphae

Stippling - small, white flecking injuries produced by certain insects (some leafhoppers, lacebugs) and spider mites resulting from removal of plant sap

Stroma (pl. **stromata**) - compact mass of vegetative fungal hyphae in which fruiting bodies and spores are usually produced

Susceptible - lacking the inherent ability to resist disease or attack by a given pathogen or insect

Symptom - the external and internal reactions or alterations of a plant as a result of a disease - what the sick plant looks like

Systemic - spreading internally throughout the plant body; said of a pathogen or a chemical.

T

Tents - protective shelter constructed of silk, spun by certain caterpillars

Terminal growth - typically new growth or buds at the end of a branch or twig

Thorax - the middle section of an insect body where the legs and wings are attached

Toxin - poisonous secretion produced by a living organism.

Tubercles - a rounded protuberance found on many insects, particularly caterpillars

Tylosis - an overgrowth of the protoplast of a parenchyma cell into an adjacent xylem vessel or tracheid

V

Vector - a living organism (ie, insect, bird, higher animal, etc.) able to carry and transmit a pathogen

Ventral - the underside of the body

Virulent - capable of causing a severe disease; strongly pathogenic

Virus - a submicroscopic obligate parasite consisting of nucleic acid and protein.

W

Wingspan - the measurement between tips of the extended forewings of an insect

Witches' broom - broomlike growth or massed proliferation caused by the dense clustering of branches of woody plants

X

Xylem - the complex supporting, water- and mineral-conducting tissue of vascular plants that makes up sapwood and heartwood

diagnostic key to common *woody plant disorders*

ALDER (*Alnus* spp.)

Affecting leaves
- Leaves mined
  - Alder leafminer (*Fenusa dorhnii*), 38
- Small pits chewed in leaves
  - Alder flea beetle (*Altica ambiens*), 33
- Beadlike galls on both sides of leaf surface
  - An eriophyid mite (*Phytoptus laevis*), 100
- Velvety patches
  - An erineum (eriophyid) mite,100

Affecting twigs
- Producing frothy spittlemass
  - Spittlebugs, 55
- Cottony material
  - Alder aphid (*Oestlundiella flava*)

Affecting trunk and larger branches
- Discolored areas, dead bark containing small pimple-like fruiting bodies (pycnidia)
  - Cytospora canker (*Cytospora* sp.), 146
- Borer at base of plant
  - Alder borer (*Saperda obliqua*)

APPLE (See **Pome Fruits**)

APRICOT (See **Stone Fruits**)

ASH (*Fraxinus* spp.)

Affecting leaves
- Leaves chewed
  - Normal chewing injuries, primarily confined to leaf edges
    - Brownheaded ash sawfly, 24
    - Cankerworms, 18
    - Forest tent caterpillar, 11
    - Fruittree leafroller, 7
    - Great ash sphinx, Twinspot sphinx, 20
    - Western tiger swallowtail, 227
  - Smooth semicircular cuts made on leaf edge
    - Leafcutter bees, 36
  - "Shothole" feeding wounds in leaves
    - Brownheaded ash sawfly (young larvae), 24

Leaves spotted
- Blackened spotting
  - Ash anthracnose (*Apiognomonia* sp.), 72
- Rust or orange color spots
  - Leaf rust (*Puccinia sparganiodes*), 77
- Leaflets being distorted and/or killed back
  - Ash plant bug, 57
- Flecking wounds on leaves
  - Ash plant bug, 57
  - Lacebug, 59

Leaves yellowing
- Witches brooming
  - Ash Yellows, 68

Leaves with internal yellowing
- Excess water

Leaves bronzed
- "Brittle-leaf" condition
  - Eriophyid mites, 99

Leaves curled
- Leaflets thickened and curled at midrib
  - Ash midrib gall midge, 91
- Leaves tightly curled, thickened
  - Leafcurl ash aphid, 50
- General distortion, with thickened veins
  - Phenoxy herbicide injury, 66

White powdery material on upper or lower surface of leaf
  - Powdery mildew, 82

Wilting of portions of tree
- Originating from roots
  - Verticillium wilt, 142

Affecting twigs and smaller branches
- Twig dieback
  - With small exit holes or ventilation holes visible
    - Ash bark beetles, 157
  - With small pimple-like fruiting bodies (pycnidia) in the bark
    - Cytospora canker (*Cytospora* sp.), 146
  - Tunneling young twigs
    - Unidentified caterpillar (Tortricidae)
    - Hunting wasps (*Pemphredon* species), nesting in pith, 138
- Witches' brooming symptoms
  - Ash yellows, 68
- Scales
  - Oystershell scale, 120
  - Walnut scale, 123
  - Common falsepit scale, 125
  - European fruit lecanium, 126

Affecting trunk and larger branches
Tunneling trunk
Lilac/Ash borer, Banded ash clearwing, 169
Ash bark beetles, 157
Carpenterworm, 172
Redheaded ash borer, 168
Masses of caterpillars resting on bark
Forest tent caterpillar, 11
Discolored areas and dead bark containing small pimple-like fruiting bodies (pycnidia)
Cytospora canker (*Cytospora* sp.), 146
Clear to white oozing or frothy malodorous liquid exiting from wounds
Bacterial wetwood/slime flux, 189
Large, dead areas of bark on southwest side of trunk
Winter sunscald, 67
Open wounds, internal decay, swollen areas on stem
Stem decay fungi (*Perenniporia fraxinophila, Phellinus punctatus* and various fungal genera), 152
Fungal fruiting bodies (mushrooms, conks) present
Stem decay fungi (*Perenniporia fraxinophila, Phellinus punctatus* and various fungal genera),152
Wilting and dieback of portions of tree
Originating from roots
Verticillium wilt, 142
Distorting flowers
Ash flower gall mite, 98
Chewing seeds, grubs drop to ground in fall
Ash seed weevil, 202

ASPEN (*Populus tremuloides*)

Affecting leaves
Leaves chewed
Leaves curled
Large aspen tortrix, 6
Tent of silk produced
Western tent caterpillar, 10
No curling or silk associated with injury
Forest tent caterpillar, 11
Nevada buck moth, 3
Redhumped caterpillar, 4
Poplar dagger moth, 16
Cottonwood leaf beetle, 30
Sawflies, 22
Masses of dark, spiny caterpillars on leaves
Spiny elm caterpillar, 20
Leaves spotted
Young leaves blackened
Shoot blight (*Venturia tremulae*), 117
Frost injury, 113
Black irregular spot
Septoria leaf spot and canker (*Septoria populicola*), 83
Dark round spot which drops out of leaf, leaving shothole appearance
Ink spot (*Ciborinia whetzelii*), 75

Black spots with yellow margins
Marssonina blight (*Marssonina populi*), 78
Rust or orange colored spots
Conifer-aspen rust (*Melampspora* spp.), 77
Underside with small depressions and patches of brown leaf hairs
Eriophyid mites, 99
Leaves generally distorted, thickened
Whole leaf or set of leaves so distorted
Poplar vagabond aphid, 87
Edge of leaf folded into a series of ridges
Eriophyid mites, 99
Red, thickened folds along leaf veins
Gall midge (unknown species)
Leaves with serpentine, silvery tunneling
Aspen leafminer, 38
Sucking insects on leaves
Clear-winged aspen aphid, 45
Leafhoppers
General dark gray-black growth on upper leaves
Sooty mold,190
White powdery material on upper or lower surface of leaf
Powdery mildew (*Erysiphe cichoracearum*), 82
Eriophyid mites, 98
Yellowed leaves
Root damage caused by under or over watering, 197
Iron chlorosis, 64
Leaves with marginal burn, discoloration
Drought
Excess soil salts
Affecting smaller branches and twigs
Twig dieback
Shoot curled to a shepherd's crook, 117
Shoot blight (*Venturia tremulae*)
With small, closely spaced, pimple-like fruiting bodies (pycnidia)
Dothiora canker (*Dothiora polyspora*), 147
Meandering tunnels under bark
Agrilus spp. of flatheaded borer, 168
With small pimple-like fruiting bodies (pycnidia) in the bark
Cytospora canker (*Cytospora* sp.), 146
Twigs with rounded swellings
Poplar twiggall fly, 93
Scales on bark
Oystershell scale, 120
Scurfy scale, 121
Affecting branches, trunks
Areas of dead bark
With discoloration and small pimple-like fruiting bodies (pycnidia)
Cytospora canker (*Cytospora* sp.), 146
Small (1/6-in) gray or brown oystershell shaped objects on bark
Oystershell scale, 120

Dark, ridged callus areas around a target-type canker on trunk
Black canker (*Ceratocystis fimbriata*), 144
Black with white flecks and/or small pimple-like fruiting bodies (perithecia)
Cryptosphareria canker (*Cryptosphareria lignyota*), 145
Darkened bark with large target-type canker and leopard patterns on wood
Sooty-bark canker (*Encoelia pruinosa*), 149
Black bark with checkered patterning
Hypoxylon canker (*Hypoxylon mammatum*), 148
Flexuous-rubbery, pendulous branches
Droopy aspen, 114
Rounded and often rough swellings on branches
Aspen gall (*Diplodia tumefaciens*), 149
Oozing liquid from wounds
Clear to white oozing or frothy malodorous liquid
Bacterial wetwood/slimeflux, 189
Orange staining ooze
Poplar borer, 180
Rough, fissured bark
Rough bark of aspen (*Cucurbitaria staphola* and other fungi), 149
Shallow tunneling under bark
Ambrosia beetles, addendum
Cottony growth on branches, trunk
Woolly aphid (species unknown)
Swellings in branches, trunk
Poplar twiggall fly, 93
Poplar gall saperda, 181
Tunnelling with coarse sawdust often forced from opening
Poplar borer, 180
Regular rows of holes in trunk
Sapsuckers, 211
Meandering tunnels under bark
Agrilus species, 168
Open wounds, internal decay, swollen areas in stem
Stem decay fungi (*Phellinus tremulae, Ganoderma applanatum*, and other fungal genera), 152
Fungal fruiting bodies (mushrooms, conks) present
Stem decay fungi (*Phellinus tremulae, Ganoderma applanatum*, and other fungal genera), 152
Root and trunk galls
Swelling of lower trunk or root tissue
Aspen gall (*Diplodia tumifaciens*), 149
Root decay
White root decay with white mycelial fans between bark and wood
Armillaria root disease (*Armillaria mellea*), 194
Ganoderma root decay (*Armillaria mellea*), 153

BIRCH (*Betula* spp.)
Affecting leaves
Leaves generally chewed
Cecropia moth, 2
Polyphemus moth, 2
Acronicta leporina, 16
Twinspot sphinx, 20

Leaves with dark, blotchy mines
Birch leafminer, 37
Aphids (several species, unknown)
Affecting branches, trunk
Tunneling of cambium, dieback
Bronze birch borer, 182
Areas of dead bark
With discoloration and small pimple-like fruiting bodies (pycnidia)
Cytospora canker (*Cytospora* sp.), 146

BOXELDER (*Acer negundo*)

Affecting leaves
Leaves chewed
American dagger moth, 16
Leaves curled, chewed
Boxelder leafroller, 7
Boxelder leafminer, 40
Leaves mined
Boxelder leafminer, 40
Leaves thickened around midrib
Gouty veingall midge, 91
New growth small, distorted
Eriophyid mites, 99
Sucking on leaves, honeydew often present
Psylla negundinis, 52
Boxelder and maple aphids, 45
Small cottony indentations on leaf underside
Eriophyid mites, 99
Masses of reddish eggs on leaves
Boxelder bug, 206
Leaves spotted
Dark spots on leaves
Phyllosticta leaf spot (*Phyllosticta* sp.), 81
Raised, black spots on leaves
Tar spot (*Rhytisma* sp.), 84
Affecting branches, trunk
Discolored areas, dead bark containing small pimple-like fruiting bodies (pycnidia)
Cytospora canker (*Cytospora* sp.), 146
Open wounds, internal decay, swollen areas in stem
Stem decay fungi (*Pholiota* sp., *Pleurotus* sp., and various fungal genera), 152
Fungal fruiting bodies (mushrooms, conks) present
Stem decay fungi (*Pholiota* sp., *Pleurotus* sp., and various fungal genera), 152
Borers
Flatheaded appletree borer, 184
Carpenterworm, 172
Affecting seeds
Feeding on seeds
Boxelder bug, 206

CACTUS (*Opuntia* spp., *Cylindropuntia* spp.)
- Insects on pads
  - Cottony masses, wilting may be associated
    - Cochineal, 54
  - Brown to grayish 1/2-inch bugs, often in large groups
    - Opuntia bug, 59
    - A leaffooted bug, 59
- Borers
  - Cholla-type
    - Cactus longhorn (*Monoleima* spp.), 179
  - Prickly pear-type
    - Blue cactus borer, 174

CATALPA (*Catalpa* sp.)
- Affecting leaves
  - Dark spots on leaves
    - Various leaf spot fungi
    - Sooty mold gowing on honeydew of mealybugs, aphids, 190
- Affect large branches and trunk
  - Wilting and dieback of portions of tree
    - Originating from roots
      - Verticillium wilt, 142
  - Masses of cottony insects
    - Grape mealybug, 53

CHOKECHERRY (See **STONE FRUITS**)

CHERRY (See **STONE FRUITS**)

COTTONWOOD, POPLAR (*Populus* spp., excluding aspen)
- Affecting leaves
  - Leaves chewed
    - Cottonwood leaf beetle, 30
    - Spiny elm caterpillar, 20
    - Fall webworm, 14
    - Nevada buck moth, 3
    - Dagger moth, 16
    - Tussock moth, 16
    - Twinspot sphinx, 20
  - Webbing/Tents produced
    - Fall webworm, 14
    - Western tent caterpillar, 10
    - A tent caterpillar, 11
  - Small holes chewed in leaves
    - Flea beetles, 32
    - Poplar blackmine beetle (adult feeding), 38
  - Leaves tunnelled
    - Poplar blackmine beetle, 38
    - Aspen leafminer, 38
    - Tentiform leafminer, 39

Leaves spotted
- Young leaves blackened
  - Shoot blight (*Venturia populina*), 117
  - Frost injury, 113
- Black spots with yellow margins
  - Marssonina blight (*Marssonina populi*), 78
- Black, irregular spots
  - Septoria leaf spot and canker (*Septoria popolicole*), 83
- Blackish-brown round spot, which drops out of, leaf, leaving shothole-like appearance
  - Ink spot (*Ciborinia whetzelii*), 75
- Rust to orange colored spots
  - Conifer-aspen rust (*Melampsora* spp.), 77

Leaf margins "burned"
- Drought

Yellowed leaves
- Root damage caused by under or over watering, 197
- Iron chlorosis, 64

White powdery material on upper or lower surface of leaf
- Powdery mildew (*Erysiphe cichoracerum*), 82

Leaf petioles, veins with swelling
- Petiole-gall aphids, 87

Leaves folded along edge
- Poplar leaffolding sawfly, 26

Leaves generally distorted and thickened
- Poplar vagabond aphid, 87

Affecting twigs, small branches
- Hollow swellings form on new shoots
  - Petiole-gall aphids, 87
- Terminal leaves distorted into thickened mass
  - Poplar vagabond aphid, 87
- Tunneling twig terminal
  - Cottonwood twig borer, 133
- Twigs shredded in irregular row
  - Cicada oviposition injury, 131
- Branch girdled, bark removed
  - Squirrel damage, 213
- Swellings in twigs, small branches
  - Poplar twiggall fly, 93
  - Poplar gall borer, 181
  - Hail injury (upper surface only), 113
- Scales on twigs, branches
  - Oystershell scale, 120
  - Scurfy scale, 121
- Buds distorted
  - Poplar budgall mite, 93
- Catkins grossly distorted and enlarged
  - Cottonwood catkin gall mite, 98

Affecting trunk, large branches
- Masses of caterpillars resting on bark
  - Forest tent caterpillar, 11
- Tunneling trunk, often with orange staining ooze
  - Cottonwood borer, 177
  - Poplar borer, 180
  - Carpenterworm, 172
  - American hornet moth, 172
- Clear to white oozing or frothy malodorous liquid exiting from wounds
  - Bacterial wetwood /slime flux, 189
- Areas of dead bark with
  - Discoloration, and small pimple-like fruiting bodies (pycnidia)
    - Cytospora canker (*Cytospora* sp.), 146
- Areas of black bark with
  - White flecks and/or small pimple-like fruiting structures (perithecia)
    - Cryptosphareria canker (*Cryptosphareria populina*), 145
- Insects visiting oozing sap from trunk
  - Bumble flower beetle, 228
  - Sap beetles
  - Flies (various families)
- Fungal fruiting bodies (mushrooms, conks) present
  - Stem decay fungi (various species), 152
- Open wounds, internal decay, swollen areas in stem
  - Stem decay fungi (various species), 152

Affecting roots or ground line area
- White root decay with white mycelial fans between bark and wood
  - Armillaria root disease (*Armillaria mellea*), 194
- Gall at ground line
  - Crown gall (*Agrobacterium tumefaciens*), 101

CRABAPPLE (See **Pome Fruits**)

CURRANT, GOOSEBERRY (*Ribes* spp.)

Affecting leaves
- White flecks in leaves
  - Leafhoppers, 55
  - Twospotted spider mite, 60
- White powdery material on upper or lower surface of leaf
  - Powdery mildew (*Mycosphaerella* sp.), 82
- Leaves puckered, often thickened and discolored
  - Currant aphid, 47
  - Psyllid (unidentified)
- Leaves chewed
  - Imported currantworm, 25
  - Currant spanworm, 18
  - Apple flea beetle, addendum
  - Western tent caterpillar, 10
- Hollow swelling in leaves
  - Gall-making sawfly, 23
- Orange to yellow colored spots on underside of leaf
  - Leaf rusts (various species), 77

Affecting stems, branches
Tunneling stems, branches
Currant borer, 169
Bronze cane borer, 183
Affecting fruit
Maggot in fruit
Gooseberry maggot

DOGWOOD (*Cornus* species)

Affecting leaves
Leaves chewed
Polyphemus moth, 2
Cecropia moth, 2
Redhumped caterpillar, 4
Sawfly (*Macremphytus* spp.), 23
Leaves with small white flecks
Dogwood leafhopper (*Edwardsiana commisuralis*), 56
New leaves curled
Sunflower aphid, 45
Black spots on leaves
Septoria leaf spot and canker (*Septoria cornicola*), 83
White powdery material on upper or lower surface of leaf
Powdery mildew, 82
Affecting branches and twigs
Scales
Oystershell scale, 120
Borers
Dogwood borer, 169
Dead areas on stem
Nectria canker (*Nectria* sp.), 151
Root collar rot (*Phytophthora* sp.), 196
Feeding in or on berries
Conifer seed bugs, 206
Rhagoletis tabellaria

DOUGLAS-FIR (*Pseudotsuga menzeisii*)

Affecting buds
Buds tunneled
Western spruce budworm, 5
Affecting needles
Newer needles being chewed
Douglas-fir tussock moth, 15
Western spruce budworm, 5
Needles with discolored spotting
Various color spots
Needle casts (various fungi), 78
Rust or orange color spots
Conifer-aspen rust (*Melampsora* spp.), 77
Needles to exterior of tree bleached or brown, developing late winter
Winter desiccation, 67

  - Reddish-brown needles
    - Bank beetles, 162
    - Salt damage, 64
  - Brown felt-like material on needles, branches (high elevations)
    - Brown felt blight (*Herpotricha juniperi*), 75
  - Needles bent, twisted
    - Cooley spruce gall adelgid, 86
    - Frost injury, 113
  - Woolly aphids on needles
    - Cooley spruce gall adelgid, 86
  - Whole tree fades, reddens
    - Douglas-fir beetle, 162
- Affecting cones
  - Flowers tunneled
    - Western spruce budworm, 5
  - Woolly aphids on cones
    - Cooley spruce gall adelgid, 86
  - Sucking on developing cones
    - Conifer seed bugs, 206
  - Cones tunneled
    - Douglas-fir cone moth
- Affecting smaller branches
  - Bark beetle
    - Douglas-fir pole beetle, 163
- Affecting trunk and larger branches
  - Witches' brooms on branches, small shoots emerging from branch
    - Dwarf mistletoe (*Arceuthobium douglasii*), 187
  - Large galls
    - Bacteria-like gall
    - Burl
  - Aphids
    - Giant conifer aphids, 47
  - Tunneling trunk, branches
    - Douglas-fir pole beetle, 163
    - Pine sawyers, 179
    - Douglas-fir beetle, 162
    - Ponderous borer, 229
  - Open wounds, internal decay, swollen areas in stem
    - Stem decay fungi (*Fomitopsis pinocola, Cryptoporus volvatus* and various fungal genera), 152
  - Fungal fruiting bodies (mushrooms, conks) present
    - Stem decay fungi (*Fomitopsis pinocola, Cryptoporus volvatus* and various fungal genera), 152
- Affecting roots
  - White root decay with white mycelial fans between bark and wood
    - Armillaria root disease (*Armillaria mellea*), 194
  - White flecks in bark with shelf-like fruiting bodies (conks) at ground line
    - Annosus root disease (*Heterobasidion annosum*), 193

ELM (*Ulmus* spp.)

- Affecting leaves
  - Leaves skeletonized, primarily on leaf underside
    - Elm leaf beetle (larvae), 31

- Holes chewed through leaves
  - Elm leaf beetle, 31
  - Cankerworms, 18
- Leaves irregularly chewed
  - Spiny elm caterpillar, 20
  - Fruittree leafroller, 7
  - Cecropia moth, 2
  - Forest tent caterpillar, 11
  - Cankerworms, 18
  - Fall webworm, 14
  - Elm sphinx/Twinspot sphinx, 20
- Masses of dark, spiny caterpillars on leaves
  - Spiny elm caterpillar, 20
- Black, irregular spots
  - Black spot of elm (*Gnomonia ulmea*), 73
- Meandering mines in leaves
  - Elm leafminer, 38
- Dark mold on upper leaf surface
  - Sooty mold, 190
- New leaves small, twisted
  - Eriophyid mites, 99
  - Phenoxy herbicide injury, 66
- Leaves curled, thickened
  - Woolly aphids (*Eriosoma* spp.), 51
- Leaves with white flecks
  - Leafhoppers, 56
  - Elm spider mite
- Pale green or yellow leaves
  - Iron chlorosis, 64
- Sucking insects on leaves, often with associated honeydew
  - European elm scale, 129
  - Elm leaf aphid, 49
- Yellowing, wilting foliage
  - Scale "flagging" (European elm scale), 129
  - Dutch elm disease (*Ophiostoma ulmi*), 141
  - Verticillium wilt, 142
  - Squirrel girdling, 213

Affecting twigs

- Scales
  - European fruit lecanium, 126
  - European elm scale, 129
- Twigs chewed
  - Fox squirrels, 213
- Twigs abnormally dark
  - Sooty mold, 190

Affecting branches, trunk

- Borers
  - Carpenterworm, 172
  - Pole borer, 177
  - Elm borer
  - Flatheaded appletree borer, 184
  - Pigeon tremex, 176

- Bark beetles in branches
    - Smaller European elm bark beetle, 159
    - Banded elm bark beetle, addendum
- Upper bark surface abnormally dark, crusty
    - Sooty mold, 190
- Cellophane-like silk late in season
    - Spider mites
- Areas of dead bark
    - With discoloration and small pimple-like fruiting bodies (pycnidia)
        - Cytospora canker (*Cytospora* sp.), 146
    - Tan to black pimple-like structures *on* bark
        - Tubercularia canker (*Tubercularia ulmea*), 151
    - Closely spaced pimples protruding from bark
        - Siberian elm canker (*Botryodiplodia hypodermia*), 144
    - Open wounds, internal decay, swollen areas on stem
        - Stem decay fungi (*Collybia velutipes, Phellinus igniarius*and various genera), 152
    - Fungal fruiting bodies (mushrooms, conks) present
        - Stem decay fungi (*Collybia velutipes, Phellinus igniarius* and various genera), 152
- Clear to white oozing or frothy malodorous liquid exiting from wounds
    - Bacterial wetwood/slime flux, 189
- Visiting oozing sap from trunk
    - Bumble flower beetle, 228
    - Sap beetles
    - Flies (various families)

EUONYMUS (*Euonymus* spp.)

- Affecting leaves
    - Notches cut in edge of leaves
        - Black vine weevil, 198
    - Leaves mined, edge curled with webbing
        - Lilac leafminer, 40
    - New leaves curled, thickened
        - Bean aphid, 45
    - White scales on leaves
        - Euonymus scale, 122
    - Flecking, discoloration
        - Twospotted spider mite, 60
- Affecting trunk
    - Bark chewed at base
        - Voles, 214
- Affecting roots
    - Chewing roots
        - Black vine weevil, 198
    - Gall at ground line
        - Crown gall (*Agrobacterium tumifaciens*), 101

FIR (*Abies* spp.)

- Affecting needles
    - New needles being chewed
        - Douglas-fir tussock moth, 15
        - Western spruce budworm, 5

Needles spotted
- Needles to exterior of tree bleached or brown, developing late winter
  - Winter desiccation, 67
- Various colored spots
  - Needlecast diseases, 78
- Orange colored pustules on needles
  - Fir broom rust (*Melampsorella caryophyllacearum*), 107

Brown felt-like material on needles, branches
- Brown felt blight (*Herpotrichia juniperi*), 75

Needles being mined
- White fir needleminer, 40

Tent-making associated
- *Lophocampa ingens*, 14

Aphids on needles
- Giant conifer aphids, 47
- *Pineus* species, 52

New needles distorted
- Balsam twig aphid
- Frost injury, 113

Affecting buds
- Buds tunneled
  - Western spruce budworm, 5

Affecting twigs and smaller branches
- Large aphids on branches, twigs
  - Giant conifer aphids, 47

Affecting trunk and larger branches
- Bark beetles
  - Fir engraver, 160
  - Balsam bark beetle
- Borers
  - Flatheaded fir borer (*Melanophila drummondi*)
  - *Buprestis* spp. (A flatheaded borer), 185
  - *Dicerca tenebrosa* (A flatheaded borer), 168
  - Black spruce borer (*Asemum striatum*)
  - Pine sawyers, 179
- Witches' brooms on branches, orange to yellow colored pustules on needles
  - Fir broom rust (*Melampsorella caryophyllacearum*), 107
- Large aphids on branches, trunk
  - Giant conifer aphids, 47
- Fungal fruiting bodies (mushrooms, conks) present
  - Stem decay fungi (various species), 152
- Open wounds, internal decay, swollen areas in stem
  - Stem decay fungi (various species), 152

Affecting roots and ground line area
- White root decay with white mycelial fans between bark and wood
  - Armillaria root disease (*Armillaria mellea*), 194
- White flecks in bark with shelf-like fruiting bodies (conks) at ground line
  - Annosus root disease (*Heterobasidion annosum*), 193
- General decline of subalpine fir
  - Dryocetes and Armillaria root disease complex, 194

GRAPE (See **VIRGINIA CREEPER/GRAPE**)

HACKBERRY (*Celtis* spp.)
Affecting leaves
Leaves being chewed
Fruittree leafroller, 7
Spiny elm caterpillar (usually in masses), 20
Fall cankerworm, 18
Ashgray blister beetle, 33
Leaves with blotchy leaf mines
Lithocolletis sp. (a leafminer)
Leaves with white flecking (net-leaf hackberry)
Lacebugs, 59
Leaves yellowed
Iron chlorosis or other deficiency, 64
Root damage from under or over watering, 197
Leaves with large, conspicuous raised areas
Hackberry nipplegall maker, 88
Leaves with small, raised areas
Hackberry blistergall maker, 88
Petiole folded, swollen with hollow area (net-leaf hackberry)
Pachypsylla venusta, 88
Hackberry "tatters" (irregular holes in some leaves)
Unknown, wind injury and/or frost injury to developing leaves is suspected, 67
Affecting buds
Buds irregularly swell
Hackberry budgall psyllid, 89
Affecting branches
Small raised area on branches
Pachypsylla species, 88
Cellophane-like webbing in late summer
Spider mites
Twigs deformed into dense witches' broom
Eriophyid mites/Powdery mildew complex, 104
Affecting large branches, trunks
Borers/Bark beetles
Flatheaded appletree borer, 184
Redheaded ash borer, 168
Scolytus fagi (a bark beetle)
Clear to white oozing or frothy malodorous liquid exiting from wounds
Bacterial wetwood and slime flux, 189

HAWTHORN (*Crataegus* spp.)
Affecting leaves
Blackened and wilting leaves
Fireblight, 114
Pale green or yellow leaves
Iron chlorosis, 64
Caterpillars associated with webbing
Fall webworm, 14
Western tent caterpillar, 10

Chewing leaves
Pearslug, 22
Forest tent caterpillar, 10
Leaves curled in spring
Apple aphid, 45
Reddish spotting on upper surface/orange pustules, spiny eruption on underside
Cedar-knotgall (*Gymnosporangium* spp.), 102
Blotchy leafmine
Hawthorn leafminer, addendum
Affecting twigs, branches
Blackened, wilting, and crooked tips
Fireblight, 114
White, wax-covered insects
Hawthorn mealybug, 53
Woolly hawthorn aphid (*Eriosoma crateagi*), 51
Scale
San Jose scale, 124
Affecting large branches, trunks
Clear to white oozing or frothy malodorous liquid exiting from wounds
Bacterial wetwood and slime flux, 189
Affecting fruit
Maggot in fruit
Apple maggot, 204

HONEYLOCUST (*Gleditschia triacanthos*)
Affecting leaves
Caterpillars chewing on leaves
Fall cankerworm, 18
Fruittree leafroller, 7
Leaflets killed back
Honeylocust plant bug, 58
Leaflets distorted into thickened pods
Honeylocust podgall midge, 89
Leaves chewed by masses of elongate, gray beetles
Ashgray blister beetle, 33
Honeydew appearing on leaves
Honeylocust leafhopper, 57
Micrutalis calva (a treehopper), 57
Cottony maple scale (nymph stage on leaves), 126
Leaves turn bronze, may prematurely drop
Honeylocust spider mite, 61
Honeylocust rust mite, 63
Yellowed leaves
Root damage caused by under or over watering, 197
Affecting twigs and smaller branches
Tips of twigs thickened, may have dieback
Honeylocust podgall midge (severe injury killing growing points), 89
Neolasioptera brevis (a gall midge), 89
Scales
Cottony maple scale, 126
Common falsepit scale, 125

Twigs with small splintering wounds
Putnam's cicada (egg laying wounds), 131
Twigs tunneled
Apple twig borer, 168
Chewing injuries on branches
Fox squirrels, 213
Affecting trunk and larger branches
Areas of dead bark (cankers)
Discoloration and small pimple-like fruiting bodies (pycnidia) in the bark
Cytospora canker (*Cytospora* sp.), 146
Small, round pinhead-sized fruiting bodies under bark
Tubercularia canker (*Tubercularia ulmea*), 151
Irregular-shaped brown/black fruiting bodies
Thyronectria canker (*Thyronectria austro-americana*), 150
Amber colored gummy exudate (Gummosis)
Stress response from drought, sunscald, collar rot, cankers or other causes
Wood borers
Honeylocust borer, 184
Bark pulled off in strips
Fox squirrel, 213
Affecting roots, ground line area
Loose bark at collar, discolored tissue under bark
Root collar rot, 196
Affecting seedpods
Feeding on seeds
Amblycerus robiniae (a bruchid/seed weevil)

HONEYSUCKLE (*Lonicera* spp.)
Affecting leaves
Leaves tightly curled, associated bunchy growth
Honeysuckle witches' broom aphid, 49
Early season leaves curled, yellow
Aphids (unidentified), 45
Leaves chewed
Common clearwing sphinx, 20
White powdery material on upper or lower surface of leaf
Powdery mildew, 82
Affecting stems, branches
Borers
Agrilus species, 168
Affecting roots, ground line area
Root and crown rot (Phytophthora and various soil fungi), 196

JUNIPER (*Juniperus* spp.)
Affecting needles
Larvae chewing on needles
Juniper sawfly, 27
Tiger moth, 14
Frothy masses on needles
Juniper spittlebug, 55

- Honeydew present
    - Fletcher scale, 128
    - Giant conifer aphids, 47
- Needles chewed and fragments tied with webbing
    - Juniper webworm, 8
- Webbing covering crown of tree
    - Tiger moth, 14
- Aphids on needles
    - Giant conifer aphids, 47
- Pale green or yellow needles
    - Iron chlorosis, 64
- Needles bleached or brown
    - Winter desiccation, 67
    - Vole injury, 214
- Needles become grayish, with small flecks
    - Spruce spider mite, 62
    - *Platytetranychus libocedri, 62*
- Scales on needles
    - Juniper scale, 123
    - Fletcher scale, 128

Affecting twigs and smaller branches

- Twig galls
    - Juniper rusts (*Gymnosporangium* spp.), 102
    - Juniper tip midge, 90
- Twig dieback
    - Kabatina blight (*Kabatina juniperi*), 76
    - Phomopsis blight (*Phomopsis juniperovora*), 76
    - Western cedar bark beetles, 166
    - Grasshoppers, 35
- Large aphids on branches, twigs
    - Giant conifer aphids, 47
- Brown felt-like material on needles and branches (higher elevations only)
    - Brown felt-blight (*Herpotricha juniperi*), 75

Affecting trunk and larger branches

- Witches' brooms or branch galls
    - Juniper broom rust (*Gymnosporangium nidus-avis*), 102
    - Juniper mistletoe (*Phorodendron juniperinum*), 188
    - Orange gall rust (*Gymnosporangium speciosum*), 102
- Bark beetles
    - Western cedar bark beetles, 166
- Large aphids on branches, trunk
    - Giant conifer aphids, 47
- Wood under bark is deeply sculptured by tunnels
    - *Callidium texanum*, 178
    - *Chrysobothris texana*, addendum
    - Juniper borer, 168
- Stem decay and/or hoof-shaped fruiting structures (conks)
    - Juniper pocket rot (*Pyrofomes demidoffii*), 156
- Dieback of branches that have been chewed externally at base
    - Voles, 214

Affecting roots
White root decay with white mycelial fans between bark and wood
Armillaria root disease (*Armillaria mellea*), 194
Gall at ground line
Crown gall (*Agrobacterium tumifaciens*), 101
Affecting berries
Berries infested with maggots
Rhagoletis juniperina

LILAC/PRIVET (*Syringa* spp., *Ligustrum* spp.)

Affecting leaves
Leaves chewed
Cecropia moth , 2
Great ash sphinx, 20
Edge of leaves cut in semicircle
Leafcutter bees, 36
Edge of leaves notched
Black vine weevil, 198
Leaves with blotchy mines
Lilac leafminer, 40
Leaves blacken and wilted
Bacterial blight (*Pseudomonas syringae* pv. *syringae*), 116
White powdery material on upper or lower surface of leaf
Powdery mildew (*Microsphaera penicillata*), 82
Leaves turn a rusty color
Rust (eriophyid) mites, 59
Affecting stems, branches
Scales
Oystershell scale, 120
Borers
Lilac/ash borer, 169
Wilting and die back at tips
Bacterial blight (*Pseudomonas syringae* pv. *syringae*), 116

LINDEN/BASSWOOD (*Tilia* spp.)

Affecting leaves
Leaves being chewed
Fruittree leafroller, 7
Linden looper, 18
Speckled green fruitworm, 18
Small pouch galls form on leaf surface
Linden fingergall mite, 100
Velvety patches on underside of leaves (littleleaf linden)
Eriophyid mites, 100
Pale green or yellow leaves
Iron chlorosis, 64
Leaves strongly cupped
Herbicide injury, 66
Aphids
Linden aphid, 45

Wilting and dieback of portions of tree
Originating from roots
Verticillium wilt, 142
Affecting twigs, small branches
Large cottony insect develops in late spring
Cottony maple scale, 126
Small crust-like growth on twigs
Walnut scale, 123
Cellophane-like webbing on trunk in late summer
Eotetranychus tilarium (a spider mite)
Affecting stems, branches
Areas of dead bark with
Discoloration and small pimple-like fruiting bodies (pycnidia)
Cytospora canker (*Cytospora* sp.), 146
Small, round pinhead sized fruiting bodies
Tubercularia canker (*Tubercularia* sp.), 151
Wilting and dieback of portions of tree
Originating from roots
Verticillium wilt, 142

MAPLE (Excluding BOXELDER) (*Acer* spp.)

Affecting leaves
Leaves chewed
American dagger moth, 16
Fruittree leafroller, 7
Cecropia/Polyphemus moth, 2
Linden looper, 18
Leaves cut at petiole in late summer
American dagger moth, 16
White flecks on leaves
Leafhopper
Black to brown spots on leaves
Anthracnose (*Kabatiella* sp.), 72
Septoria leaf spot and canker (*Septoria* sp.), 83
Phyllosticta leaf spot (*Phyllosticta* sp.), 81
Tar spot (*Rhytisma* sp.), 84
Patches of reddish hairs (Rocky Mountain maple)
Eriophyid mites, 100
Sucking insects on leaves, often honeydew
Boxelder and maple aphids, 45
Cottony maple scale (nymphs), 126
Leaves yellowed (particularly silver maple)
Iron chlorosis, 64
Affecting twigs, small branches
Twigs with row of irregularly shredded punctures
Cicada injury, 131
Large cottony insect on twigs
Cottony maple scale, 126
Affecting trunks, branches
Crusty scale
Walnut scale, 123

- Clear to white oozing or frothy malodorous liquid exiting from wounds
    - Bacterial wetwood and slime flux, 189
- Areas of dead bark
    - Discoloration and small pimple-like fruiting bodies (pycnidia) in the bark
        - Cytospora canker (*Cytospora* sp.), 146
    - Tan to black pimple-like structures *on* bark
        - Tubercularia canker (*Tubercularia* sp.), 151
- Elongated dead and discolored areas on southwest side of trunk
    - Winter sunscald, 113
- Borers
    - Flatheaded appletree borer, 184
    - Pole borer, 177
    - Carpenterworm, 172
    - Pigeon tremex, 176
    - Maple wood wasp, addendum
- Wilting and dieback of portions of tree
    - Originating from roots
        - Verticillium wilt, 142

MOUNTAIN-ASH (*Sorbus* spp.)

- Affecting leaves
    - Orange spots
        - Juniper rusts (*Gymnosporangium* spp.), 102
    - Rusty scabs on leaf
        - Pearleaf blister mite, 98
    - Chewing leaves
        - Pearslug, 22
    - Aphids
        - Apple aphid, 45
    - Small pouch galls
        - *Phytoptus sorbi, 99*
    - Blackened and wilting leaves
        - Fireblight, 114
- Affecting twigs
    - Woolly aphids (*Eriosoma sp.*), 51
    - Scales
        - Oystershell scale, 120
        - Scurfy scale, 121
        - Walnut scale, 123
    - Hawthorn mealybug, 53
- Affecting branches, trunk
    - Blackened, wilting, and crooked tips
        - Fireblight, 114
    - Discolored areas and dead bark containing small pimple-like fruiting bodies (pycnidia)
        - Cytospora canker (*Cytospora* sp.), 146
    - Elongated dead and discolored areas on southwest side of trunk
        - Winter sunscald, 113
    - Clear to white oozing or frothy malodorous liquid exiting from wounds
        - Bacterial wetwood and slime flux, 189
    - Borers
        - Flatheaded appletree borer, 194
        - Shothole borer, 158

MULBERRY (*Morus* spp.)

On leaves
- Chewing leaves
  - Io moth, 3
  - Small mulberry borer, 168
- Small black insect sucking sap
  - Mulberry whitefly, 55

On twigs
- White scale on twigs
  - Common falsepit scale, 125
- Borer in twigs, small branches
  - Small mulberry borer, 168

Affecting large branches, trunk
- Clear to white oozing or frothy malodorous liquid exiting from wounds
  - Bacterial wetwood and slime flux, 189
- Areas of dead bark
  - Discoloration and small pimple-like fruiting bodies (pycnidia) in the bark
  - Cytospora canker (*Cytospora* sp.), 146
- Borer
  - Carpenterworm, 172
  - Small mulberry borer, 168
- Tan to black pimple-like structures *on* bark
  - Tubercularia canker (*Tubercularia ulmea*), 151
  - Nectria canker (*Nectria* sp.), 151

OAK (*Quercus* spp.)

Affecting leaves
- Leaves chewed
  - Sonoran tent caterpillar, 11
  - Oak leafroller, 7
  - Speckled green fruitworm, 18
  - *Hemileuca diana*, 3
  - Linden looper/Oak looper, 18
- Leaves tied together with silk, chewed
  - Oak leafroller, 7
- Leaf blistered, cupped
  - Oak leaf blister (*Taphrina caerulesecens*), 81
- Black/brown spotting on leaves
  - Anthracnose (*Apiognomonia quercina*), 72
- Frothy mass on leaf vein
  - Spittlebugs, 55
- Insects sucking on leaves
  - Aphids, 45
  - Treehoppers
- Spider mites
  - *Oligonychus platani*
- Small balls on leaves
  - Gall wasps, 94
- Cottony growth on leaves, near midrib
  - Gall wasps, 94
- Pale green or yellow leaves
  - Iron chlorosis, 64

Affecting twigs
- Woody, dark rounded objects on twigs
  - Rough bulletgall wasp, 95
- Twigs dieback in late summer
  - Kermes scale, 130
- Oozing sap at terminal
  - Kermes scale, 130
- Scales settled in small depressions
  - Golden oak scale, 125
- Twigs shredded by multiple punctures
  - Cicada oviposition injury, 131
  - Oak treehoppers (*Platycottis vittata* and other species)
- Borers
  - Gambel oak borer, addendum

Affecting branches
- Witches' brooms on Gambel oak
  - Oak witches' broom (*Articularia quercina*), 104
- Areas of dead bark
  - Discoloration and small pimple-like fruiting bodies (pycnidia) in the bark
    - Cytospora canker (*Cytospora* sp.), 146
  - Large pustule fruiting body (stromata), blackish, breaking to rust red
    - Endothia canker (*Endothia singularis*), 147

Affecting large branches or trunk
- Clear to white oozing or frothy malodorous liquid exiting from wounds
  - Bacterial wetwood / slime flux, 189
  - Gambel oak borer, addendum
- Stem decay and /or hoof-shaped fruiting bodies (conks) on Gambel oak
  - White trunk rot of oak (*Phellinus everhartii*), 153
- Borers
  - Gambel oak borer, addendum
  - Flatheaded apple tree borer, 184

Affecting acorns
- Grubs feeding in acorn
  - Acorn weevils, 202

PEACH (See **Stone Fruits**)

PEAR (See **Pome Fruits**)

PINES (excluding PINYON) (*Pinus* spp.)

Affecting needles
- Older needles being chewed
  - Tiger moth, 14
  - Conifer sawfly, 28
  - Bull pine sawfly, 28
  - Pandora moth, 4
  - Pine butterfly, 21
- Older needles chewed and mixed with webbing, pellets
  - Web-spinning sawflies, 28
  - *Tetralopha* sp. (pine webworm), 13

Newer needles being chewed
Ponderosa pine budworm, 5
Mottled yellowing of needles on a branch
Winter exposure injury, 113
Needles to exterior of tree bleached or brown, developing late winter
Winter desiccation, 113
Discoloration, needle drop of current or previous years growth
Needle casts (*Bifusella* sp., *Davisomycella* sp., *Elytoderma* deormans, *Lophodermella* sp., *Lophoedermium* sp.), 78
Giant conifer aphids, 47
Brown felt like material on needles (high elevations)
Brown felt blight (*Neopeckia coulteri*), 75
Webbing covering crown of tree (ponderosa pine)
Tiger moth, 14
Needles being mined
Pine needle sheath miner, 41
Ponderosa pine budworm (young larvae), 5
Ponderosa pine needleminer, 41
Needles stunted, swollen
Stubby needlegall midge (ponderosa pine, 91
Needles twisted, stunted
Eriophyid mites (*Trisecatus* spp.), 100
Aphids on needles
Pineus spp., *Essigella* spp., *Eulachnus* spp., 52
Giant conifer aphids, 47
Scales on needles
Pine needle scale, 121
Pine tortoise scale (nymphs), 127
Striped pine scale (males), 127
Black pineleaf scale, 123
Honeydew present
Striped pine scale, 127
Pine tortoise scale, 127
Conspicuous blue/black weevil observed
Magdalis leconti, 137
Whole tree fades, reddens
Mountain pine beetle, 161
Affecting twigs and smaller branches
Twigs tunneled
Twig beetles, 137
Southwestern pine tip moth, *Rhyacionia bushnelli*, 133
Petrova metallica, 134
Magdalis lecontei, 137, addendum
Terminal killed back (lodgepole pine)
Pissodes terminalis, 136
Oozing small pockets of sap
Pitch midge
Cottony aphids on twigs
Pineus spp., 52

- Scales on twigs
    - Striped pine scale, 127
    - Pine tortoise scale, 127
- Galls on small branches
    - Western gall rust (*Endocronartium harknessii*), 110
- Stunted, twisted witches' brooms
    - Eriophyid mites (*Trisecatus* spp.), 100
- Swollen and/or twisted terminal growth
    - Herbicide injury, 66

Affecting trunk and larger branches

- Oozing pink/yellow pitch from trunk
    - Pinyon "pitch mass" borer, 174
- Tunnels oozing popcorn-like white pitch, often near crotches
    - Zimmerman pine moth, 175
    - Limb rust, ponderosa pine only (*Peridermium filamentosum*), 108
- Galls on large branches or trunk
    - Western gall rust (*Endocronartium harknessii*), 110
- Cankers
    - With resin and squirrel chewing
        - Comandra rust (lodgepole pine) (*Cronartium comandrae*), 108
    - Roughened bark and resin production
        - White pine blister rust (on five needle pines), 108
- Fungal fruiting bodies (mushrooms, conks) present
    - Stem decay fungi (various species), 154
- Open wounds, internal decay, swollen areas in stem
    - Stem decay fungi (various species), 154
- Regular rows of holes
    - Woodpeckers, 212
- Sawdust associated with tunneling of trunk, branches
    - Pine sawyers, 179
    - Blackhorned pine borer, 178
    - *Chalcophora* spp./*Buprestis* spp./Ponderous borer, 168
- Bark beetles
    - Engraver (ips) beetles, 165, addendum
    - Mountain pine beetle, 161
    - Red turpentine beetle, 163
    - Twig beetles, 137
    - Ambrosia beetles, addendum
- Chewing off twigs
    - Abert's squirrel (ponderosa pine), red squirrel (lodgepole pine), 213
- Small shoots emerging from the branch
    - Dwarf mistletoe (*Arceuthobium* sp.), 187

Affecting roots

- White root decay with white mycelial fans between bark and wood
    - Armillaria root disease (*Armillaria mellea*), 194
- General dieback of tree (ponderosa pine)
    - Black stain root disease, 195

Affecting cones
    Cones tunneled by caterpillars
        Coneworms (*Dioryctria* spp.)
        Cone beetles (*Conophthorus* spp.)
        Cone weevil
    Sucking on developing cones
        Conifer seed bugs, 206

PINYON (*Pinus edulis*)
Affecting needles
    Needles being chewed
        Tiger moth, 14
        Conifer sawfly, 28
        Pine butterfly, 21
    Webbing covering crown of tree
        Tiger moth, 14
    Needles being mined
        Pinyon needleminer, 41
    Discoloration, needle drop of current or previous year's growth
        Needle cast (various fungi), 78
        Needle rust (Coleosporium jonesii), 77
    Needles to exterior of tree bleached or brown, developing late winter
        Winter desiccation, 113
    Needles galled, swollen
        Pinyon spindlegall midge, 91
        Pinyon stunt needlegall midge, 92
    Aphids
        Pineus spp., 52
        Giant conifer aphids, 47
    Scales
        Pine needle scale, 121
        Pine tortoise scale (nymphs), 127
        Pinyon needle scale (bean stage), 30
    Row of small dark objects on needle
        Aphid eggs, 52
Affecting twigs and smaller branches
    Twigs tunneled
        Twig beetles, 137
        Pinyon tip moth, 134
        Pinyon pitch nodule moth, 134
    Twigs chewed
        Red squirrel, 213
    Cottony aphids on twigs
        Pineus spp., 52
    Scales on twigs
        Pine tortoise scale, 127
        Striped pine scale, 127
    Large sucking insects feeding on twig
        Cicadas, 131

Affecting trunk and larger branches
- Oozing pink/yellow pitch from trunk
  - Pinyon "pitch mass" borer, 174
  - *Dioryctria* spp. (unidentified)
- Orange spore-forming bodies between bark cracks
  - Pinyon blister rust (*Cronartium occidentale*), 108
- Galls on branches or stem
  - Western gall rust (*Endocronartium harknessii*), 110
- Sawdust tunneling of trunk, branches
  - Engraver beetles, 165
  - Pine sawyers, 179
- Cottony masses on trunk in spring
  - Pinyon needle scale, 130
- Sapwood decay
  - Pouch fungus (*Cryptoporus volvatus*), 154
- Small (1/2-in to 1-in) stems emerging from the branch
  - Dwarf mistletoe (*Arceuthobium divaricatum*), 187

Affecting roots
- General die back of tree
  - Black stain root disease (*Leptographium wageneri*), 195
  - Pinyon decline (multiple abiotic and biotic factors)
- White root decay with white mycelial fans between bark and wood
  - Armillaria root disease (*Armillaria mellea*), 194

Affecting cones
- Cones tunneled by caterpillars
  - Coneworms (*Dioryctria* spp.)
- Sucking injuries, poor seed fill
  - Insufficient water after pollination
  - Conifer seed bugs, 206

PLUM (See **Stone Fruits**)

POME FRUITS [Crabapple, Apple (*Malus* spp.); Pear (*Pyrus* spp.)]

Affecting leaves
- Chewing large holes in leaves
  - Speckled green fruitworm, 18
  - Redhumped caterpillar, 4
  - Forest tent caterpillar, 11
  - Cankerworms, 18
- Silken tents produced
  - Eastern tent caterpillar, 11
  - Western tent caterpillar, 10
  - Fall webworm, 14
- Caterpillar living within a case
  - Casebearer, 10
  - Snailcase bagworm, 9
- Pale yellow to cream color areas on leaves in various patterns
  - Apple mosaic virus (ApMV), 71
- Terminal leaves curled and tied together with silk
  - Fruittree leafroller, 7

- Skeletonizing leaves
    - Apple flea beetle, 32
    - Pearslug (pear only), 22
- "Shothole" feeding wounds in leaf (usually sucker growth)
    - Apple flea beetle, 32
- Producing raised leafmines
    - Spotted tentiform leafminer, 39
- White powdery material on upper or lower surface of leaf
    - Powdery mildew (*Podosphaera leucotricha*), 82
- Pale green or yellow leaves
    - Iron chlorosis, 64
- Rust or orange spotting
    - Juniper rusts (*Gymnosporangium* spp.), 102
- Black mold on surface of leaf
    - Apple scab (*Venturia inaequalis*), 73
- Producing rusty blisters or scabby patches on leaves
    - Blister mites, 98
- Bronzing of leaves (spider mites)
    - Twospotted spider mite, 60
    - McDaniel spider mite, 60
- Curling distortions of new growth in spring
    - Rosy apple aphid, 45
    - Apple aphid, 45
- Blackened and wilting leaves
    - Fireblight (*Erwinia amylovora*), 114
- Large droplets of honeydew on pear leaves
    - Pear psylla, 52
- Small scales
    - San Jose scale, 124

Affecting twigs, small branches

- Twigs shredded by a line of multiple punctures
    - Cicada oviposition wounds, 131
    - Buffalo treehopper, 131
    - Hail damage (restricted to upper surface), 113
- Blackened, wilting, and crooked tips
    - Fireblight (*Erwinia amylovora*), 114
- Bark beetle tunneling, small exit holes
    - Shothole borer, 158
- Scales on twigs
    - San Jose scale, 124
    - Oystershell scale, 120
    - European fruit lecanium, 126
- Cottony insects on twigs
    - Woolly apple aphid, 51
    - Grape mealybug, 53
- Discolored areas, dead bark containing small pimple-like fruiting bodies (pycnidia)
    - Cytospora canker (*Cytospora* sp.), 146

Affecting trunk

- Boring into trunk
    - Flatheaded appletree borer, 184
    - Redheaded ash borer, 168

Bark beetle tunneling, small exit holes
Shothole borer, 158
Cottony insects on trunk and or roots
Woolly apple aphid, 51
Internal decay and/ or shelf-like fruiting structures (conks)
Decay fungus (*Trametes hirsuta*), 154
Wilting and die back of portions of tree
Originating from roots
Verticillium wilt, 142
Affecting ground line area of trunk
Discolored tissue under bark at ground line
Phytophthora root rot (*Phytophthora* sp.), 196
Gall at ground line
Crown gall (*Agrobacterium tumefaciens*), 101
Pitting at graft union
Apple union necrosis virus (TmRSV), 71
Affecting roots
White root decay and white mycelial fans between bark and wood
Armillaria root disease (*Armillaria mellea*), 194
Affecting fruit
Tunneling fruit
Codling moth, 208
Apple maggot, 204
Scales, with reddened area on fruit
San Jose scale, 124
Scarring or scabbing of fruit
Speckled green fruitworm, 18
Fruittree leafroller, 7
Hail injury, 113
Apple scab (*Venturia inaequalis*), 73

PYRACANTHA

Affecting foliage
Spider mites, 60
Affecting twigs
Scale
Hawthorn mealybug, 53
San Jose scale, 124
Twig blight/dieback
Fireblight, 114

RABBITBRUSH (*Chrysothamnus* spp.)

Affecting leaves
Leaves chewed
Rabbitbrush beetle, 32
Rabbitbrush webbing moth, 8
Snailcase bagworm, 9
White powdery material on upper or lower surface of leaf
Powdery mildew, 82
Orange pustules on leaf
Leaf rust (*Pucciniastrum* sp.), 77

Affecting stems
Frothy mass on stems
Spittlebug, 55
Cottony balls on stems
Rabbitbrush gall flies, 94
Green flowerlike swellings
Rabbitbrush gall flies, 94

ROSE (*Rosa* spp.)

Affecting leaves
White powdery material on upper or lower surface of leaf
Powdery mildew (*Sphaerotheca pannosa*), 82
Leaves with white spotting
Rose leafhopper, 55
Twospotted spider mite, 60
Rust (*Phragmidium mucronatum*), 77
Small ball-like growths on leaves, usually reddish
Gall wasps, 96
Leaves mottled with yellowish areas
Iron chlorosis, 64
Rose mosaic complex (various viruses are responsible), 71
Orange-red colored patches on lower leaf surface
Rose rust (Phragmidium *mucronatum*), 77
Angular brown spots on leaves
Anthracnose, 72
Interior areas of leaves chewed
Roseslug and other sawflies, 27
Even, semicircular cuts in leaf edge
Leafcutter bees, 36
Leaves curled
Powdery mildew, 82
Dark spots on leaves
Black spot (*Diplocarpon rosae*), 74
Affecting canes
Insects tunneling into pith of cane
Hunting wasps, 138
Small carpenter bee
Leafcutter bee, 36
Exterior area of cane girdled, sometimes with associated diebck
Stemboring sawfly, 177
Bronze cane borer/rose stem girdler, 183
Mossy or ball-like growths on stem
Gall wasps, 96
Woody, tumor-like growth, usually on lower stem
Crown gall (*Agrobacterium tumefaceins*), 101
Swellings in canes, often with associated vertical cracking
Bronze cane borer/rose stem girdler, 183
Purple to tan, sometimes sunken spots on canes
Rose canker (*Cryptosporella* or *Leptosphaeria* sp., 148

Affecting flowers
Flower buds killed, fail to emerge ("blind" shoots)
Rose midge, 203
Fluctuating, cool temperatures
Rose curculio, 202
Flower petals scarred
Flower thrips
Flower petals chewed or tunneled
Earwigs, 225
Rose curculio, 202
Tobacco budworm
Speckled green fruitworm, 18
Flowers produced are "off-type" after winter (Grandiflora and hybrid tea roses)
Die back to the grafted rootstock

RUSSIAN-OLIVE (*Eleagnus angustifolia*)

Insects associated with foliage
Aphids, 45
Grasshoppers, 35
Thrips
On trunks and branches
Oozing gum
Russian olive decline (gummosis), 190
Botryodiplodia canker (*Botryodiplodia hypodermia*), 144
Tubercularia canker (*Tubercularia* sp.), 151
Small, round pinhead-sized fruiting bodies under bark
Tubercularia canker (*Tubercularia* sp.), 151
Borers
Redheaded ash borer, 168
Branch borer (*Anelaphus vellosus*)
Branch dieback with bark girdled
Fox squirrel, 213
Affecting roots and ground line
Discolored tissue under bark at ground line
Phytophthora root rot (*Phytophthora* sp.), 196

SPIREA (*Spirea* spp.)

On leaves, sometimes associated with leaf curls
Spirea aphid (*Aphis citricola*), 45

SPRUCE (*Picea* spp.)

Affecting needles
New needles being chewed
Douglas-fir tussock moth, 15
Western spruce budworm, 5
Needles chewed and fragments tied with webbing
Spruce needleminer, 40
Web-spinning sawfly, 28
Brown felt-like material on needles, branches (high elevation)
Brown felt blight (snow mold) (*Herpotrichia juniperi*), 75

Needles being mined
- Spruce needleminer, 40
- *Coleotechnites piceaella*, 41

Aphids on needles
- Giant conifer aphids, 47
- Balsam twig aphid

Needles become grayish, with small flecks
- Spruce spider mite, 62

White scales on needles
- Pine needle scale, 121

Yellow witches' broom
- Spruce broom rust (*Chrysomyxa arctistaphyli*), 109

Brown needles with dark fruiting bodies
- Needlecasts (*Lophodermium* sp. and others), 78

Needles to exterior of tree bleached or brown, developing late winter
- Winter desiccation, 113

Needles on branch turn reddish brown
- Cytospora canker, 146
- Root injury

Mottled yellowing of needles on a branch
- Winter exposure injury, 113

Affecting twigs and smaller branches

Twigs distorted into cone-like gall
- Cooley spruce gall adelgid, 86
- *Pineus pinifoliae*, 86

Woolly aphid on underside of twigs in spring
- Cooley spruce gall adelgid, 86

Twigs tunneled
- Twig beetles, 137

Small bud-like scales on twigs
- Spruce bud scale

Large aphids on branches, twigs
- Giant conifer aphids, 47

Resinous canker on branch
- Cytospora canker (*Leucostoma kunzei*), 146

Affecting trunk and larger branches

Bark beetles
- Spruce beetle, 164
- Spruce Ips, 165, addendum

Borers
- Pine sawyers, 179
- *Neoclytus muricatulus*, 168
- *Synanthedon novaroensis*, 169
- Horntails, 176

Aphids
- Giant conifer aphids, 47

Resinous canker on branch
- Cytospora canker (*Leucostoma kunzei*), 146

Open wounds, internal decay, swollen areas in stem
- Stem decay fungi (*Phellinus pini, Inonotus circinatus, Fomitopsis pinocola* and various fungal genera), 154

Fungal fruiting bodies (mushrooms, conks) present
Stem decay fungi (*Phellinus pini, Inonotus circinatus, Fomitopsis pinocola* and various fungal genera), 154
Affecting top of tree
Dieback restricted to terminal
White pine weevil, 136
Upper crown defoliated
Douglas-fir tussock moth, 15
Upper crown dieback
Spruce ips, 165
Root damage
Affecting roots
White root decay with white mycelial fans between bark and wood
Armillaria root disease (*Armillaria mellea*), 194

STONE FRUITS (Cherry, Chokecherry, Plum, Peach, Apricot, etc.) (*Prunus* spp.)

Affecting whole plant
Trees decline in vigor
Peach tree borer, 171
Phytophthora root rot (*Phytophthora* sp.), 196
X-disease, 68
Cytospora canker (*Cytospora* sp.), 146
Verticillium wilt, 142
Virus disease (various), 71
Affecting leaves
Leaves chewed, no associated webbing
Redhumped caterpillar, 4
Twotailed swallowtail, 227
Pearslug, 26
Upper surface of the leaf skeletonized
Pearslug, 26
Webbing associated with chewing
Fall webworm, 14
Uglynest caterpillar, 7
Western tent caterpillar/Eastern tent caterpillar, 10
Upper surface of the leaf skeletonized
Pearslug, 26
Leaves with small, circular holes
Shot hole disease, 83
Prunus necrotic ring spot virus (PNRSV), 71
White flecking injuries
White apple leafhopper (on peach), 56
Lacebug (*Corythuca padi*) (on chokecherry), 59
Leaves containing abnormal color, shape or growth
Virus disease (various), 71
White powdery material on upper or lower surface of leaf
Powdery mildew, 82
Leaves yellow
Iron chlorosis, 64
X-disease, 68

- Leaves with small, pouch or finger-like projections
    - Fingergall mites, 98
- New leaves curled
    - Green peach aphid, 46
    - Leafcurl plum aphid, 46
    - Black cherry aphid, 46
- New leaves curled and thickened (plum only)
    - Plum pockets (*Taphrina communis*), 209

Affecting branches, trunk

- Galls on branches
    - Black knot of cherry (*Apiosporina morbosa*), 117
- Stem decay and /or hoof-shaped fruiting body (conk)
    - Decay fungus (*Phellinus pomaceus*), 153
- Scales
    - European fruit lecanium, 126
    - San Jose scale, 124
    - Oystershell scale, 120
- Small pinhead-sized exit holes in branches
    - Shothole borer, 158
- Tips of branches dieback
    - Peach twig borer, 135
- Branches die back,
    - Cytospora canker (*Cytospora* sp.), 146
    - Shothole borer, 158
- Gumming
    - Peach tree borer, 171
    - Shothole borer, 158
    - Cytospora canker (*Cytospora* sp.), 146
- Dead bark with discoloration and small pimple-like fruiting bodies (pycnidia)
    - Cytospora canker (*Cytospora* sp.), 146
- Borers
    - Peach tree borer, 171
    - Redheaded ash borer, 168
- Wilting and dieback of portions of tree
    - Originating from roots
        - Verticillium wilt, 142

Affecting ground line area of trunk

- Girdling wounds made at or near soil line
    - Peach tree borer, 171
- Discolored tissue under bark at ground line
    - Phytophthora root rot (*Phytophthora* sp.), 196
    - Prune brown line virus (TmRSV), 71
- Pitting at graft union
    - Cherry stem pitting virus, 71
- Gall at ground line
    - Crown gall (*Agrobacterium tumefaciens*), 101

Affecting fruit
- Fruit tunneled
  - Peach twig borer, 135
  - Oriental fruit moth
  - Walnut husk fly, 204
  - European earwig, 225
- Fruit with sunken, corky areas
  - Plant bug, 207
  - Boxelder bug, 206
  - Hail injury, 113
- Puncture wounds in fruit
  - Cherry curculio, 201
  - Bird damage (Robins, finches, etc.)
- Chokecherry fruit enlarged, hollow
  - Chokecherry gall midge, 203
- Maggots in cherry fruit
  - Western cherry fruit fly, 204
  - Chokecherry gall midge, 203
- Grub in pit
  - Cherry curculio, 201
  - Plum gouger, 201
- Ooze from fruit (plum)
  - Plum gouger, 201
- Reduced fruit size
  - Viral disease (various), 71

SUMAC/SKUNKBRUSH (*Rhus* spp.)

Affecting leaves
- Leaves chewed
  - *Hemileuca neumoegeni*, 3
  - Sumac leaf beetle, 33
  - *Coleothorpa dominicana*, addendum
- Small reddish swellings
  - Eriophyid mite gall, 100
- Sucking insect
  - Sumac psyllid, 52

Affecting twigs
- Dark insect on twigs
  - Sumac psyllid (overwintering nymphs), 52
  - Oystershell scale, 120

Affecting stems, branches
- Discolored areas, dead bark containing small pimple-like fruiting bodies (pycnidia)
  - Cytospora canker (*Cytospora* sp.), 146
- Wilting and dieback of portions of plant originating from roots
  - Verticillium wilt (*Verticillium* sp.), 142

SYCAMORE (*Platanus* spp.)

Affecting leaves
- Leaves with blackened spotting
  - Anthracnose (Apiognomonia veneta), 72

Affecting twigs, small branches
- Discolored areas, dead bark containing small pimple-like fruiting bodies (pycnidia)
  - Cytospora canker (*Cytospora* sp.), 142
- Witches' broom with dead small branches
  - Anthracnose (*Apiognomonia veneta*), 72

Affecting branches and trunks
- Discolored areas, dead bark containing small pimple-like fruiting bodies (pycnidia)
  - Cytospora canker (*Cytospora* sp.), 142
- Clear to white oozing or frothy malodorous liquid
  - Bacterial wetwood/slimeflux, 189

VIBURNUM (*Viburnum* spp.)

Affecting leaves
- Leaves chewed
  - Cecropia moth, 2
- Leaves curled in spring
  - Snowball (viburnum) aphid, 50
- Leaves blacken
  - Bacterial leaf blight (*Pseudomonas* sp.), 116
- Flecking, bronzing
  - Twospotted spider mite, 60

Affecting stems
- Scoring bark near base of plant
  - Viburnum borer, 168
  - Dogwood borer, 168

VIRGINIA CREEPER (*Parthenocissus quinquefolia*)

GRAPE (*Vitis* spp.)

Affecting leaves
- Leaf edge cut in semicircle
  - Leafcutter bees, 36
- Leaves chewed in general manner
  - Achemon sphinx, 20
  - Whitelined sphinx, 20
- Shotholes chewed in leaf interior
  - Apple leaf beetle, 32, addendum
- Leaves chewed in skeletonizing pattern
  - Western grapeleaf skeletonizer, 19
- Leaves with white flecking
  - Zic-zac leafhopper, 56
  - Grape leafhopper, 56
- Leaves with meandering silver tunnels
  - *Antispila isabella* (Larva of a leafmining moth)

Affecting branches, stems, or crown
- Tunneling crown
  - *Albuna fraxini* (Clearwing borer)
- Borer in small twigs
  - Apple twig borer, 168

WALNUT (*Juglans* spp.)
- Affecting leaves
  - Leaves chewed
    - Polyphemus moth, 2
    - Cecropia moth, 2
  - Aphids
    - American walnut aphid, 45
  - New growth killed or distorted in patches
    - Plant bugs, 207
- Affecting twigs, branches
  - Scales
    - Oystershell scale, 120
- Tunneling husk of nuts
  - Walnut husk fly, 204

WILLOW (*Salix* spp.)
- Affecting leaves
  - Leaves chewed
    - Spiny elm caterpillar, 20
    - Redhumped caterpillar, 4
    - Nevada buck moth, 3
    - Cecropia moth/Glover's silk moth, 2
    - Polyphemus moth, 2
    - Speckled green fruitworm, 18
    - Western tent caterpillar/Eastern tent caterpillar, 10
    - Fall webworm, 14
    - *Acronicta leporina*, 16
    - Twinspot sphinx/Columbia Basin sphinx/Giant poplar sphinx, 20
    - Willow leaf beetle (*Chrysomela aeneicollis*), 30
    - Willow flea beetles (*Disonycha* sp.), 32
    - Apple flea beetle, addendum
  - Masses of dark, spiny caterpillars on leaves
    - Spiny elm caterpillar, 20
  - Edges of leaves curled over
    - Poplar leaffolding sawfly, 26
  - Leaves with reddish, hollow swellings
    - Willow redgall sawfly, 97
    - Psyllid (unidentified) gall
  - Leaves with black spotting
    - Tar spot (*Rhytisma salicinium*), 84
  - Rust to orange colored leaf spots
    - Willow leaf rust (*Melamspora epita*), 77
  - Leaves with irregular raised pouch galls
    - *Aculops tetanothrix* (eriophyid mite), 100
  - Leaves skeletonized by beetle larvae
    - Willow flea beetles, 32
    - Cottonwood leaf beetle, 30
    - Willow leaf beetle, *Chrysomela aeneicollis*, 30

Sticky honeydew on leaves
- Black willow aphids, 45
- Carrot-willow aphid, 45
- Little green and yellow willow aphids, 45

White powdery material on upper or lower surface of leaf
- Powdery mildew, 82

Affecting twigs

Thickened swellings in twigs
- Willow stemgall sawflies (*Euura* spp.), 96

Large aphids on twigs
- Giant willow aphid, 48
- Black willow aphid, 45

Scales
- Oystershell scale, 120
- Scurfy scale, 121

Cone-like gall at end of twig
- Willow cone gall midges, 90

Large beetles chewing bark
- Cottonwood borer, 177

Twigs with closely spaced pimple-like fruiting bodies (pycnidia)
- Dothiora canker, 147

Affecting trunk and branches

Tunneling into wood
- Poplar and willow borer, 168
- Flatheaded appletree borer, 184
- Cottonwood borer, 177
- Bronze cane borer and related species, 183

Areas of dead bark
- Discoloration and small pimple-like fruiting bodies (pycnidia) in the bark
  - Cytospora canker (*Cytospora* sp.), 146

Regular rows of holes in trunk
- Sapsucker, 212

Open wounds, internal decay, swollen areas in stem
- Stem decay fungi (*Collybia velutipes, Phellinus igniarius* and various fungal genera), 152

Fungal fruiting bodies (mushrooms, conks) present
- Stem decay fungi (*Collybia velutipes, Phellinus igniarius* and various fungal genera), 152

Clear to white oozing or frothy malodorous liquid exiting from wounds
- Bacterial wetwood and slime flux, 189

Visiting oozing sap from trunk
- Bumble flower beetle, 228
- Sap beetles
- Flies (fruit flies and other families)

index

Q

R

S

T

U

V

W

X

Y

Z